AF775308

MATHEMATICAL SURVEYS AND MONOGRAPHS SERIES LIST

Volume

1 **The problem of moments,** J. A. Shohat and J. D. Tamarkin

2 **The theory of rings,** N. Jacobson

3 **Geometry of polynomials,** M. Marden

4 **The theory of valuations,** O. F. G. Schilling

5 **The kernel function and conformal mapping,** S. Bergman

6 **Introduction to the theory of algebraic functions of one variable,** C. C. Chevalley

7.1 **The algebraic theory of semigroups, Volume I,** A. H. Clifford and G. B. Preston

7.2 **The algebraic theory of semigroups, Volume II,** A. H. Clifford and G. B. Preston

8 **Discontinuous groups and automorphic functions,** J. Lehner

9 **Linear approximation,** Arthur Sard

10 **An introduction to the analytic theory of numbers,** R. Ayoub

11 **Fixed points and topological degree in nonlinear analysis,** J. Cronin

12 **Uniform spaces,** J. R. Isbell

13 **Topics in operator theory,** A. Brown, R. G. Douglas, C. Pearcy, D. Sarason, A. L. Shields; C. Pearcy, Editor

14 **Geometric asymptotics,** V. Guillemin and S. Sternberg

15 **Vector measures,** J. Diestel and J. J. Uhl, Jr.

16 **Symplectic groups,** O. Timothy O'Meara

17 **Approximation by polynomials with integral coefficients,** Le Baron O. Ferguson

18 **Essentials of Brownian motion and diffusion,** Frank B. Knight

19 **Contributions to the theory of transcendental numbers,** Gregory V. Chudnovsky

20 **Partially ordered abelian groups with interpolation,** Kenneth R. Goodearl

21 **The Bieberbach conjecture: Proceedings of the symposium on the occasion of the proof,** Albert Baernstein, David Drasin, Peter Duren, and Albert Marden, Editors

22 **Noncommutative harmonic analysis,** Michael E. Taylor

23 **Introduction to various aspects of degree theory in Banach spaces,** E. H. Rothe

24 **Noetherian rings and their applications,** Lance W. Small, Editor

25 **Asymptotic behavior of dissipative systems,** Jack K. Hale

26 **Operator theory and arithmetic in H^∞,** Hari Bercovici

27 **Basic hypergeometric series and applications,** Nathan J. Fine

28 **Direct and inverse scattering on the lines,** Richard Beals, Percy Deift, and Carlos Tomei

29 **Amenability,** Alan L. T. Paterson

30 **The Markoff and Lagrange spectra,** Thomas W. Cusick and Mary E. Flahive

MATHEMATICAL SURVEYS AND MONOGRAPHS SERIES LIST

Volume

31 **Representation theory and harmonic analysis on semisimple Lie groups,** Paul J. Sally, Jr. and David A. Vogan, Jr., Editors

32 **An introduction to CR structures,** Howard Jacobowitz

Spectral Theory and Analytic Geometry Over Non-Archimedean Fields

Mathematical
Surveys
and
Monographs

Volume 33

Spectral Theory and Analytic Geometry Over Non-Archimedean Fields

Vladimir G. Berkovich

American Mathematical Society
Providence, Rhode Island

Translated by Neal I. Koblitz

2000 *Mathematics Subject Classification.* Primary 12J25, 32-XX, 14G20, 20G25, 46-XX, 46J99; Secondary 47-XX.

Library of Congress Cataloging-in-Publication Data

Berkovich, Vladimir G.

Spectral theory and analytic geometry over non-Archimedean fields/Vladimir G. Berkovich.

p. cm.—(Mathematical surveys and monographs, ISSN 0076-5376; no. 33)

Includes bibliographical references and index.

ISBN 0-8218-1534-2 (alk. paper)

1. Spectral theory (Mathematics). 2. Geometry, Analytic. I. Title. II. Series.

QA320.B44 1990 90-830

515′.7222—dc20 CIP

AMS softcover ISBN 978-0-8218-9020-2

∞ The paper used in this book is acid-free and falls within the guidelines
established to ensure permanence and durability.
This publication was typeset using AMS-TeX,
the American Mathematical Society's TeX macro system.
Visit the AMS home page at http://www.ams.org/

12 11 10 9 8 7 6 5 4 3 27 26 25 24 23 22

Contents

Preface

According to A. Grothendieck one really does not need a space to do geometry, all one needs is a category of sheaves on this would-be space. A nice illustration of this idea is the introduction by J. Tate of the notion of rigid analytic space, which is a p-adic analog of the notion of complex analytic space. Rigid analytic spaces are good enough for the construction of the category of sheaves, but they don't allow direct application of the geometrical intuition that is so valuable over the complex field. And so they are called spaces only conditionally. Several years ago I found that p-adic analytic spaces really exist. They are quite elegant objects possessing many topological properties of complex analytic spaces that are sufficient, for example, for applying to them the homotopy and singular homology notions in the usual sense. This book is devoted to the study of these analytic spaces. We try to show that geometrical considerations are relevant and useful over p-adic fields too. Of course, time will show better whether the lost p-adic spaces were worth finding.

In the early stages of this work, the interest and encouragement of M. Vishik were a necessary stimulus for me. I am very grateful to him for this. I am also grateful to M. Gromov for his interest, encouragement, and valuable remarks. I would like to thank V. Hinich for useful discussion.

I currently hold the Reiter Family Career Development Chair at the Weizmann Institute.

Introduction

A non-Archimedean field is a field complete with respect to a non-Archimedean valuation (see §1.1). The most important example is the field of p-adic numbers $\mathbf{Q}_p$ discovered by K. Hensel in 1905. The simplest example is an arbitrary field endowed with the trivial valuation. Many classical notions—such as the notions of a Banach space, a bounded linear operator, and a convergent power series over $\mathbf{R}$ or $\mathbf{C}$—have their non-Archimedean counterparts. But there are some "pathological" properties of non-Archimedean fields which make them, at least on the face of it, essentially different from $\mathbf{R}$ and $\mathbf{C}$. First of all, non-Archimedean fields are totally disconnected. This means, for example, that the intuitive idea of continuity will not work in the same way as in the classical situation. And, in fact, the construction of a reasonable theory of analytic functions, in which the local properties of functions determine their global properties, is highly nontrivial. Furthermore, for every non-Archimedean field k there exists a non-Archimedean field K that contains k and does not coincide with it. This does not correspond to the classical Gel′fand-Mazur Theorem, which asserts that any commutative Banach field over $\mathbf{C}$ coincides with $\mathbf{C}$. As a consequence of this fact, one finds that there exist bounded linear operators with empty spectrum (if one defines the spectrum in the same way as in the classical situation). Indeed, if α is an element of K not in k, then multiplication by α in K (which is a Banach space over k) is such an operator.

As for the theory of analytic functions, it was constructed to some extent by M. Krasner in the 1940s ([**Kr**]) and in full generality by J. Tate in 1961 ([**Ta**]). The approach of Krasner is elementary. (The ground field is assumed to be algebraically closed and have a nontrivial valuation.) First, he defines an analytic element on a subset $U \subset k$ as a function $f\colon U \to k$ which is a limit of rational functions without poles in U. Next, he introduces a class of subsets of k (quasiconnected subsets) that have the property of analytic continuation. The latter means that, for any analytic element f on such a subset U, if f is zero on an open subset of U, then f is zero on all of U. For example, the standard sets (closed discs minus a finite number of open discs) are quasiconnected. Finally, a function f defined on an open subset $U \subset k$ is said to be analytic if there is a covering $\{U_i\}_{i \in I}$ of U

by quasiconnected subsets such that the restriction of f to each U_i is an analytic element.

The approach of Tate, while less elementary, in principle generalizes the approach of Krasner (the ground field is no longer assumed to be algebraically closed). First, he introduces a class of commutative Banach algebras over k (affinoid algebras), which are quotient algebras of $k\{T_1, \dots, T_n\}$ (the algebra of convergent power series on the n-dimensional unit disc). The maximal spectrum $X = \mathrm{Max}(\mathscr{A})$ has a canonical topology induced by the topology of k. But this topology is not interesting, because X is also totally disconnected with respect to it. Instead of this, Tate introduces a class of open subsets of X and a class of coverings of these open sets which gives rise to a Grothendieck topology on X; then he constructs a sheaf of rings $\mathscr{O}_X$ in this topology. Such a space is called an affinoid space. An analytic space (or, as one often says, a rigid analytic space) is a set X endowed with a Grothendieck topology and a sheaf of rings on it which is locally isomorphic to an affinoid space. Rigid analytic spaces have the property of local connectedness, and they also have a rich cohomology theory of coherent sheaves which is very similar to the theory for complex analytic spaces (see **[Ki1]**, **[Ki2]**, **[BGR]**). The most beautiful result in this subject is the uniformization of curves with totally degenerate reduction, which was discovered by Tate for elliptic curves and generalized by Mumford to curves of higher genus (**[M2]**). However, one cannot apply to rigid analytic spaces the usual topological properties of complex analytic spaces, such as local compactness or local arcwise connectedness. Thus, one cannot consider singular homology and homotopy groups, although in some sense they certainly should exist. (Some people believe that, in principle, singular homology and homotopy groups of rigid analytic spaces can be defined, but it is clear that such a definition would not be natural from the geometrical point of view.)

The spectral theory of bounded linear operators was constructed by M. Vishik (**[Vi]**) for a class of operators that he called analytic operators with compact spectrum (the ground field is assumed to be algebraically closed and have nontrivial valuation). These are operators whose resolvent is a Krasner analytic function on the complement of the spectrum (defined in the usual sense). Actually, our work arose as a result of an attempt to extend Vishik's theory to arbitrary bounded linear operators. We shall return to this topic at the end of the introduction.

The purpose of this book is to introduce non-Archimedean analytic spaces that have the topological properties mentioned previously, to use them to construct the basics of the spectral theory of bounded linear operators, and to give some applications. Our approach, which is closely related to the approaches of Krasner and Tate, generalizes the definitions of analytic functions and analytic spaces over $\mathbf{C}$ in a natural way. Moreover, the basic definitions work over arbitrary commutative Banach rings, including the fields

$\mathbf{C}$ and $\mathbf{R}$, fields with the trivial valuation, and the ring of integers $\mathbf{Z}$ with the Archimedean absolute value (although we shall restrict ourselves to non-Archimedean fields). We will now give some motivation and explanation of the idea.

Consider the following naive definition of an analytic function in one variable over a non-Archimedean field k. A function $f: k \to k$ is analytic if, for any $a \in k$, there exists $r > 0$ such that the restriction of f to $E(a, r)$ (the closed disc of radius r with center a) is equal to a power series $\sum_{i=0}^{\infty} a_i(T-a)^i$ which is convergent on $E(a, r)$. It is clear that this is not a good definition because the field k is totally disconnected, and therefore the class of functions obtained is too large. But it is also clear that with any good definition the set of functions analytic on the unit disc $E(0, 1)$ should coincide with the algebra $k\{T\} = \{\sum_{i=0}^{\infty} a_i T^i \mid |a_i| \to 0 \text{ as } i \to \infty\}$.

We now remark that we would have a similar phenomenon if we replaced all entries of $\mathbf{R}$ by $\mathbf{Q}$ in the definition of complex valued continuous functions on $\mathbf{R}$, as follows. A function $f: \mathbf{Q} \to \mathbf{C}$ is continuous if, for any $a \in \mathbf{Q}$ and any $\varepsilon > 0$, there exists $\delta > 0$ such that $|f(a') - f(a)| < \varepsilon$ for all $a' \in \mathbf{Q}$ with $|a' - a| < \delta$. As above, the class of functions given by this definition is much larger than the class of continuous functions on $\mathbf{R}$, because the field $\mathbf{Q}$ is also totally disconnected with respect to the topology of $\mathbf{R}$. But in this situation we know what to do in order to obtain the usual continuous functions on $\mathbf{R}$. Namely, we have to complete $\mathbf{Q}$ with respect to the absolute value.

Let us try to do something like this with our non-Archimedean field k; let us try to find holes in it. The field k is complete with respect to its valuation, and so we need a completion of another type, but one which would still give $\mathbf{R}$ if one starts with $\mathbf{Q}$. Recall that we would like to have the algebra $k\{T\}$ as the set of functions analytic on the unit disc $E(0, 1)$. Once again let us come back to the classical situation and make a model of the same problem. Suppose that we have an algebra $\mathscr{C}$ which should be the algebra of continuous functions on the unit interval $[0, 1]$. How can we recover the interval $[0, 1]$? The answer is clear. We must take the Gel′fand space of maximal ideals of the algebra $\mathscr{C}$ (which is a commutative Banach algebra over $\mathbf{C}$). This gives us an idea of what to do in the non-Archimedean situation, because $k\{T\}$ is also a commutative Banach algebra over k. Before continuing, we should make a preliminary observation. In the last procedure we actually tried to recover the real affine line as a topological manifold. (It was pure coincidence that the real line coincides with the field of real numbers.) Our attempt to do something similar in the non-Archimedean situation is none other than an attempt to recover the non-Archimedean affine line as an analytic space. In this situation we cannot hope that the affine line $\mathbf{A}^1$ over k coincides with k as a set.

Thus, our first goal is to understand what the non-Archimedean analog of the Gel′fand space of maximal ideals is. Our second goal is to understand what the non-Archimedean affine line $\mathbf{A}^1$ is.

The answer to the first problem is given in Chapter 1 and is in fact very simple. Namely, the spectrum $\mathcal{M}(\mathcal{A})$ of a commutative Banach algebra $\mathcal{A}$ over k with identity is the set of all bounded multiplicative seminorms on $\mathcal{A}$ endowed with the obvious topology (see §1.2). This definition works for any commutative Banach ring, and in the classical situation it gives the space of maximal ideals. The space $\mathcal{M}(\mathcal{A})$ is always nonempty, compact, and Hausdorff. In §1.4.4 it is computed for the algebra $k\{T\}$. We note that every point $x \in \mathcal{M}(\mathcal{A})$ determines a character $\mathcal{A} \to \mathcal{H}(x): f \mapsto f(x)$ from $\mathcal{A}$ to a certain non-Archimedean field $\mathcal{H}(x)$ over k. The multiplicative seminorm corresponding to x is extended to the fraction field of the quotient ring of $\mathcal{A}$ by its kernel; $\mathcal{H}(x)$ is the completion of this field.

Now we explain what the affine line $\mathbf{A}^1$ is, and what the analytic functions on it are. First of all, $\mathbf{A}^1$ is the set of all multiplicative seminorms on the ring of polynomials $k[T]$ which extend the valuation of k. As above, every point $x \in \mathbf{A}^1$ determines a character $k[T] \to \mathcal{H}(x): f \mapsto f(x)$. Next, $\mathbf{A}^1$ is endowed with the weakest topology with respect to which the real valued functions $\mathbf{A}^1 \to \mathbf{R}$ of the form $x \mapsto |f(x)|$ for $f \in k[T]$ are continuous. Finally, an analytic function φ on a subset $\mathcal{U} \subset \mathbf{A}^1$ is a rule which to each point $x \in \mathcal{U}$ assigns an element $\varphi(x) \in \mathcal{H}(x)$ so that φ is a local limit of rational functions. This means that, for any $x \in \mathcal{U}$, there exists an open neighborhood $\mathcal{U}'$ of x in $\mathcal{U}$ such that, for any $\varepsilon > 0$, there exist polynomials $f, g \in k[T]$ with $g(x') \neq 0$ and $|\varphi(x') - \frac{f(x')}{g(x')}| < \varepsilon$ for all $x' \in \mathcal{U}'$. It turns out that $\mathbf{A}^1$ is a one-dimensional, locally compact, contractible and locally contractible space (see §4). The field k is naturally embedded in $\mathbf{A}^1$ (an element $a \in k$ goes to the multiplicative seminorm $f \mapsto |f(a)|$), and the topology of $\mathbf{A}^1$ induces the canonical topology on k. Furthermore, the set of analytic functions on the closed unit disc $\mathrm{E}(0, 1) = \{x \in \mathbf{A}^1 \,|\, |T(x)| \leq 1\}$ coincides with $k\{T\}$, and $\mathcal{M}(k\{T\}) = \mathrm{E}(0, 1)$. It is clear that the construction generalizes naturally to the multidimensional case, not only over k but over any commutative Banach ring. We note that multiplicative seminorms were already used in rigid analytic geometry by M. van der Put, but not for the same purposes (see **[P3]**, **[P4]**).

We now give a summary of the material that follows. In §1.1 we recall some basic notions, such as seminorm, Banach ring, admissible homomorphism. In §1.2 we introduce the spectrum $\mathcal{M}(\mathcal{A})$ of a commutative Banach ring with identity $\mathcal{A}$, and we prove that $\mathcal{M}(\mathcal{A})$ is a nonempty compact Hausdorff space. In §1.3 we establish basic properties of the spectrum, such as the formula for the spectral radius $\rho(f)$ of an element $f \in \mathcal{A}$, which is a natural generalization of the corresponding classical facts (**[Bou2]**, **[GRS]**). In §1.4 we consider several examples, including the algebra $k\{T\}$. In §1.5 we define analytic spaces over an arbitrary commutative Banach ring.

Beginning with the second section, the ground Banach ring is assumed to be a non-Archimedean field k (we also consider the case of the trivial valuation). Although the definitions in §1.5 are elementary, it is not so easy to deduce various properties of analytic functions from them. For example, the fact that a function analytic on the affine line over k is a power series that is convergent on all of k. In the classical situation, for such purposes one uses real analysis and, in particular, Cauchy's integral formula. These things do not work in the non-Archimedean situation. Instead, one uses (in rigid analytic geometry and here) the apparatus of affinoid algebras, which makes many problems purely algebraic.

In Chapter 2 we introduce and study a class of affinoid algebras which is larger than the class considered in rigid analytic geometry. Namely, an affinoid algebra is a commutative Banach algebra $\mathscr{A}$ for which there exists an admissible epimorphism

$$k\{r_1^{-1}T_1, \dots, r_n^{-1}T_n\} \to \mathscr{A},$$

where $k\{r_1^{-1}T_1, \dots, r_n^{-1}T_n\}$ is the algebra of power series convergent on the n-dimensional closed disc of radius $(r_1, \dots, r_n)$. The algebras which are affinoid in the usual sense, i.e., for which such an epimorphism can be found with $r_i = 1$, $1 \le i \le n$, are said to be strictly affinoid. For example, the algebra $k\{r_1^{-1}T_1, \dots, r_n^{-1}T_n\}$ is strictly affinoid if and only if all of the r_i belong to the set $\sqrt{|k^*|} = \{\alpha \in \mathbf{R}_+ | \alpha^n \in |k^*| \text{ for some } n \ge 1\}$ (**[BGR]**, 6.1.5/4). One can already see that the consideration of this more general class of algebras makes it possible to work with continuous families of analytic spaces. Every affinoid algebra becomes strictly affinoid after tensoring with an appropriate non-Archimedean field. In this way various properties of affinoid algebras and affinoid spaces (the spectra of affinoid algebras) are deduced from the properties of strictly affinoid algebras. Given a morphism of affinoid spaces $\varphi: Y \to X$, in §2.5 we define the interior $\mathrm{Int}(Y/X)$ and the boundary $\partial(Y/X)$. These are subsets of Y. The first is open, and the second is closed. One has $\partial(Y/X) = \varnothing$ if and only if φ is finite. If Y is an affinoid subdomain of X, then $\mathrm{Int}(Y/X)$ coincides with the topological interior of Y in X.

In Chapter 3 we study analytic spaces. In §3.1 we consider their basic properties. For example, an analytic space X is separated if and only if its underlying topological space $|X|$ is Hausdorff. The interior and the boundary are defined for any morphism of analytic spaces. A morphism $\varphi: Y \to X$ is said to be proper if it is proper in the usual sense (the inverse image of a compact is a compact) and $\partial(Y/X) = \varnothing$. In §3.2 we show that analytic spaces are locally arcwise connected and that the topological dimension of an analytic space is equal to its dimension. In §3.3 we associate a rigid analytic space X_0 with every separated analytic space X. The correspondence $X \mapsto X_0$ is a fully faithful functor. A morphism $\varphi: Y \to X$ is a closed immersion (resp. proper) if and only if the corresponding morphism $\varphi_0: Y_0 \to X_0$ has the same property. In the general case, however, open immersions do not go to open immersions. Furthermore, the functor $X \mapsto X_0$ preserves the categories of coherent (resp. locally constant) sheaves and their cohomology groups. This implies that all of the cohomological results from rigid analytic geometry are applicable to our analytic spaces. We list several of these results. In §3.4 (for the nontrivial valuation case) and §3.5 (for the trivial valuation case) to every scheme X of locally finite type over k we associate an analytic space X^{an}, and we study the properties of the correspondence $X \mapsto X^{an}$. For example, a scheme X is connected (resp. separated, resp. proper) if and only if the underlying topological space of X^{an} is arcwise connected (resp. Hausdorff, resp. Hausdorff and compact). The topological dimension of X^{an} is equal to the dimension of X. We list some cohomological theorems which are verified in the same way as in complex analysis (and, of course, are also known in rigid analytic geometry). In §3.6 we give a proof of some facts on meromorphic functions on proper analytic spaces.

In Chapter 4 we give a topological description of one-dimensional analytic spaces. Let X be a smooth projective curve of genus g. If the valuation on k is trivial, then X^{an} is contractible (for trivial reasons). So we assume that the valuation on k is nontrivial. In §4.2 we show that if $g = 0$, then X^{an} is contractible. In §4.3 we consider the case $g \geq 1$. We construct a closed subset $\Delta^{an}(X) \subset X^{an}$ (the analytic skeleton of X) which is a finite graph with Betti number $\leq g$ and is a strong deformation retract of X^{an}. As a first application, we show that any one-dimensional analytic space is locally contractible and, in particular, has a universal covering which is a contractible space. Furthermore, we show that, for any finite connected graph Σ, there exists a commutative Banach field over k whose spectrum is homeomorphic to Σ. In §4.4 we give an account of the Mumford theory of p-adic uniformization. According to this theory, a smooth projective curve of genus $g \geq 1$ has split totally degenerate reduction if and only if the universal covering Ω of X^{an} is an open subset of the projective line $\mathbf{P}^1$ whose complement lies in $\mathbf{P}^1(k)$. Moreover, in this case Ω is the set of nonlimits points of some Schottky group $\Gamma \subset \mathrm{PGL}_2(k)$ of rank g, and $\pi_1(X^{an}) = \Gamma$. (Such a curve is called a Mumford curve.) In the rest of the section we give applications. In §4.5 we prove a non-Archimedean analog of Picard's Theorem, which asserts that every holomorphic map from an affine line to a smooth algebraic curve of genus ≥ 1 is constant. (I am indebted to M. Gromov for suggesting that Picard's Theorem could be proved using topological considerations, as in complex analysis.) In §4.6 we describe the smooth projective curves X of genus $g \geq 1$ for which the Betti number of X^{an} is equal to g. Namely, X is such a curve if and only if X is a Mumford curve or a principal homogeneous space over a Tate elliptic curve. We note that both results can be formulated within the framework of rigid analytic geometry. The second result is well known if the ground field is algebraically closed.

In Chapter 5 we study analytic groups (group objects in the category of analytic spaces) and give an interpretation of the Bruhat-Tits buildings in terms of analytic spaces. Let G be an analytic group. As a set, G is not a group, because the canonical map $p: |G \times G| \to |G| \times |G|$ is not bijective ($|X|$ is the underlying topological space of an analytic space X). Nevertheless, for two elements $g, h \in G$ we define the set $gh = m(p^{-1}(g, h))$ (m is the morphism of multiplication). Furthermore, one introduces the notion of a subgroup H of G (this is a subset with $HH^{-1} \subset H$) and the space of cosets G/H. If a subgroup H is compact, then G/H is a locally compact space, and the natural action of the group of k-points $G(k)$ on G/H is proper. Now suppose that G is a Chevalley group over k, $\mathbf{P}$ is a maximal compact subgroup of G^{an} (which comes from a model of G over the ring of integers of k), and T is a maximal torus of G. Then the image of T^{an} in $G^{\text{an}}/\mathbf{P}$ is naturally homeomorphic to the $\mathbf{R}$-vector space $\Lambda = \text{Hom}(\widehat{T}, \mathbf{R}_+^*)$ ($\widehat{T}$ is the character group of T), and the orbit of Λ with respect to the group $G(k)$ coincides with the Bruhat-Tits building $\mathcal{B}(G)$ of G (Λ and its images are the apartments of the building). If the field is local or finite, then $\mathcal{B}(G)$ is closed in $G^{\text{an}}/\mathbf{P}$. In fact, the construction works over arbitrary commutative Banach rings, and in the case of $\mathbf{C}$ or $\mathbf{R}$ it gives the usual symmetric spaces. But in our situation one has the following purely non-Archimedean construction.

In §5.2 we introduce the notion of a peaked point of an analytic space. It turns out that if one of the points $g, h \in G$ is peaked, then the set gh contains a certain canonical element, which is denoted $g * h$. Next, if G is a Stein group and H is a compact subgroup that coincides with the holomorphically convex envelope of a peaked point, then all of the cosets of H contain certain canonical points. The map $G/H \to G$ constructed in this way is continuous and commutes with the action of $G(k)$. All of the conditions necessary for the existence of such a construction are satisfied in the case of the subgroup $\mathbf{P} \subset G^{\text{an}}$, where G is a Chevalley group. Therefore, one has a section $G^{\text{an}}/\mathbf{P} \to G^{\text{an}}$ and, in particular, a $G(k)$-equivariant embedding $\Theta: \mathcal{B}(G) \to G^{\text{an}}$ ($G(k)$ acts on G^{an} by left multiplication). Furthermore, if G is semisimple, one can derive from Θ another $G(k)$-equivariant embedding $\theta: \mathcal{B}(G) \to G^{\text{an}}$ but with the action of $G(k)$ on G^{an} given by inner automorphisms; and, given a parabolic subgroup $H \subset G$, one has an embedding $\Theta_H: \mathcal{B}(G) \to (G/H)^{\text{an}}$. In the case when the field k is local or finite, we construct a $G(k)$-invariant open subset $\Omega^H \subset (G/H)^{\text{an}}$ that contains $\mathcal{B}(G)$ as a closed subset. It is very likely that the étale cohomology of Ω^H is related to the representations of the group $G(k)$.

In Chapter 6 we use the constructions in Chapter 5 to find the homotopy type of certain analytic spaces. In §6.1 we show that the projective space $\mathbf{P}^n$ is contractible. Moreover, any subset of $\mathbf{P}^n$ whose complement lies in a union of (proper) Zariski closed subsets is contractible. Furthermore, we show that for any differentiable manifold there exists a smooth analytic space homotopy equivalent to it. For example, the analytic space $\mathbf{A}^{n+1}\backslash\{x \in \mathbf{A}^{n+1} \| |T_1(x)| = \cdots = |T_{n+1}(x)| = 1\}$ is homotopy equivalent to the n-dimensional sphere S^n. In §6.2 we show that the analytic space associated with a Chevalley group G or a compact homogeneous space G/H is contractible. In §6.3–6.5 we consider abelian varieties. Given an abelian variety A over k, we construct a closed subset $\Delta(A) \subset A^{\mathrm{an}}$ (the skeleton of A) which is a deformation retract of A^{an} and is homeomorphic to the quotient space of the real torus $(S^1)^n$ by an action of a finite group G. Moreover, the action of G on $(S^1)^n$ comes from a linear action of G on the universal covering $\mathbf{R}^n$ of $(S^1)^n$. We show that, for any quotient space $(S^1)^n/G$ of the above type, there exists an abelian variety A over k whose skeleton $\Delta(A)$ is homeomorphic to $(S^1)^n/G$. For example, if E is a twisted Tate elliptic curve, then E^{an} is contractible, but $(E \times E)^{\mathrm{an}}$ is homotopy equivalent to the two-dimensional sphere S^2. It follows from the construction that A^{an} has a universal covering. Furthermore, the first Betti number of A^{an} is equal to the dimension of A if and only if A^{an} is a split analytic torus. Ideally, it would be nice to show that all analytic spaces are locally contractible (as in the classical situation or in the one-dimensional case). In §6.6 we show that the two-dimensional closed disc $X = \mathcal{M}(k\{r_1^{-1}T_1, r_2^{-1}T_2\})$ is weakly locally contractible, i.e., for any point $x \in X$ and any open neighborhood $\mathcal{U}$ of x, there exists an open neighborhood $\mathcal{V}$ of x in $\mathcal{U}$ that can be contracted in $\mathcal{U}$ to a point.

In Chapter 7 we introduce and study some basic notions of spectral theory which are very similar to the classical ones (**[Bou2]**, **[GRS]**). Let $\mathcal{A}$ be a Banach algebra with identity. The spectrum of an element $f \in \mathcal{A}$ is the set Σ_f of points $x \in \mathbf{A}^1$ such that element $1 \otimes T(x) - f \otimes 1$ is noninvertible in the Banach algebra $\mathcal{A}\widehat{\otimes}\mathcal{H}(x)$. The function $x \mapsto R_f(x) = (1 \otimes T(x) - f \otimes 1)^{-1} \in \mathcal{A}\widehat{\otimes}\mathcal{H}(x)$ on $\mathbf{A}^1\backslash\Sigma_f$ is called the resolvent of f. Furthermore, if A is a bounded linear operator on a Banach space B over k, then the spectrum Σ_A and the resolvent R_A are the spectrum and resolvent of A considered as an element of the Banach algebra of bounded linear operators $\mathrm{End}(B)$. We prove that (i) the spectrum Σ_f is nonempty and compact in $\mathbf{A}^1$; (ii) the radius of the minimal closed disc with center at zero which contains Σ_f is equal to the spectral radius $\rho(f)$; (iii) the resolvent $x \mapsto R_f(x)$ is an analytic function on the set $\mathbf{P}^1\backslash\Sigma_f$ which is equal to zero at infinity. If the algebra $\mathcal{A}$ is commutative, the spectrum Σ_f coincides with the image of the map $\mathcal{M}(\mathcal{A}) \to \mathbf{A}^1$ induced by the homomorphism $k[T] \to \mathcal{A}$ which sends T to f, and the canonical map $\mathcal{M}(\mathcal{A}) \to \prod_{f \in \mathcal{A}} \Sigma_f$ induces a homeomorphism of $\mathcal{M}(\mathcal{A})$ with its image. In §7.3 we construct the holomorphic functional calculus (in a slightly more general form than its classical counterpart). The construction is parallel to the construction of the classical holomorphic functional calculus but, instead of real analysis, the apparatus of affinoid algebras is used. In §7.4 the holomorphic functional calculus is used to prove the Shilov Idempotent Theorem, which asserts that, for any open and closed subset $\mathscr{U}$ of the spectrum $\mathscr{M}(\mathscr{A})$ of a commutative Banach k-algebra with identity $\mathscr{A}$, there exists a unique idempotent $e \in \mathscr{A}$ which is equal to 1 on $\mathscr{U}$ and 0 outside $\mathscr{U}$. In §7.5 the discrete spectrum and the essential spectrum of a bounded linear operator are introduced.

In Chapter 8 we generalize Vishik's results from **[Vi]**. The first result is the description of distributions with compact support. Let Σ be a nonempty compact subset of $\mathbf{A}^1$, and let B be a Banach space over k. A B-valued distribution with support in Σ is a bounded linear operator $\Gamma(\Sigma, \mathscr{O}_{\mathbf{A}^1}) \to B$. Let $M: \Gamma(\Sigma, \mathscr{O}_{\mathbf{A}^1}) \to B$ be a distribution. Then the image of the function $x \mapsto (1 \otimes T(x) - T \otimes 1)^{-1}$ with respect to M belongs to the space $\Gamma_0(\mathbf{P}^1 \backslash \Sigma, \mathscr{O}_{\mathbf{P}^1}(B))$ of analytic functions on $\mathbf{P}^1 \backslash \Sigma$ with values in B which vanish at infinity. In this way one obtains a k-linear map

$$\mathscr{K}: \mathrm{Hom}(\Gamma(\Sigma, \mathscr{O}_{\mathbf{A}^1}), B) \to \Gamma_0(\mathbf{P}^1 \backslash \Sigma, \mathscr{O}_{\mathbf{P}^1}(B))$$

(the Cauchy transform). Theorem 8.1.1 asserts that $\mathscr{K}$ is an isomorphism. Theorem 8.1.3 describes the functions in $\Gamma_0(\mathbf{P}^1 \backslash \Sigma, \mathscr{O}_{\mathbf{P}^1}(B))$ which correspond to distributions that factor through $\mathscr{H}(\Sigma)$ (completion of $\Gamma(\Sigma, \mathscr{O}_{\mathbf{A}^1})$ with respect to the maximum modulus norm on Σ). Both facts are proved by Vishik in the case $\Sigma \subset k$. For the construction of the inverse map

$$\mathscr{E}: \Gamma_0(\mathbf{P}^1 \backslash \Sigma, \mathscr{O}_{\mathbf{P}^1}(B)) \to \mathrm{Hom}(\Gamma(\Sigma, \mathscr{O}_{\mathbf{A}^1}), B)$$

he used the Shnirelman integral. Our construction is more elementary. Vishik's second result is his perturbation theory, which in §8.3–8.4 is generalized to arbitrary bounded linear operators. Our account of the theory closely follows **[Vi]**. It consists of a theorem asserting that the essential spectrum of an operator does not change under perturbations by completely continuous operators (Theorem 8.3.2) and a trace formula (Theorem 8.4.1).

In Chapter 9 we study the dimension and stable dimension of a commutative Banach algebra. These notions give, in particular, some numerical characteristics of points of analytic spaces. One of the corollaries of the results obtained is the Stone-Weierstrass Theorem proved by Kaplansky (**[Ka]**). Here is another one. If $\mathscr{A}$ is a uniform commutative Banach algebra such that all of its characters take values in k, then the space $\mathscr{M}(\mathscr{A})$ is totally disconnected, and the Gel′fand transform induces an isomorphism of $\mathscr{A}$ onto the algebra of k-valued continuous functions on $\mathscr{M}(\mathscr{A})$.

Our interest in the subject was stimulated by the nice book of Yu. I. Lyubich **[Ly]** devoted to classical functional analysis and by Vishik's paper **[Vi]**. After reading **[Vi]**, we started investigating subsets of k on which the (usual) resolvent R_f of an element f of a Banach algebra $\mathscr{A}$ is Krasner analytic. It turned out that if R_f is an analytic element on an infraconnected set—this is a subset of k that is a closed disc minus a (possibly infinite) set of open discs and has the property of analytic continuation ([**Es**], [**Ro**])—then R_f is an analytic element on a strictly larger standard set. Therefore, every element $f \in \mathscr{A}$ determines a cofilter of subsets of k which are finite unions of standard sets (on which R_f is an analytic element) or, equivalently, the complementary filter of subsets of the same type. One can regard this filter as the spectrum of f. Furthermore, the maximal elements in the family of filters (or ultrafilters) form a locally compact space which can be regarded as the affine line (and, in fact, it coincides with the affine line $\mathbf{A}^1$ of this book). Using these objects, we extended Vishik's results to arbitrary bounded linear operators. After that, we found that they can be understood within the framework of non-Archimedean Gel′fand theory. Furthermore, the spectrum of a commutative Banach algebra given by this theory can be used to introduce analytic spaces endowed with a good topology.

CHAPTER 1

The Spectrum of a Commutative Banach Ring

1.1. Banach rings. Let M be an abelian group. A *seminorm* on M is a function $\| \ \|: M \to \mathbf{R}_+$ with values in the set of non-negative real numbers possessing the following properties: $\|0\| = 0$ and $\|f - g\| \le \|f\| + \|g\|$ for all $f, g \in M$. A seminorm is called *non-Archimedean* if $\|f - g\| \le \max(\|f\|, \|g\|)$ for all $f, g \in M$. Each seminorm $\| \ \|$ determines a topology on M. This topology is Hausdorff if and only if $\| \ \|$ is a *norm*, i.e., if the equality $\|f\| = 0$ holds only when $f = 0$. For any seminorm on M one can in the obvious way construct the completion of M with respect to this seminorm.

Two seminorms $\| \ \|$ and $\| \ \|'$ on M are called *equivalent* if there exist $C, C' > 0$ such that $C\|f\| \le \|f\|' \le C'\|f\|$ for all $f \in M$. Equivalent seminorms determine the same topology. If N is a subgroup of M, then one can define the *residue seminorm* on the quotient group M/N as follows: $\|f\| = \inf\{\|g\| | g \in \pi^{-1}(f)\}$, where π denotes the canonical homomorphism $M \to M/N$. The residue seminorm is a norm if and only if N is closed in M. If M is complete with respect to a norm and N is closed, then M/N is complete with respect to the residue norm.

Let $\varphi: M \to N$ be a homomorphism of seminormed groups. It is said to be *bounded* if there exists $C > 0$ such that $\|\varphi(f)\| \le C\|f\|$ for all $f \in M$. It is said to be *admissible* if the residue seminorm on $M/\mathrm{Ker}(\varphi)$ is equivalent to the restriction to $\mathrm{Im}(\varphi)$ (which is isomorphic to $M/\mathrm{Ker}(\varphi)$) of the seminorm of N.

Now let $\mathcal{A}$ be a ring with identity 1. A *seminorm* on $\mathcal{A}$ is a seminorm $\| \ \|$ on the additive group of $\mathcal{A}$ possessing the following properties: $\|1\| = 1$ and $\|fg\| \le \|f\| \cdot \|g\|$ for all $f, g \in \mathcal{A}$. A seminorm is called *power-multiplicative* (resp. *multiplicative*) if $\|f^n\| = \|f\|^n$ (resp. $\|fg\| = \|f\| \cdot \|g\|$) for all $f, g \in \mathcal{A}$, and $n \ge 1$. A multiplicative norm is called a *valuation.* A ring with a fixed seminorm (resp. norm) is called a *seminormed* (resp. *normed*) *ring.* A *Banach ring* is a normed ring $\mathcal{A}$ that is complete with respect to its norm.

1.1.1. *Examples.* (i) Any ring may be regarded as a Banach ring with respect to the *trivial norm* $| \ |_0$: $|f|_0 = 1$ for all $f \ne 0$.

(ii) The ring of integers $\mathbf{Z}$ is a Banach ring with respect to the *absolute value* $| \ |_\infty$.

(iii) If $\mathbf{a}$ is a proper closed two-sided ideal of a Banach ring $\mathcal{A}$, then the quotient ring $\mathcal{A}/\mathbf{a}$ is Banach with respect to the residue norm. Note that all maximal two-sided ideals of $\mathcal{A}$ are closed.

(iv) For a family of Banach rings $\{\mathcal{A}_i\}_{i\in I}$, let $\prod_{i\in I}\mathcal{A}_i$ denote the set of all elements $f = (f_i)_{i\in I}$, $f_i \in \mathcal{A}_i$, for which there exists $C > 0$ with $||f_i|| \le C$, $i \in I$. Then $\prod_{i\in I}\mathcal{A}_i$ is a Banach ring with respect to the norm $||f|| = \sup_{i\in I}||f_i||$.

(v) For a Banach ring $\mathcal{A}$ and a positive number r, let $\mathcal{A}\langle r^{-1}T\rangle$ denote the set of all power series $f = \sum_{i=0}^{\infty} a_i T^i$ such that $\sum_{i=0}^{\infty}||a_i||r^i < \infty$. Then $\mathcal{A}\langle r^{-1}T\rangle$ is a Banach ring with respect to the norm $||f|| = \sum_{i=0}^{\infty}||a_i||r^i < \infty$. Note that an element $1 - aT$, $a \in \mathcal{A}$, is invertible in $\mathcal{A}\langle r^{-1}T\rangle$ if and only if $\sum_{i=0}^{\infty}||a^i||r^i < \infty$.

A seminorm $|\ |$ on a Banach ring $\mathcal{A}$ is said to be *bounded* if there exists $C > 0$ such that $|f| \le C||f||$ for all $f \in \mathcal{A}$. If a bounded seminorm $|\ |$ is power-multiplicative, then one can set $C = 1$. Indeed, since $|f^n| \le C||f^n||$, we have $|f| \le \sqrt[n]{C}||f||$ for all $n \ge 1$ and hence, $|f| \le ||f||$.

A *valuation field* is a commutative Banach field whose norm is a valuation. A *non-Archimedean field* is a valuation field with non-Archimedean valuation. If a valuation field is not non-Archimedean, it is isomorphic to $\mathbf{R}$ or to $\mathbf{C}$ with the norm $|\ |_{\infty,\varepsilon}$, where $0 < \varepsilon \le 1$, $|f|_{\infty,\varepsilon} = |f|_\infty^\varepsilon$ and $|\ |_\infty$ is the absolute value. Such fields are said to be *Archimedean*. If $|\ |$ is the valuation of a normed field and $|\ |'$ is a valuation inducing the same topology, then there exists $a > 0$ such that $|f|' = |f|^a$ for all f. It is clear that if $|\ | \ne |\ |_0$, the number a is uniquely determined.

Let $\mathcal{A}$ be a normed ring. A *seminormed* (resp. *normed*) $\mathcal{A}$*-module* is an $\mathcal{A}$-module M with a seminorm (resp. norm) $||\ ||$ possessing the following property. There exists $C > 0$ such that $||fm|| \le C||f|| \cdot ||m||$ for all $f \in \mathcal{A}$ and $m \in M$. Note that in this situation there exists an equivalent seminorm (resp. norm) $||\ ||'$ for which the inequality is valid with $C = 1$. A *Banach* $\mathcal{A}$*-module* is a complete normed $\mathcal{A}$-module.

Finally, we recall the notion of *complete tensor product.* Let M and N be seminormed $\mathcal{A}$-modules. One can define the following seminorm on $M\otimes_{\mathcal{A}}N$: $||f|| = \inf\sum_i||m_i||\cdot||n_i||$, where the infimum is taken over all representations $f = \sum m_i\otimes n_i$. The completion of $M\otimes_{\mathcal{A}}N$ with respect to this seminorm is denoted by $M\widehat{\otimes}_{\mathcal{A}}N$. Note that if $M' \to M$ and $N' \to N$ are admissible homomorphisms of seminormed $\mathcal{A}$-modules, then the induced homomorphism $M'\otimes_{\mathcal{A}}N' \to M\otimes_{\mathcal{A}}N$ is admissible. Beginning with Chapter 2, only non-Archimedean Banach rings and modules are considered. For them we use another definition. Namely, $M\widehat{\otimes}_{\mathcal{A}}N$ will denote the completion of $M\otimes_{\mathcal{A}}N$ with respect to the seminorm $||f|| = \inf\max_i||m_i||\cdot||n_i||$.

1.2. The spectrum. Let $\mathcal{A}$ be a commutative Banach ring with identity. The *spectrum* $\mathcal{M}(\mathcal{A})$ is the set of all bounded multiplicative seminorms on

$\mathscr{A}$ provided with the weakest topology with respect to which all real valued functions on $\mathscr{M}(\mathscr{A})$ of the form $|\ | \mapsto |f|$, $f \in \mathscr{A}$, are continuous.

1.2.1. THEOREM. *The spectrum $\mathscr{M}(\mathscr{A})$ is a nonempty, compact Hausdorff space.*

1.2.2. REMARKS. (i) Let x be a point of $\mathscr{M}(\mathscr{A})$ and let $|\ |$ be the corresponding seminorm. The kernel $\wp_x$ of $|\ |$ is a closed prime ideal of $\mathscr{A}$. The value $|f|$ depends only on the residue class of f in $\mathscr{A}/\wp_x$. The resulting valuation on the integral domain $\mathscr{A}/\wp_x$ extends to a valuation on its fraction field F. The closure of F with respect to the valuation is a valuation field denoted by $\mathscr{H}(x)$. The image of an element $f \in \mathscr{A}$ in $\mathscr{H}(x)$ will be denoted by $f(x)$. The homomorphism

$$\wedge : \mathscr{A} \to \prod_{x \in \mathscr{M}(\mathscr{A})} \mathscr{H}(x)$$

which sends f to the element $\hat{f} = (f(x))_{x \in \mathscr{M}(\mathscr{A})}$ is called the *Gel'fand transform.*

(ii) A nonzero bounded homomorphism χ from $\mathscr{A}$ to a valuation field K is called a *character* of $\mathscr{A}$. Two characters $\chi' : \mathscr{A} \to K'$ and $\chi'' : \mathscr{A} \to K''$ are said to be *equivalent* if there exist a character $\chi : \mathscr{A} \to K$ and embeddings $K \to K'$ and $K \to K''$ such that the diagram

$$\begin{array}{ccccc} & & \mathscr{A} & & \\ & \swarrow & \downarrow & \searrow & \\ K' & \longleftarrow & K & \longrightarrow & K'' \end{array}$$

is commutative. Using the construction in (i), it is easy to show that $\mathscr{M}(\mathscr{A})$ coincides with the set of equivalence classes of characters of $\mathscr{A}$.

(iii) Any bounded homomorphism of commutative Banach rings $\varphi : \mathscr{A} \to \mathscr{B}$ sending the identity to the identity induces a continuous map $\varphi^* : \mathscr{M}(\mathscr{B}) \to \mathscr{M}(\mathscr{A})$. If the set of elements of the form $\varphi(f)/\varphi(g)$, where $f, g \in \mathscr{A}$ and $\varphi(g)$ is invertible in $\mathscr{B}$, is everywhere dense in $\mathscr{B}$, the map φ^* is injective.

PROOF OF THEOREM 1.2.1. First we have to show that the space $\mathscr{M}(\mathscr{A})$ is nonempty. Replacing $\mathscr{A}$ by the quotient ring $\mathscr{A}/\mathbf{m}$, where $\mathbf{m}$ is a maximal ideal of $\mathscr{A}$, we may assume that $\mathscr{A}$ is a field. Let S denote the set of all nonzero bounded seminorms on $\mathscr{A}$. The set S is nonempty because the norm of $\mathscr{A}$ belongs to S. It is partially ordered with respect to the ordering for which $|\ |' \le |\ |''$ if $|f|' \le |f|''$ for all $f \in \mathscr{A}$. This ordering satisfies the conditions of Zorn's Lemma; therefore, there exist minimal elements in S. Let $|\ |$ be a minimal element in S. Replacing $\mathscr{A}$ by the completion with respect to $|\ |$, we may assume that $|\ |$ is the norm of $\mathscr{A}$. We claim that the norm $|\ |$ is multiplicative.

Suppose that there exists an element $f \in \mathscr{A}$ with $|f^n| < |f|^n$ for some $n > 1$. We claim that the element $f - T$ is noninvertible in the Banach ring $\mathscr{A}\langle r^{-1}T\rangle$ with $r = \sqrt[n]{|f^n|}$. By Remark 1.1.1(v), it suffices to show that the series $\sum_{i=0}^{\infty} |f^{-i}|r^i$ does not converge. If $i = pn + q$, where $0 \le q \le n-1$, then $|f^i| \le |f^n|^p |f^q|$ and

$$|f^{-i}|r^i \ge |f^i|^{-1} |f^n|^{p+\frac{q}{n}} \ge \frac{|f^n|^{\frac{q}{n}}}{|f^q|}.$$

Hence, $|f^{-i}|r^i \ge \varepsilon > 0$ for all $i \ge 0$, where $\varepsilon = \min\{|f^n|^{q/n}/|f^q| \,|\, 0 \le q \le n-1\}$. Consider the homomorphism $\varphi: \mathscr{A} \to \mathscr{A}\langle r^{-1}T\rangle/(f-T)$. It is injective, since $\mathscr{A}$ is a field, and $\|\varphi(f)\| = \|T\| \le r < |f|$. But this is impossible because $|\ |$ is a minimal bounded seminorm on $\mathscr{A}$.

Now suppose that there exists an element $f \in \mathscr{A}\backslash\{0\}$ with $|f|^{-1} < |f^{-1}|$. We claim that the element $f - T$ is noninvertible in $\mathscr{A}\langle r^{-1}T\rangle$ with $r = |f^{-1}|^{-1}$. Indeed, since $|f^{-i}|r^i = |f^{-1}|^i |f^{-1}|^{-i} = 1$, the series $\sum_{i=0}^{\infty} |f^{-i}|r^i$ does not converge. As above, for the homomorphism $\varphi: \mathscr{A} \to \mathscr{A}\langle r^{-1}T\rangle/(f-T)$ we have $\|\varphi(f)\| = \|T\| \le r < |f|$. Hence, the norm $|\ |$ is multiplicative.

Let x' and x'' be two different points of $\mathscr{M}(\mathscr{A})$. Then there exists an element $f \in \mathscr{A}$ with $|f(x')| \ne |f(x'')|$. Suppose that $|f(x')| < |f(x'')|$. We choose a real number r with $|f(x')| < r < |f(x'')|$ and put $\mathscr{U}' = \{x \in \mathscr{M}(\mathscr{A}) \,|\, |f(x)| < r\}$, $\mathscr{U}'' = \{x \in \mathscr{M}(\mathscr{A}) \,|\, |f(x)| > r\}$. The sets $\mathscr{U}'$ and $\mathscr{U}''$ are disjoint open neighborhoods of the points x' and x'', respectively. Hence, the space $\mathscr{M}(\mathscr{A})$ is Hausdorff.

To prove that the space $\mathscr{M}(\mathscr{A})$ is compact, consider the Gel′fand transform:

$$\wedge : \mathscr{A} \to \mathscr{B} = \prod_{x \in \mathscr{M}(\mathscr{A})} \mathscr{H}(x).$$

It is clear that the induced map $\mathscr{M}(\mathscr{B}) \to \mathscr{M}(\mathscr{A})$ is surjective. Hence, the compactness of $\mathscr{A}$ follows from following

1.2.3. Proposition. *Let $\{K_i\}_{i \in I}$ be a family of valuation fields. Then the spectrum $\mathscr{M}(\mathscr{A})$ of the Banach ring $\mathscr{A} = \prod_{i \in I} K_i$ is homeomorphic to the Stone-Čech compactification of the* (*discrete*) *set I.*

Proof. Recall that a *filter* on I is a family Φ of nonempty subsets of I such that: (a) if $A \in \Phi$ and $A \subset B$, then $B \in \Phi$; (b) if $A, B \in \Phi$, then $A \cap B \in \Phi$. The set of filters is partially ordered by inclusion. Its maximal elements are called *ultrafilters*. The *Stone-Čech compactification* of I is the set of ultrafilters endowed with the topology whose open basis is formed by subsets of the form $\{\Phi | J \in \Phi\}$, $J \subset I$. This space is Hausdorff and compact.

First we construct a one-to-one correspondence between the set of proper closed ideals of $\mathscr{A}$ and the set of filters on I. For a subset $J \subset I$, we denote by a_J the element of $\mathscr{A}$ with $a_{J,i} = 0$ if $i \in J$, and $a_{J,i} = 1$ if $i \notin J$. Let $\mathbf{a}$ be a proper closed ideal of $\mathscr{A}$. Then the set $\Phi_{\mathbf{a}} = \{J \subset I | a_J \in \mathbf{a}\}$ is a filter on I. Conversely, given a filter Φ on I, we denote by $\mathbf{a}_\Phi$ the closed ideal of $\mathscr{A}$ generated by the elements a_J for $J \in \Phi$. It is easy to verify that the maps $\mathbf{a} \mapsto \Phi_{\mathbf{a}}$ and $\Phi \mapsto \mathbf{a}_{\Phi}$ are inverse to one another and that maximal ideals correspond to ultrafilters.

To prove the proposition, it suffices to verify the following facts:

(1) all prime ideals of $\mathscr{A}$ are maximal;

(2) if $\mathbf{m}$ is a maximal ideal of $\mathscr{A}$, then the residue norm on the field $\mathscr{A}/\mathbf{m}$ is multiplicative;

(3) the topology of $\mathscr{M}(\mathscr{A})$ corresponds to that of the Stone-Čech compactification of I.

Let $\wp$ be a prime ideal of $\mathscr{A}$, and suppose that there exists a maximal ideal $\mathbf{m}$ with $\wp \subset \mathbf{m}$ and $\wp \neq \mathbf{m}$. Let $J \in \Phi_{\mathbf{m}} \backslash \Phi_{\mathbf{a}}$. Then $a_J \in \mathbf{m} \backslash \wp$. Since $I \backslash J \notin \Phi_{\mathbf{m}}$, it follows that $a_{I\backslash J} \notin \mathbf{m}$. We have $a_J \cdot a_{I\backslash J} = 0$; but $a_J \notin \wp$ and $a_{I\backslash J} \notin \wp$.

Let $\mathbf{m}$ be a maximal ideal of $\mathscr{A}$. It is easy to verify that for $f \in \mathscr{A}$

$$||\pi(f)|| = \inf_{J\in\Phi_{\mathbf{m}}} \sup_{i\in J} |f_i|,$$

where π denotes the canonical map $\mathscr{A} \to \mathscr{A}/\mathbf{m}$. From this it follows that $||\pi(fg)|| = ||\pi(f)|| \cdot ||\pi(g)||$.

To verify (3) it suffices to note that, if Φ is an ultrafilter and $\mathbf{m} = \mathbf{a}_{\Phi}$, then $||\pi(a_J)|| = 0$ if $J \in \Phi$, and $||\pi(a_J)|| = 1$ if $J \notin \Phi$. ∎

1.2.4. COROLLARY. *An element f of a commutative Banach ring $\mathscr{A}$ is invertible if and only if $f(x) \neq 0$ for all $x \in \mathscr{M}(\mathscr{A})$.* ∎

1.2.5. REMARKS. (i) There is a natural continuous map

$$\mathscr{M}(\mathscr{A}) \to \mathrm{Spec}(\mathscr{A}) : x \mapsto \wp_x,$$

where $\wp_x$ is the kernel of the seminorm corresponding to the point x. For example, if the norm of $\mathscr{A}$ is trivial, this map is surjective.

(ii) The proof of the nonemptiness of the spectrum gives the following slightly more general fact. Let $\mathscr{B}$ be a Banach ring, and let $\mathscr{A}$ be a closed subring containing the identity of $\mathscr{B}$ and lying in its center. Then there exists a bounded seminorm on $\mathscr{B}$ whose restriction to $\mathscr{A}$ is multiplicative.

1.3. Properties of the spectrum. Let $\mathscr{A}$ be a Banach ring and $f \in \mathscr{A}$. The *spectral radius* of f is the number

$$\rho(f) = \lim_{n\to\infty} \sqrt[n]{||f^n||} = \inf_n \sqrt[n]{||f^n||}.$$

The existence of the limit and its equality with the infimum are well known (see **[Bou2]**, 1.2.3). Now suppose that $\mathscr{A}$ is commutative. Recall that $\widehat{f}$ denotes the image of f under the Gel′fand transform. One has:

$$||\widehat{f}|| = \max_{x\in\mathscr{M}(\mathscr{A})} |f(x)|.$$

1.3.1. THEOREM. *For any element $f \in \mathscr{A}$ of a commutative Banach ring $\mathscr{A}$, one has the equality $\rho(f) = ||\widehat{f}||$.*

PROOF. The inequality $\|\widehat{f}\| \le \rho(f)$ is obvious. To verify the reverse inequality it suffices to show that if $|f(x)| < r$ for all $x \in \mathcal{M}(\mathcal{A})$, then $\rho(f) < r$.

Consider the Banach ring $\mathcal{B} = \mathcal{A}\langle rT\rangle$. Since $\|T\| = r^{-1}$, we have $|T(x)| \le r^{-1}$ for all $x \in \mathcal{M}(\mathcal{B})$. Therefore, $|(fT)(x)| < 1$. In particular, $(1 - fT)(x) \ne 0$ for all $x \in \mathcal{M}(\mathcal{B})$. By Corollary 1.2.4, the element $1 - fT$ is invertible in $\mathcal{B}$. From Remark 1.1.1(v) it follows that the series $\sum_{i=0}^{\infty} \|f^i\| r^{-i}$ is convergent. Hence, $\rho(f) < r$. ∎

1.3.2. COROLLARY. (i) *The kernel of the Gel'fand transform coincides with the set of quasinilpotent elements* (*i.e. elements* f *with* $\rho(f) = 0$).

(ii) *The Gel'fand transform is isometric if and only if* $\mathcal{A}$ *is uniform.* ∎

Recall that $\mathcal{A}$ is called *uniform* if its norm is power multiplicative. This is equivalent to the relation $\|f^2\| = \|f\|^2$ for all $f \in \mathcal{A}$.

1.3.3. COROLLARY. *For any two commuting elements* f *and* g *of a Banach ring, one has* $\rho(f+g) \le \rho(f) + \rho(g)$ *and* $\rho(fg) \le \rho(f)\rho(g)$. ∎

From Corollary 1.3.3 it follows that the function $f \mapsto \rho(f)$ on a commutative Banach ring $\mathcal{A}$ is a bounded seminorm. Let $\mathcal{A}^u$ denote the completion of $\mathcal{A}$ with respect to this seminorm. This is a uniform ring, and any bounded homomorphism $\mathcal{A} \to \mathcal{B}$ from $\mathcal{A}$ to a uniform ring $\mathcal{B}$ factors through a homomorphism $\mathcal{A}^u \to \mathcal{B}$. Moreover, $\mathcal{M}(\mathcal{A}^u) \xrightarrow{\sim} \mathcal{M}(\mathcal{A})$.

1.3.4. COROLLARY. (i) *The spectrum* $\mathcal{M}(\mathcal{A})$ *consists of one point if and only if* $\mathcal{A}^u$ *is a valuation field.*

(ii) $\mathcal{A}^u$ *is a non-Archimedean ring if and only if the fields* $\mathcal{H}(x)$ *are non-Archimedean for all* $x \in \mathcal{M}(\mathcal{A})$. ∎

The situation of Corollary 1.3.4(ii) can occur, for example, if the ring $\mathcal{A}$ contains a non-Archimedean field whose norm is induced by the norm of $\mathcal{A}$.

1.3.5. PROPOSITION. *Suppose that* $\mathcal{A}$ *is a commutative Banach algebra over a valuation field* k, *and let* K *be a finite normal extension of* k.

(i) *If* K *is separable over* k, *then* $\mathcal{M}(\mathcal{A} \otimes K)/\mathrm{Gal}(K/k) \xrightarrow{\sim} \mathcal{M}(\mathcal{A})$.

(ii) *If* K *is purely inseparable over* k, *then* $\mathcal{M}(\mathcal{A} \otimes K) \xrightarrow{\sim} \mathcal{M}(\mathcal{A})$.

PROOF. The inverse image of a character $\chi: \mathcal{A} \to \mathcal{H}(x)$ in $\mathcal{M}(\mathcal{A} \otimes K)$ coincides with the image of $\mathcal{M}(\mathcal{H}(x) \otimes K)$ in $\mathcal{M}(\mathcal{A} \otimes K)$. If K is separable over k, the group $\mathrm{Gal}(K/k)$ acts transitively on the set $\mathcal{M}(\mathcal{H}(x) \otimes K)$, which coincides with the set of maximal ideals in $\mathcal{H}(x) \otimes K$. If K is purely inseparable over k, then $\mathcal{H}(x) \otimes K$ is a local algebra. ∎

Let $\widehat{k^a}$ denote the completion of the algebraic closure k^a of k. This is an algebraically closed field, and the Galois group $\mathrm{Gal}(k^a/k)$ acts on it.

1.3.6. COROLLARY. $\mathcal{M}(\mathcal{A} \widehat{\otimes} \widehat{k^a})/\mathrm{Gal}(k^a/k) \xrightarrow{\sim} \mathcal{M}(\mathcal{A})$. ∎

1.4. Examples

1.4.1. Consider the ring of integers $\mathbf{Z}$ endowed with the absolute value $|\ |_\infty$. (Note that $|\ |_\infty$ is the maximal seminorm which can be defined on $\mathbf{Z}$.) By Ostrowski's Theorem, any multiplicative seminorm on $\mathbf{Z}$ is one of the following types: $|\ |_{\infty,\varepsilon}$ $(0 < \varepsilon \le 1)$, $|\ |_0$, $|\ |_{p,\varepsilon}$ $(0 < \varepsilon < 1)$ and $|\ |_{p,0}$ (p is a prime). Here $|\ |_{\infty,\varepsilon}$ is the ε-power of $|\ |_\infty$, $|\ |_{p,\varepsilon}$ is the p-adic norm for which $|p| = \varepsilon$, and $|\ |_{p,0}$ is the seminorm induced by the trivial norm on $\mathbf{Z}/p\mathbf{Z}$. The spectrum $\mathcal{M}(\mathbf{Z})$ has the following form:

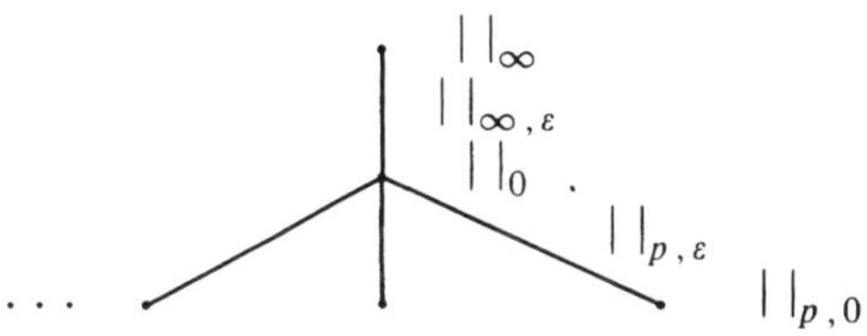

Each interval is embedded in $\mathcal{M}(\mathbf{Z})$ homeomorphically, and open neighborhoods of the point $|\ |_0$ contain almost all intervals. The map $\mathcal{M}(\mathbf{Z}) \to \operatorname{Spec}(\mathbf{Z})$ sends the lower ends of the prime number intervals to the corresponding closed points and all other points to the generic point of $\operatorname{Spec}(\mathbf{Z})$.

One similarly describes the spectrum of the ring of integers $\mathcal{A}$ in a number field K endowed with the norm $\|f\| = \max|\sigma(f)|_\infty$, where σ runs through the Archimedean places $\sigma: K \to \mathbf{C}$ (this is the maximal power multiplicative seminorm which can be defined on $\mathcal{A}$). The space $\mathcal{M}(\mathcal{A})$ is a union of intervals connecting the point $|\ |_0$ with the Archimedean places of K and with the seminorms $|\ |_{\wp,0}$ for all maximal ideals $\wp \subset \mathcal{A}$.

1.4.2. Let $\mathcal{A}$ be an arbitrary commutative ring endowed with the trivial norm. The map $\mathcal{M}(\mathcal{A}) \to \operatorname{Spec}(\mathcal{A})$ in Remark 1.2.5(i) has a canonical section $\operatorname{Spec}(\mathcal{A}) \to \mathcal{M}(\mathcal{A})$ sending a prime ideal $\wp \subset \mathcal{A}$ to the multiplicative seminorm which is induced by the trivial norm of $\mathcal{A}/\wp$.

Suppose that $\mathcal{A}$ is a Dedekind domain. Then the space $\mathcal{M}(\mathcal{A})$ is a union of intervals connecting the generic point of $\operatorname{Spec}(\mathcal{A})$ with all points in the maximal spectrum $\operatorname{Max}(\mathcal{A})$. More generally, let $\mathcal{A}$ be a one-dimensional domain whose integral closure $\mathcal{A}'$ is a Dedekind ring. Then $\mathcal{M}(\mathcal{A})$ is obtained from $\mathcal{M}(\mathcal{A}')$ by identifying the points of $\operatorname{Max}(\mathcal{A}')$ which have common images in $\operatorname{Max}(\mathcal{A})$. In particular, if n_x denotes the number of inverse images of a point $x \in \operatorname{Max}(\mathcal{A})$, then $\mathcal{M}(\mathcal{A})$ is homotopy equivalent to the wedge of $\sum_{x\in\operatorname{Max}(\mathcal{A})}(n_x - 1)$ circles. The structure of $\mathcal{M}(\mathcal{A})$ in the general case is complicated.

1.4.3. Let k be a valuation field, and let $\mathcal{A} = k\langle r^{-1}T\rangle$. The spectral radius of an element $f = \sum_{i=0}^\infty a_i T^i \in \mathcal{A}$ is described as follows: (1) $\rho(f) = \max\{|f(z)| \,|\, z \in \mathbf{C} \text{ with } |z| \le r\}$ if k is Archimedean, and (2) $\rho(f) = \max_{i\ge0}|a_i|r^i$ if k is non-Archimedean. Indeed, let $|f|$ denote the number defined on the right side. Since the function $f \mapsto |f|$ is a bounded power-multiplicative norm on $\mathcal{A}$, we have $|f| \le \rho(f)$. It suffices to verify the reverse inequality for polynomials. Let $f = \sum_{i=0}^d a_i T^i$. In both cases one has $|a_i| \le r^{-i}|f|$, $0 \le i \le d$. (This is obvious in case (2) and follows from Cauchy's formula in case (1).) Hence, $\|f\| \le (d+1)|f|$. Applying this inequality for f^n, we obtain $\sqrt[n]{\|f^n\|} \le \sqrt[n]{nd+1}|f|$. Therefore, $\rho(f) \le |f|$. The uniform completion $\mathscr{A}^u$ is described as follows. If $k = \mathbf{C}$, $\mathscr{A}^u$ is the algebra of continuous functions on the closed disc of radius r with center 0 that are analytic in the open disc. If $k = \mathbf{R}$, $\mathscr{A}^u$ is the subalgebra of the algebra for $k = \mathbf{C}$ consisting of functions with real Taylor coefficients. If k is non-Archimedean, $\mathscr{A}^u = k\{r^{-1}T\} = \{f = \sum_{i=0}^{\infty} a_i T^i \mid \sup |a_i| r^i < \infty\}$ with the norm $\|f\| = \sup |a_i| r^i$.

1.4.4. Let $\mathscr{A} = k\{r^{-1}T\}$, where k is a non-Archimedean field, and set $X = \mathscr{M}(\mathscr{A})$. If the valuation on k is trivial, the space X has the same structure as the spectrum of $\mathbf{Z}$ when $r \ge 1$ and is homeomorphic to the unit interval when $r < 1$. So we suppose that the valuation on k is nontrivial. To describe X it suffices to suppose that k is algebraically closed (see Corollary 1.3.6). For $a \in k$ and $\rho \in \mathbf{R}$ with $|a| \le r$ and $0 < \rho \le r$, we set $E(a, \rho) = \{x \in X \mid |(T-a)(x)| \le \rho\}$ (closed disc) and $D(a, \rho) = \{x \in X \mid |(T-a)(x)| < \rho\}$ (open disc). For example, $E(0, r) = X$.

Every element $a \in k$ with $|a| \le r$ determines a multiplicative seminorm $k\{T\} \to \mathbf{R}_+ : f \mapsto |f(a)|$. The corresponding point of X is called a point of *type (1)*.

Furthermore, every closed disc $E = E(a, \rho) \subset X$ determines a multiplicative seminorm $f \mapsto |f|_E = \max_n |a_n| \rho^n$, where $\sum_{n=0}^{\infty} a_n (T-a)^n$ is the expansion of f with center a. The corresponding point of X is denoted by $p(E)$ and is called a point of *type (2)* if $\rho \in |k|$ and of *type (3)* if $\rho \notin |k|$.

Finally, let $\mathscr{E} = \{E^{(\rho)}\}$ be a family of embedded closed discs in X. That is, $E^{(\rho)}$ is a closed disc of radius ρ and $E^{(\rho)} \supset E^{(\rho')}$ if $\rho > \rho'$. Then $\mathscr{E}$ determines a multiplicative seminorm $f \mapsto |f|_{\mathscr{E}} = \inf |f|_{E^{(\rho)}}$. The corresponding point of X is denoted by $p(\mathscr{E})$. Let σ denote the intersection of all $E^{(\rho)} \cap k$ in k. If σ is nonempty, then σ is a point $a \in k$ and $p(\mathscr{E}) = a$, or else σ is equal to $E \cap k$ for some closed disc $E \subset X$ and $p(\mathscr{E}) = p(E)$. If σ is empty, we obtain a new point of X which is called a point of *type (4)*.

Points of type (4) occur only in the case when the field k is not maximally complete. We recall ([**Bou1**], Ch. IV, §10, ex. 2) that a non-Archimedean field is said to be *maximally complete* if every family of embedded discs has a nonempty intersection. This condition is equivalent to the property that the field has no extensions with the same residue field and the same set of valuation values in $\mathbf{R}_+$ (*immediate extensions*). For example, if a field contains a countable everywhere dense subset and its valuation is not discrete, then it is not maximally complete.

We claim that every point of X can be obtained in the manner described above. Indeed, let $|\ |$ be a multiplicative seminorm on $k\{r^{-1}T\}$. Consider the family of closed discs $\mathscr{E} = \{E(a, |T-a|) \mid a \in k \text{ with } |a| \le r\}$. It is easy to verify that if $|T-a| \le |T-b|$, then $E(a, |T-a|) \subset E(b, |T-b|)$. Hence, $\mathscr{E}$ is a family of embedded discs. By construction, $|T-a| = |T-a|_{\mathscr{E}}$ for all $a \in k$. But multiplicative seminorms on $k\{r^{-1}T\}$ are completely determined by their values on the polynomials of the type $T-a$, $a \in k$. Therefore, $|\ | = |\ |_{\mathscr{E}}$.

We list some properties of the fields $\mathscr{H}(x)$ for $x \in X$.

(1) If x is of type (1), then $\mathscr{H}(x) = k$.

(2) If x is of type (2), then $\widetilde{\mathscr{H}(x)} \approx \tilde{k}(T)$ and $|\mathscr{H}(x)| = |k|$.

(3) If $x = p(E(a, \rho))$ with $\rho \notin |k|$, then $\widetilde{\mathscr{H}(x)} = \tilde{k}$ and $|\mathscr{H}(x)^*|$ is generated by $|k^*|$ and ρ.

(4) If x is of type (4), $\mathscr{H}(x)$ is an immediate extension of k. Here $\tilde{k}$ and $\widetilde{\mathscr{H}(x)}$ denote the residue fields of k and $\mathscr{H}(x)$, respectively.

1.5. Analytic spaces over a commutative Banach ring. Let k be a commutative Banach ring.

1.5.1. DEFINITION. The *n-dimensional affine space* $\mathbf{A}^n = \mathbf{A}^n_k$ *over* k is the set of all multiplicative seminorms on the ring of polynomials $k[T_1, \dots, T_n]$ whose restriction to k is bounded. The set $\mathbf{A}^n$ is provided with the weakest topology with respect to which all real valued functions on $\mathbf{A}^n$ of the form $|\ | \mapsto |f|$, $f \in k[T_1, \dots, T_n]$, are continuous.

1.5.2. REMARKS. (i) The space $\mathbf{A}^n$ is locally compact. Indeed, one has $\mathbf{A}^n = \cup E(0, r)$, where

$$r = (r_1, \dots, r_n) \quad \text{and} \quad E(0, r) = \mathscr{M}(\mathscr{A}\langle r_1^{-1}T_1, \dots, r_n^{-1}T_n\rangle).$$

(ii) Each point $x \in \mathbf{A}^n$ determines a valuation field $\mathscr{H}(x)$. The image of an element $f \in k[T_1, \dots, T_n]$ in $\mathscr{H}(x)$ is denoted by $f(x)$.

(iii) There is a canonical projection $\pi: \mathbf{A}^n \to \mathscr{M}(k) = \mathbf{A}^0$. If $x \in \mathscr{M}(k)$, then $\pi^{-1}(x) = \mathbf{A}^n_{\mathscr{H}(x)}$.

Further, let $\mathscr{K}_n$ denote the ring of fractions of $k[T_1, \dots, T_n]$. We say that an element $f \in \mathscr{K}_n$ is *defined* on a subset $\mathscr{U} \subset \mathbf{A}^n$ if f can be represented in the form $f = g/h$ with $g, h \in k[T_1, \dots, T_n]$ and $h(x) \neq 0$ for all $x \in \mathscr{U}$. In this case one can define the *value* of f at x: $f(x) = g(x)/h(x) \in \mathscr{H}(x)$.

1.5.3 DEFINITION. An *analytic function* on a subset $\mathscr{U} \subset \mathbf{A}^n$ is a rule which associates to every point $x \in \mathscr{U}$ an element $f(x) \in \mathscr{H}(x)$ such that f is a local limit of rational functions. That is, for each $x \in \mathscr{U}$ there exists an open neighborhood $\mathscr{U}'$ of x in $\mathscr{U}$ such that, for any $\varepsilon > 0$, there exists an element $g \in \mathscr{K}_n$ defined on $\mathscr{U}'$ with $|f(x') - g(x')| < \varepsilon$ for all $x' \in \mathscr{U}'$.

It is easy to see that the correspondence $\mathscr{U} \mapsto \mathscr{O}(\mathscr{U})$ (the set of analytic functions on $\mathscr{U}$) is a sheaf of local rings.

1.5.4. EXAMPLES. (i) If $k = \mathbf{C}$, $\mathbf{A}^n$ is the complex analytic affine space $\mathbf{C}^n$.

(ii) If $k = \mathbf{R}$, $\mathbf{A}^n$ is the quotient of $\mathbf{C}^n$ by complex conjugation (the quotient is taken in the category of locally ringed spaces).

(iii) If k is a non-Archimedean field, $\mathbf{A}^n$ is the affine space considered in this paper.

(iv) Let $k = \mathbf{Z}$ be provided with the usual absolute value. The space $\mathbf{A}^0 = \mathscr{M}(\mathbf{Z})$ is described in 1.4.1. The local rings $\mathscr{O}_x$ of the sheaf $\mathscr{O}_{\mathbf{A}^0}$ are the following: $\mathscr{O}_x = \mathbf{Q}$ if $x = |\ |_0$, $\mathscr{O}_x = \mathbf{R}$ if x lies in an Archimedean interval, $\mathscr{O}_x = \mathbf{Q}_p$ if x lies in the interior of a p-adic interval, $\mathscr{O}_x = \mathbf{Z}_p$ if x is the end of a p-adic interval. The local ring of the point $x_0 \in \mathbf{A}^1$ corresponding to zero on the affine line $\mathbf{A}^1_{\mathbf{Q}} = \pi^{-1}(|\ |_0)$, consists of all power series $f \in \mathbf{Q}[[T]]$ which are p-integral for almost all primes p and, for all places of $\mathbf{Q}$, converge in a small neighborhood of zero in the completion of $\mathbf{Q}$ with respect to this place.

Having defined affine spaces as analytic spaces over k, one can define arbitrary analytic spaces by a straightforward generalization of the definition of complex analytic spaces. Namely, these spaces are locally ringed spaces with an open covering and with a system of isomorphisms between each element of the covering and a local model. The latter objects are defined as follows. Let $\mathscr{U}$ be an open subset of $\mathbf{A}^n$ and let $\mathscr{J}$ be a subsheaf of ideals in $\mathscr{O}_{\mathscr{U}} = \mathscr{O}_{\mathbf{A}^n}|_{\mathscr{U}}$ which is of finite type over $\mathscr{O}_{\mathscr{U}}$. The local model associated with these data is the support X of the sheaf $\mathscr{O}_{\mathscr{U}}/\mathscr{J}$ with sheaf $\mathscr{O}_X$, which is the restriction of $\mathscr{O}_{\mathscr{U}}/\mathscr{J}$ to X.

1.5.5. Remark. In the case when k is a non-Archimedean field the previous definition does not give all reasonable analytic spaces (for example, affinoid spaces). So we slightly improve the definition by extending the class of local models as follows (k is arbitrary). We note that Definition 1.5.3 gives the notion of an analytic function on an arbitrary subset of $\mathbf{A}^n$. In particular, one has a sheaf of analytic functions on a closed disc $E(0, r) \subset \mathbf{A}^n$. We define the local models as above, but proceeding from the open subsets of closed discs. In the case of a non-Archimedean field one obtains all of the analytic spaces studied in this paper. But in the case $k = \mathbf{C}$ (and also $k = \mathbf{R}$) the analytic spaces defined in this way include real manifolds. (For example, if $k = \mathbf{C}$ and $n = 2$, the subsheaf of ideals of $\mathscr{O}_{E(0,1)}$ generated by the function $T_1T_2 - 1$ determines the local model $X = \{x \in \mathbf{C}^2 \,|\, |T_1(x)| = |T_2(x)| = 1\}$ homeomorphic to the unit circle with the sheaf of complex valued continuous functions.) It seems that such analytic spaces are worth studying.

CHAPTER 2

Affinoid Spaces

Beginning with this section, we fix a non-Archimedean field k with a valuation $| \ |$ which we do not assume to be nontrivial. One sets $\sqrt{|k^*|} = \{\alpha \in \mathbf{R}_+ | \alpha^n \in |k^*| \text{ for some } n \geq 1\}$.

2.1. Affinoid algebras. For $r_1, \dots, r_n > 0$, we set:

$$k\{r_1^{-1}T_1, \dots, r_n^{-1}T_n\} = \{f = \sum_{\nu=0}^{\infty} a_\nu T^\nu | a_\nu \in k \text{ and } |a_\nu| r^\nu \to 0 \text{ as } |\nu| \to \infty\}$$

(here $\nu = (\nu_1, \dots, \nu_n)$, $|\nu| = \nu_1 + \cdots + \nu_n$, $T^\nu = T_1^{\nu_1} \dots T_n^{\nu_n}$ and $r^\nu = r_1^{\nu_1} \dots r_n^{\nu_n}$). This is a commutative Banach k-algebra with respect to the multiplicative norm $||f|| = \max_\nu |a_\nu| r^\nu$. For brevity this algebra will also be denoted by $k\{r^{-1}T\}$.

For example, suppose that the valuation on k is trivial. If $r_i \geq 1$ for all $1 \leq i \leq n$, $k\{r^{-1}T\}$ coincides with the ring of polynomials $k[T_1, \dots, T_n]$. If $r_i < 1$ for all $1 \leq i \leq n$, then $k\{r^{-1}T\}$ coincides with the ring of formal power series $k[[T_1, \dots, T_n]]$.

2.1.1. DEFINITION. A commutative Banach k-algebra $\mathcal{A}$ is said to be *k-affinoid* if there exists an admissible epimorphism $k\{r^{-1}T\} \to \mathcal{A}$. If such an epimorphism can be found with $r = 1$, $\mathcal{A}$ is said to be *strictly k-affinoid*. By an *affinoid k-algebra* we mean a K-affinoid algebra for some non-Archimedean field K over k.

For example, if the valuation on k is trivial, any strictly k-affinoid algebra is isomorphic to a finitely generated k-algebra endowed with the trivial valuation (and conversely). The algebra $k\{r^{-1}T\}$ is strictly k-affinoid if and only if $r_i \in \sqrt{|k^*|}$ for all $1 \leq i \leq n$ (**[BGR]**, 6.1.5/4). For strictly k-affinoid algebras, the Noether Normalization Lemma holds. Namely, for any $\mathcal{A}$ of this type there exists a finite admissible monomorphism $k\{T_1, \dots, T_n\} \hookrightarrow \mathcal{A}$.

Here is an important example of a k-affinoid algebra. For $r > 0$, let K_r denote the set of all formal series $f = \sum_{i=-\infty}^{+\infty} a_i T^i$, $a_i \in k$, with $|a_i| r^i \to 0$ as $|i| \to \infty$. This is a commutative Banach k-algebra with respect to the multiplicative norm $||f|| = \max_i |a_i| r^i$. Moreover, it is a k-affinoid algebra, because there is an isometric isomorphism $k\{r^{-1}T_1, rT_2\}/(T_1T_2 - 1) \tilde{\to} K_r$: $T_1 \mapsto T$, $T_2 \mapsto T^{-1}$. Now suppose that $r \notin \sqrt{|k^*|}$. In this case K_r is a non-

Archimedean field with nontrivial valuation, because the group $|K_r^*|$ is generated by $|k^*|$ and r. If the valuation on k is nontrivial, K_r coincides with the field $\mathcal{H}(x)$ of the point $x = p(E(0, r))$ of type (3) (see 1.4.4). The construction can be iterated. Namely, for any set of positive numbers $r_1, \dots, r_n$ whose images in the $\mathbf{Q}$-vector space $\mathbf{R}_+^*/\sqrt{|k^*|}$ are linearly independent, one can construct the non-Archimedean field $K_{r_1,\dots,r_n} = K_{r_1}\widehat{\otimes}\dots\widehat{\otimes}K_{r_n}$, which is a k-affinoid algebra whose group $|K_{r_1,\dots,r_n}^*|$ is generated by $|k^*|$ and $r_1, \dots, r_n$.

For a k-Banach space X, the tensor product $X\widehat{\otimes}K_r$ is isomorphic to the space of series $f = \sum_{i=-\infty}^{+\infty} x_i T^i$, $x_i \in X$, with $||x_i||r^i \to 0$ as $|i| \to \infty$. This is a K_r-Banach space. Tensoring with K_r is a useful tool to reduce many questions to the case of strictly affinoid algebras over fields with nontrivial valuation. Here are some simple properties of this tensor product.

2.1.2. Proposition. *Let X be a k-Banach space, and let $r \notin \sqrt{|k^*|}$.*

(i) *The canonical map $X \to X\widehat{\otimes}K_r$ is an isometric embedding.*

(ii) *A sequence of bounded homomorphisms of k-Banach spaces $X \to Y \to Z$ is exact and admissible if and only if the corresponding sequence of K_r-spaces $X\widehat{\otimes}K_r \to Y\widehat{\otimes}K_r \to Z\widehat{\otimes}K_r$ is exact and admissible.* ■

For example, for any k-affinoid algebra $\mathcal{A}$, there exists a field $K = K_{r_1,\dots,r_n}$ with nontrivial valuation such that the algebra $\mathcal{A}\ \widehat{\otimes}\ K$ is strictly K-affinoid.

2.1.3. Proposition. *Any k-affinoid algebra $\mathcal{A}$ is Noetherian, and all of its ideals are closed.*

Proof. (see **[P1]**). Strictly k-affinoid algebras possess the required properties. So it suffices to show that if the ring $\mathcal{A}\widehat{\otimes}K_r$ is Noetherian for some $r \notin \sqrt{|k^*|}$, and if all of its ideals are closed, then the ring $\mathcal{A}$ possesses the same properties. Indeed, let $\mathbf{a}$ be an ideal of $\mathcal{A}$. Then the ideal $\mathbf{a}(\mathcal{A}\widehat{\otimes}K_r)$ is generated by elements $f_1, \dots, f_n \in \mathbf{a}$. Any $f \in \mathbf{a}$ can be written in the form $f = \sum_{i=1}^n f_i g_i$, where $g_i = \sum_{j=-\infty}^{+\infty} g_{i,j}T^j \in \mathcal{A}\widehat{\otimes}K_r$. We have $f = \sum_{i=1}^n g_{i,0}f_i$, i.e., $f_1, \dots, f_n$ generate $\mathbf{a}$ and $\mathbf{a} = \mathcal{A} \cap \mathbf{a}(\mathcal{A}\widehat{\otimes}K_r)$. Since $\mathcal{A}$ and $\mathbf{a}(\mathcal{A}\widehat{\otimes}K_r)$ are closed in $\mathcal{A}\widehat{\otimes}K_r$, it follows that $\mathbf{a}$ is closed in $\mathcal{A}$. ■

We remark that for

$$f = \sum_{i=-\infty}^{+\infty} a_i T^i \in \mathcal{A}\widehat{\otimes}K_r,$$

one has $\rho(f) = \max_i \rho(a_i)r^i$.

2.1.4. Proposition. (i) *For any $f \in \mathcal{A}$, there exist $C > 0$ and $N \geq 1$ such that $||f^n|| \leq C\rho(f)^n$ for all $n \geq N$. In particular, if f is quasinilpotent (i.e., $\rho(f) = 0$), then f is nilpotent.*

(ii) *If $\mathcal{A}$ is reduced, then there exists $C > 0$ such that $||f|| \leq C\rho(f)$ for all $f \in \mathcal{A}$.*

(iii) *If the radical of* $\mathscr{A}$ *is prime, the same is true for the algebra* $\mathscr{A}\widehat{\otimes}K_r$, *where* $r \notin \sqrt{|k^*|}$.

PROOF. To verify (i), it suffices to assume that the valuation on k is nontrivial and that the algebra $\mathscr{A}$ is strictly k-affinoid. If $\rho(f) = 0$, then f is nilpotent, because $\mathscr{A}$ is a Noetherian Jacobson ring. Suppose that $\rho(f) > 0$. Then $\rho(f) \in \sqrt{|k^*|}$ (**[BGR]**, 6.2.1/4). Let $\rho(f)^d = |a|$ for $a \in k^*$. Then the spectral norm of the element $g = f^d/a$ is equal to 1. Hence, it is power bounded (**[BGR]**, 6.2.3/1). That is, there exists $C > 0$ such that $\|g^n\| \le C$ for all $n \ge 1$. Therefore, $\|f^{nd}\| \le C|a|^n = C\rho(f)^{nd}$ for all $n \ge 1$.

If $\mathscr{A}$ is reduced, then for any $r \notin \sqrt{|k^*|}$ the algebra $\mathscr{A}\widehat{\otimes}K_r$ also is reduced. Hence, the statement (ii) follows from the corresponding fact for strictly k-affinoid algebras (**[BGR]**, 6.2.4/1). The statement (iii) is easily obtained using the remark before the formulation of the proposition. ∎

2.1.5. COROLLARY. *Let* $\varphi: \mathscr{A} \to \mathscr{B}$ *be a bounded homomorphism between affinoid* k*-algebras. Let* $f_1, \dots, f_n \in \mathscr{B}$, *and let* $r_1, \dots, r_n$ *be positive numbers with* $r_i \ge \rho(f_i)$, $1 \le i \le n$. *Then there exists a unique bounded homomorphism* $\Phi: \mathscr{A}\{r^{-1}T\} \to \mathscr{B}$ *extending* φ *and sending* T_i *to* f_i, $1 \le i \le n$.

PROOF. For $h = \sum a_{i_1,\dots,i_n} T_1^{i_1} \dots T_n^{i_n} \in \mathscr{A}\{r^{-1}T\}$, we set

$$\Phi(h) = \sum \varphi(a_{i_1,\dots,i_n})\, f_1^{i_1} \dots f_n^{i_n}.$$

By Proposition 2.1.4(i), the series on the right side represents a well-defined element of $\mathscr{B}$. It is easy to verify that Φ is the required unique homomorphism. ∎

2.1.6. COROLLARY. *A* k*-affinoid algebra* $\mathscr{A}$ *is strictly* k*-affinoid if and only if* $\rho(f) \in \sqrt{|k^*|} \cup \{0\}$ *for all* $f \in \mathscr{A}$.

PROOF. Assume that the valuation on k is nontrivial. The direct implication follows from **[BGR]**, 6.2.1/4. Now suppose that $\rho(f) \in \sqrt{|k^*|} \cup \{0\}$ for all $f \in \mathscr{A}$. Consider an admissible epimorphism $\varphi: k\{r^{-1}T\} \to \mathscr{A}$, and set $f_i = \varphi(T_i)$, $1 \le i \le n$. Suppose that $r_1, \dots, r_m \notin \sqrt{|k^*|}$ and $r_{m+1}, \dots, r_n \in \sqrt{|k^*|}$. By hypothesis, $\rho(f_i) < r_i$ for $1 \le i \le m$. Let $r'_1, \dots, r'_m \in \sqrt{|k^*|}$ be numbers with $\rho(f_i) \le r'_i < r_i$, $1 \le i \le m$, and set $r'_i = r_i$, $m+1 \le i \le n$. Then the homomorphism $k\{r'^{-1}T\} \to \mathscr{A}$ sending T_i to f_i, $1 \le i \le n$, is well defined, surjective, and admissible. Because the algebra $k\{r'^{-1}T\}$ is strictly k-affinoid (**[BGR]**, 6.1.5/4), $\mathscr{A}$ is also strictly k-affinoid. ∎

The construction of the algebra $k\{r^{-1}T\}$ obviously extends to a construction of $\mathscr{A}\{r^{-1}T\}$ for any commutative Banach k-algebra $\mathscr{A}$. As in Definition 2.1.1, one defines $\mathscr{A}$*-affinoid algebras* and *affinoid* $\mathscr{A}$*-algebras* (the latter are affinoid k-algebras $\mathscr{B}$ with a fixed bounded homomorphism $\mathscr{A} \to \mathscr{B}$). The following statement is proved in the same way as the corresponding statement [**BGR**], 7.3.4/3.

2.1.7. Proposition. *Let $\mathscr{A}$ and $\mathscr{B}$ be k-affinoid algebras, $\varphi : \mathscr{A}\{r^{-1}T\} \to \mathscr{B}$ be an admissible epimorphism, and set $f_i = \varphi(T_i)$, $1 \le i \le n$. Then there exists $\varepsilon > 0$ such that for any set $f' = (f_1', \dots , f_n')$ of elements of $\mathscr{B}$ with $\|f_i' - f_i\| < \varepsilon$, the homomorphism $\varphi' : \mathscr{A}\{r^{-1}T\} \to \mathscr{B}$ that coincides with φ on $\mathscr{A}$ and sends T_i to f_i', is also an admissible epimorphism.* ∎

2.1.8. Corollary. *A commutative Banach k-algebra $\mathscr{A}$ is k-affinoid if and only if the algebra $\mathscr{A}\widehat{\otimes}K_r$ is K_r-affinoid for some $r \notin \sqrt{|k^*|}$.*

Proof. Let $\pi : K_r\{p_1^{-1}T_1, \dots , p_n^{-1}T_n\} \to \mathscr{A}\widehat{\otimes}K_r$ be an admissible epimorphism, and let $\pi(T_i) = \sum_{j=-\infty}^{+\infty} a_{i,j}T^j$. By Proposition 2.1.7, we may assume that there exists a large enough integer l such that $a_{i,j} = 0$ for $|j| > l$. Let $\mathscr{B} = k\{p_i^{-1}r^jT_{i,j}\}$, $1 \le i \le n$, $-l \le j \le l$. Since the algebra $\mathscr{A}$ satisfies the property (i) from Proposition 2.1.4 and $\rho(a_{i,j}) \le p_ir^{-j}$, the homomorphism $\varphi : \mathscr{B} \to \mathscr{A}$ sending $T_{i,j}$ to $a_{i,j}$ is well defined. We claim that φ is surjective and admissible. Indeed, this follows from the fact that the composition of $\varphi\widehat{\otimes}K_r$ with the homomorphism $K_r\{p_1^{-1}T_1, \dots , p_n^{-1}T_n\} \to \mathscr{B}\widehat{\otimes}K_r$ sending T_i to $\sum_{j=-l}^{l} T_{i,j}T^j$ coincides with π. ∎

Let $\mathscr{A}$ be a k-affinoid algebra. A Banach $\mathscr{A}$-module M is said to be *finite* if there exists an admissible epimorphism $\mathscr{A}^n \to M$. The category of finite Banach $\mathscr{A}$-modules with bounded $\mathscr{A}$-linear maps as morphisms is denoted by $\mathrm{Mod}_b^h(\mathscr{A})$. The category of finite $\mathscr{A}$-modules is denoted by $\mathrm{Mod}^h(\mathscr{A})$. There is an obvious functor $\theta : \mathrm{Mod}_b^h(\mathscr{A}) \to \mathrm{Mod}^h(\mathscr{A})$.

2.1.9. Proposition. *θ is an equivalence of categories.*

Proof. Let $M \in \mathrm{Mod}_b^h(\mathscr{A})$ and let $\varphi : M \to N$ be an $\mathscr{A}$-linear map to a Banach $\mathscr{A}$-module N. We claim that φ is bounded. To show this, consider an admissible epimorphism $\pi : \mathscr{A}^n \to M$. It suffices to show that the composition map $\psi : \mathscr{A}^n \to N$ is bounded. Let $e_1, \dots , e_n$ be a basis of $\mathscr{A}^n$ and $C = \max(\|\psi(e_i)\|)$. Then for an element $f = \sum a_ie_i$ one has $\|\psi(f)\| = \|\sum a_i\psi(e_i)\| \le C\|f\|$. That is, ψ is bounded.

Now let M be a finite $\mathscr{A}$-module. Consider an $\mathscr{A}$-linear epimorphism $\pi : \mathscr{A}^n \to M$. Since $\mathscr{A}^n$ has a k-affinoid algebra structure, the kernel of π is closed. Therefore, one can endow M with the residue norm, with respect to which M is complete. ∎

It is not true that a Banach $\mathscr{A}$-module which is finite as an $\mathscr{A}$-module is a finite Banach $\mathscr{A}$-module. Indeed, suppose that the valuation on k is trivial. Let $0 < p < q < 1$, $\mathscr{A} = k\{q^{-1}T\}$ and $\mathscr{B} = k\{p^{-1}T\}$. Then $\mathscr{B}$ is a Banach $\mathscr{A}$-module and is finite over $\mathscr{A}$, since the canonical map $\mathscr{A} \to \mathscr{B}$ is bijective. But $\mathscr{B}$ is not a finite Banach $\mathscr{A}$-module, because the inverse map $\mathscr{B} \to \mathscr{A}$ is not bounded.

2.1.10. PROPOSITION. *Let* M, $N \in \mathrm{Mod}_b^h(\mathscr{A})$, *and let* $\mathscr{B}$ *be an affinoid* $\mathscr{A}$*-algebra. Then*:

(i) $M \otimes_{\mathscr{A}} N \tilde{\to} M \widehat{\otimes}_{\mathscr{A}} N \in \mathrm{Mod}_b^h(\mathscr{A})$;

(ii) $M \otimes_{\mathscr{A}} \mathscr{B} \tilde{\to} M \widehat{\otimes}_{\mathscr{A}} \mathscr{B} \in \mathrm{Mod}_b^h(\mathscr{B})$;

(iii) *any* $\mathscr{A}$*-linear map* $M \to N$ *is admissible.*

PROOF. The statements (i) and (ii) easily follow from the remark at the end of §1.1. The statement (iii) is obtained from the corresponding fact for strictly affinoid algebras by applying Proposition 2.1.2(ii). ∎

2.1.11. PROPOSITION. *Let* M *be a Banach* $\mathscr{A}$*-module, and let* $r \notin \sqrt{|k^*|}$. *Then* $M \in \mathrm{Mod}_b^h(\mathscr{A})$ *if and only if* $M \widehat{\otimes} K_r \in \mathrm{Mod}_b^h(\mathscr{A} \widehat{\otimes} K_r)$.

PROOF. Let $\varphi: (\mathscr{A} \widehat{\otimes} K_r)^n = \sum_{i=1}^n (\mathscr{A} \widehat{\otimes} K_r) e_i \to M \widehat{\otimes} K_r$ be an admissible epimorphism, and set $\varphi(e_i) = \sum_{j=-\infty}^{+\infty} m_{i,j} T^j$. We may assume that for l large enough, $m_{i,j} = 0$ for all $1 \le i \le n$ and $|j| > l$. It is easy to verify that the homomorphism $\psi: (\mathscr{A})^{n(2l+1)} = \sum_{i=1}^n \sum_{j=-l}^{l} \mathscr{A} e_{i,j} \to M$ sending $e_{i,j}$ to $m_{i,j}$ is an admissible epimorphism. ∎

A commutative Banach $\mathscr{A}$-algebra $\mathscr{B}$ is said to be *finite* if $\mathscr{B}$ is a finite Banach $\mathscr{A}$-module. It is easy to see that any finite Banach $\mathscr{A}$-algebra is an $\mathscr{A}$-affinoid algebra.

2.1.12. PROPOSITION. *The category of finite Banach* $\mathscr{A}$*-algebras is equivalent to the category of finite* $\mathscr{A}$*-algebras.*

PROOF. Let $\mathscr{B}$ be a finite $\mathscr{A}$-algebra, and let $\varphi: \mathscr{A}^n \to \mathscr{B}$ be an epimorphism of $\mathscr{A}$-modules. We endow $\mathscr{B}$ with the residue norm via φ. It suffices to show that there exists a constant $C > 0$ such that $||fg|| \le C||f|| \cdot ||g||$ for all $f, g \in \mathscr{B}$. Consider the induced $\mathscr{A}$-linear epimorphism $\psi: \mathscr{A}^n \otimes_{\mathscr{A}} \mathscr{A}^n \to \mathscr{B} \otimes_{\mathscr{A}} \mathscr{B}$. By Proposition 2.1.10(iii), the map ψ is admissible. From this it follows that $||f \otimes g|| \le C_1 ||f|| \cdot ||g||$ for some $C_1 > 0$ and all $f, g \in \mathscr{B}$. By Proposition 2.1.9, the multiplication map $\mathscr{B} \otimes_{\mathscr{A}} \mathscr{B} \to \mathscr{B}$ is bounded. Therefore, there exists $C_2 > 0$ such that $||fg|| \le C_2 ||f \otimes g|| \le C_1 C_2 ||f|| \cdot ||g||$ for all $f, g \in \mathscr{B}$. ∎

2.1.13. REMARK. Any homomorphism of strictly k-affinoid algebras is bounded ([**BGR**], 6.1.3/1). This is not true for k-affinoid algebras. Indeed, if the valuation on k is trivial, the canonical homomorphism $K_p \to K_q$, where $0 < p < q < 1$, is bijective but not bounded. One can pose the following question. Suppose that $\varphi: \mathscr{A} \to \mathscr{B}$ is a homomorphism of k-affinoid algebras such that every character of $\mathscr{B}$ induces a character on $\mathscr{A}$. Is it true that φ is bounded? By Proposition 2.1.4(ii), this is the case if the algebra $\mathscr{B}$ is reduced. But it is not true in general. Here is an example. Suppose that $\mathrm{char}(k) = 0$, and let $\mathscr{A} = K_r$ and $\mathscr{B} = K_r[\varepsilon]$, where $r \notin \sqrt{|k^*|}$ and $\varepsilon^2 = 0$. Since $\mathrm{char}(k) = 0$, there exists a differentiation $d: K_r \to$

K_r over k which is not continuous. Then the homomorphism $\varphi\colon \mathcal{A} \to \mathcal{B}$ sending f to $f + (df)\varepsilon$ is not continuous.

The following proposition also extends the properties of strictly k-affinoid algebras to k-affinoid ones.

2.1.14. PROPOSITION. (i) *Every k-affinoid algebra which is an integral domain is Japanese.*

(ii) *If G is a finite group of automorphisms of a k-affinoid algebra $\mathcal{A}$, then $\mathcal{A}^G$ is a closed subalgebra of $\mathcal{A}$, which is k-affinoid. Furthermore, $\mathcal{A}$ is a finite Banach $\mathcal{A}^G$-algebra.*

PROOF. (i) Let L be a finite extension of the field of fractions $Q(\mathcal{A})$ of $\mathcal{A}$. We have to show that the integral closure of $\mathcal{A}$ in L is a finite $\mathcal{A}$-module. Let $b_1, \dots, b_m$ be elements of L which are integral over $\mathcal{A}$ and generate L over $Q(\mathcal{A})$. Then the algebra $\mathcal{B}$ generated by $b_1, \dots, b_m$ over $\mathcal{A}$ is a finite $\mathcal{A}$-module, and $L = Q(\mathcal{B})$. By Proposition 2.1.12, $\mathcal{B}$ may be regarded as a finite Banach $\mathcal{A}$-algebra. Let $\mathcal{B}'$ denote the integral closure of $\mathcal{B}$ in L. It suffices to show that there exists a nonzero element $b \in \mathcal{B}$ with $b \cdot \mathcal{B}' \subset \mathcal{B}$. Let K be a field of the form $K_{r_1, \dots, r_n}$ for which the algebra $\mathcal{B}\widehat{\otimes}K$ is strictly K-affinoid. Because our claim is valid for $\mathcal{B}\widehat{\otimes}K$ ([**BGR**], 6.1.2/4), there exists a nonzero element $b' = \sum_\nu b_\nu T^\nu \in \mathcal{B}\widehat{\otimes}K$ with $b' \cdot \mathcal{B}' \subset \mathcal{B}\widehat{\otimes}K$. It follows that $b_\nu \cdot \mathcal{B}' \subset \mathcal{B}$ for any $b_\nu \neq 0$.

(ii) $\mathcal{A}^G$ is a commutative Banach k-algebra with respect to the norm of $\mathcal{A}$. Let K be a field of the form $K_{r_1, \dots, r_n}$ for which the algebra $\mathcal{B}\widehat{\otimes}K$ is strictly K-affinoid. Because $(\mathcal{A}\widehat{\otimes}K)^G = \mathcal{A}^G\widehat{\otimes}K$, from [**BGR**], 6.3.3, and Corollary 2.1.8 it follows that the algebra $\mathcal{A}^G$ is k-affinoid. The second assertion follows from Proposition 2.1.11. ∎

Let $\mathcal{A}$ be a strictly k-affinoid algebra. Because the maximal ideals of $\mathcal{A}$ have finite codimension, there is a canonical injective map $\operatorname{Max}(\mathcal{A}) \to \mathcal{M}(\mathcal{A})$. From [**BGR**], 6.2.1/4, it follows that each element takes its maximum on $\operatorname{Max}(\mathcal{A})$, and we have the equality ([**BGR**], 6.2.3/3)

$$\rho(f) = |f|_{\sup} = \max_{x \in \operatorname{Max}(\mathcal{A})} |f(x)|.$$

Note that if the valuation on k is trivial, the topology of $\mathcal{M}(\mathcal{A})$ induces the discrete topology on $\operatorname{Max}(\mathcal{A})$. Moreover, the closure of $\operatorname{Max}(\mathcal{A})$ in $\mathcal{M}(\mathcal{A})$ coincides with the image of the map $\operatorname{Spec}(\mathcal{A}) \to \mathcal{M}(\mathcal{A})$, which sends a prime ideal $\wp \subset \mathcal{A}$ to the multiplicative seminorm induced by the trivial norm on $\mathcal{A}/\wp$.

2.1.15. PROPOSITION. *If the valuation on k is nontrivial, then for a strictly k-affinoid algebra $\mathcal{A}$ the canonical map $\operatorname{Max}(\mathcal{A}) \to \mathcal{M}(\mathcal{A})$ induces a homeomorphism of $\operatorname{Max}(\mathcal{A})$ with an everywhere dense subset of $\mathcal{M}(\mathcal{A})$.*

PROOF. Let $\mathcal{U}$ be an open neighborhood of a point $x_0 \in \mathcal{M}(\mathcal{A})$. We may assume that $\mathcal{U} = \{x \in \mathcal{M}(\mathcal{A}) \mid |f_i(x)| < 1,\ |g_j(x)| > 1,\ 1 \le i \le n,\ 1 \le$

$j \le m\}$. Choose $p_i, q_j \in |k^*|$ with $|f_i(x_0)| < p_i < 1$ and $|g_j(x_0)| > q_j > 1$. The algebra $\mathscr{B} = \mathscr{A}\{p_1^{-1}T_1, \dots, p_n^{-1}T_n, q_1^{-1}S_1, \dots, q_m^{-1}S_m\}/(T_i - f_i, g_jS_j - 1)$ is strictly k-affinoid and nonzero. Hence, $\mathrm{Max}(\mathscr{B}) \ne \varnothing$. Because the image of $\mathscr{M}(\mathscr{B})$ in $\mathscr{M}(\mathscr{A})$ lies in $\mathscr{U}$ and contains the point x_0, $\mathrm{Max}(\mathscr{A})$ is everywhere dense in $\mathscr{M}(\mathscr{A})$. Furthermore, open subsets of the type $V = \{x \in \mathrm{Max}(\mathscr{A}) \mid |f_i(x)| < 1, 1 \le i \le n\}$ for $f_1, \dots, f_n \in \mathbf{m}_{x_0}$ form a fundamental system of neighborhoods of the point x_0 in $\mathrm{Max}(\mathscr{A})$. The set $\mathscr{U} = \{x \in \mathscr{M}(\mathscr{A}) \mid |f_i(x)| < 1, 1 \le i \le n\}$ is open in $\mathscr{M}(\mathscr{A})$, and $\mathscr{U} \cap \mathrm{Max}(\mathscr{A}) = V$. ∎

2.1.16. COROLLARY. *Let $\mathscr{B}$ be a finite Banach algebra over a k-affinoid algebra $\mathscr{A}$, and suppose the canonical homomorphism $\mathscr{A} \to \mathscr{B}$ is injective. Then the map $\mathscr{M}(\mathscr{B}) \to \mathscr{M}(\mathscr{A})$ is surjective and has finite fibres.* ∎

2.2. Affinoid domains. Let $\mathscr{A}$ be a k-affinoid algebra, and let $X = \mathscr{M}(\mathscr{A})$. If the valuation on k is nontrivial and $\mathscr{A}$ is strictly k-affinoid, one defines affinoid domains in $\mathrm{Max}(\mathscr{A})$ (see **[BGR]**, 7.2.2/2). We call their closures in X *strictly affinoid domains in X*.

2.2.1 DEFINITION. A closed subset $V \subset X$ is said to be an *affinoid domain* in X if there exists a bounded homomorphism of k-affinoid algebras $\varphi: \mathscr{A} \to \mathscr{A}_V$ satisfying the following universal property. Given a bounded homomorphism of affinoid k-algebras $\mathscr{A} \to \mathscr{B}$ such that the image of $\mathscr{M}(\mathscr{B})$ in X lies in V, there exists a unique bounded homomorphism $\mathscr{A}_V \to \mathscr{B}$ making the diagram

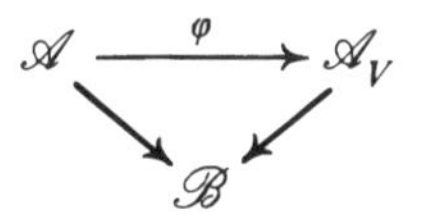

commutative.

2.2.2. REMARKS. (i) Let $f = (f_1, \dots, f_n)$ and $g = (g_1, \dots, g_m)$ be sets of elements of $\mathscr{A}$, and let $p = (p_1, \dots, p_n)$ and $q = (q_1, \dots, q_m)$ be sets of positive numbers. Then the set

$$X(p^{-1}f, qg^{-1}) := \{x \in X \mid |f_i(x)| \le p_i, |g_j(x)| \ge q_j, 1 \le i \le n, 1 \le j \le m\}$$

is an affinoid domain in X represented by the homomorphism:

$$\mathscr{A} \to \mathscr{A}\{p^{-1}f, qg^{-1}\} := \mathscr{A}\{p_1^{-1}T_1, \dots, p_n^{-1}T_n, q_1S_1, \dots, q_mS_m\}/(T_i - f_i, g_jS_j - 1).$$

This follows easily from Corollary 2.1.5. We write $X(p^{-1}f)$ or $X(qg^{-1})$ instead of $X(p^{-1}f, qg^{-1})$ in the cases $m = 1$, $g = 1$, $q = 1$ or $n = 1$, $f = 1$, $p = 1$. Affinoid domains of the form $X(p^{-1}f)$ (resp. $X(p^{-1}f, qg^{-1})$) are called *Weierstrass* (resp. *Laurent*) *domains* in X. We note that the Laurent neighborhoods of a point $x \in X$ form a basis of closed neighborhoods of x.

(ii) Let $g, f_1, \dots, f_n \in \mathcal{A}$ be elements without common zeros in X. That is, they generate the unit ideal in $\mathcal{A}$ and let $p = (p_1, \dots, p_n)$ be a set of positive numbers. Then the set

$$X(p^{-1}\frac{f}{g}) := \{x \in X \,|\, |f_i(x)| \le p_i|g(x)|,\ 1 \le i \le n\}$$

is an affinoid domain in X represented by the homomorphism:

$$\mathcal{A} \to \mathcal{A}\{p^{-1}\frac{f}{g}\} := \mathcal{A}\{p_1^{-1}T_1, \dots, p_n^{-1}T_n\}/(gT_i - f_i).$$

Affinoid domains of this type are called *rational domains*. From Remark 1.2.2(iii) it follows that $\mathcal{M}(\mathcal{A}_V) \tilde{\to} V$ for any rational domain V in X.

(iii) Let $\varphi: Y = \mathcal{M}(\mathcal{B}) \to \mathcal{M}(\mathcal{A})$ be a map induced by a bounded homomorphism of affinoid k-algebras $\mathcal{A} \to \mathcal{B}$. If V is a Weierstrass (resp. Laurent, resp. rational, resp. affinoid) domain in X, then $\varphi^{-1}(V)$ is an affinoid domain of the same type represented by the homomorphism $\mathcal{B} \to \mathcal{B}\widehat{\otimes}_{\mathcal{A}}\mathcal{A}_V$.

(iv) If U and V are affinoid domains in X, then the intersection $U \cap V$ is an affinoid domain represented by the homomorphism $\mathcal{A} \to \mathcal{A}_U\widehat{\otimes}_{\mathcal{A}}\mathcal{A}_V$. If U and V are Weierstrass (resp. Laurent, resp. rational), then $U \cap V$ is an affinoid domain of the same type. This is verified in the same way as in **[BGR]**, 7.2.3/7. In particular, any Laurent domain is also rational.

2.2.3. PROPOSITION. *Suppose that the valuation on k is nontrivial and that the algebra $\mathcal{A}$ is strictly k-affinoid.*

(i) *The strictly affinoid domains in X are affinoid domains.*

(ii) *If V is an affinoid domain in X such that the algebra $\mathcal{A}_V$ is strictly k-affinoid, then V is a strictly affinoid domain.*

(iii) *The strictly affinoid neighborhoods of a point form a basis of closed neighborhoods of the point.*

PROOF. (i) Let V be a strictly affinoid domain in X. We have to show that the canonical map $\mathcal{A} \to \mathcal{A}_V$ satisfies the universal property with respect to bounded homomorphisms $\mathcal{A} \to \mathcal{B}$ to any affinoid k-algebra $\mathcal{B}$. By the Gerritzen-Grauert Theorem (**[BGR]**, 7.3.5/3), there exists a finite covering $V = \cup V_i$ by strictly rational domains $V_i \subset X$. We set $Y = \mathcal{M}(\mathcal{B})$ and let Y_i denote the inverse image of V_i in Y. Then the Y_i are strictly rational domains of Y, and $Y = \cup Y_i$. The required extension $\mathcal{A}_V \to \mathcal{B}$ of the homomorphism $\mathcal{A} \to \mathcal{B}$ is constructed using the admissible exact sequences $0 \to \mathcal{A}_V \to \prod_i \mathcal{A}_{V_i} \to \prod_{i,j} \mathcal{A}_{V_i \cap V_j}$ and $0 \to \mathcal{B} \to \prod_i \mathcal{B}_{Y_i} \to \prod_{i,j} \mathcal{B}_{Y_i \cap Y_j}$ and the fact that strictly rational domains are affinoid domains.

(ii) First suppose that the intersection $V \cap \mathrm{Max}(\mathcal{A})$ is nonempty, and let W denote its closure in X. Then the homomorphism $\mathcal{A} \to \mathcal{A}_V$ satisfies the universal property in Definition 2.2.1 for the set W and for homomorphisms of strictly k-affinoid algebras $\mathcal{A} \to \mathcal{B}$. This means that W is a strictly affinoid domain in X. In particular, $\mathrm{Max}(\mathcal{A}_V) = W \cap \mathrm{Max}(\mathcal{A})$. By Proposition 2.1.15, $\mathrm{Max}(\mathcal{A}_V)$ is everywhere dense in $\mathcal{M}(\mathcal{A}_V)$. This implies that the image of $\mathcal{M}(\mathcal{A}_V)$ in X coincides with W. Since V is an affinoid domain, this image contains V. Therefore, $V = W$; V is a strictly affinoid domain.

In the general case, let K be a non-Archimedean field which is large enough so that $\pi^{-1}(V)$ contains a point from $\mathrm{Max}(\mathcal{A}\widehat{\otimes}K)$ (π denotes the canonical map $X\widehat{\otimes}K \to X$). Then $\pi^{-1}(V)$ is a strictly affinoid domain in $X\widehat{\otimes}K$. In particular, the image of $\mathcal{M}(\mathcal{A}_V\widehat{\otimes}K)$ in $X\widehat{\otimes}K$ coincides with $\pi^{-1}(V)$. Because the map $\mathcal{M}(\mathcal{A}_V\widehat{\otimes}K) \to \mathcal{M}(\mathcal{A}_V)$ is surjective, the image of $\mathcal{M}(\mathcal{A}_V)$ in X coincides with V. It follows that $V \cap \mathrm{Max}(\mathcal{A}) \neq \emptyset$, because the algebra $\mathcal{A}_V$ is strictly k-affinoid.

(iii) Let $x_0 \in X$. The open neighborhoods of x_0 of the type $\mathcal{U} = \{x \in X \mid |f_i(x)| < r_i, |g_j(x)| > s_j, 1 \le i \le n, 1 \le j \le m\}$ form a basis of open neighborhoods of x_0. We choose numbers $p_i, q_j \in \sqrt{|k^*|}$ with $|f_i(x_0)| < p_i < r_i$ and $|g_j(x_0)| > q_j > s_j$. Then the Laurent domain $V = X(p^{-1}f, qg^{-1})$ is strictly affinoid. Moreover, for the open subset $\mathcal{U}' = \{x \in X \mid |f_i(x)| < p_i, |g_j(x)| > q_j, 1 \le i \le n, 1 \le j \le m\}$, one has $x_0 \in \mathcal{U}' \subset V \subset \mathcal{U}$. ∎

2.2.4. Proposition. *Let V be an affinoid domain in X. Then:*

(i) $\mathcal{M}(\mathcal{A}_V) \tilde{\to} V$; *in particular, the homomorphism $\mathcal{A} \to \mathcal{A}_V$ is uniquely determined by V;*

(ii) *$\mathcal{A}_V$ is a flat $\mathcal{A}$-algebra.*

Proof. First suppose that the valuation on k is nontrivial and that $\mathcal{A}$ and $\mathcal{A}_V$ are strictly k-affinoid. By **[BGR]**, 7.2.2/1, the map $\mathrm{Max}(\mathcal{A}_V) \to \mathrm{Max}(\mathcal{A})$ is injective. If V' and V'' are Laurent domains in $\mathcal{M}(\mathcal{A}_V)$ with empty intersection, then $\mathcal{A}_{V'}\widehat{\otimes}_{\mathcal{A}}\mathcal{A}_{V''} = 0$. Hence the images of V' and V'' in $\mathcal{M}(\mathcal{A})$ also have empty intersection. It follows that the map $\mathcal{M}(\mathcal{A}_V) \to \mathcal{M}(\mathcal{A})$ is injective and that its image coincides with V. The $\mathcal{A}$-algebra $\mathcal{A}_V$ is flat, by **[BGR]**, 7.3.2/6.

In the general case, let K be a sufficiently large non-Archimedean field with nontrivial valuation such that the algebras $\mathcal{A}\widehat{\otimes}K$ and $\mathcal{A}_V\widehat{\otimes}K$ are strictly K-affinoid. Then $\pi^{-1}(V)$ is a strictly affinoid domain in $X\widehat{\otimes}K = \mathcal{M}(\mathcal{A}\widehat{\otimes}K)$ represented by the homomorphism $\mathcal{A}\widehat{\otimes}K \to \mathcal{A}_V\widehat{\otimes}K$ (π denotes the canonical surjective map $X\widehat{\otimes}K \to X$). It follows that $\mathcal{M}(\mathcal{A}_V\widehat{\otimes}K) \tilde{\to} \pi^{-1}(V)$. Because the map $\mathcal{M}(\mathcal{A}_V\widehat{\otimes}K) \to \mathcal{M}(\mathcal{A}_V)$ is surjective, the image of $\mathcal{M}(\mathcal{A}_V)$ in X coincides with V. The injectivity of the map $\mathcal{M}(\mathcal{A}_V) \to X$ follows from the universal property of the homomorphism $\mathcal{A} \to \mathcal{A}_V$.

Now let $\varphi: M \to N$ be an injective $\mathcal{A}$-homomorphism of finite $\mathcal{A}$-modules. We endow M and N with the structure of finite Banach $\mathcal{A}$-modules. By Proposition 2.1.9, the map φ is admissible. Hence the $\mathcal{A}\widehat{\otimes}K$-homomorphism of finite Banach $\mathcal{A}\widehat{\otimes}K$-modules $M\widehat{\otimes}K \to N\widehat{\otimes}K$ is injective and admissible. Since $\mathcal{A}_{\pi^{-1}(V)} = \mathcal{A}_V\widehat{\otimes}K$ is a flat $\mathcal{A}$-algebra, the map

$(M\widehat{\otimes}K) \otimes_{\mathscr{A}\widehat{\otimes}K} (\mathscr{A}_V\widehat{\otimes}K) \to (N\widehat{\otimes}K) \otimes_{\mathscr{A}\widehat{\otimes}K} (\mathscr{A}_V\widehat{\otimes}K)$ is injective. But the first module coincides with $(M\widehat{\otimes}K)\widehat{\otimes}_{\mathscr{A}\widehat{\otimes}K}(\mathscr{A}_V\widehat{\otimes}K) = (M\otimes_{\mathscr{A}}\mathscr{A}_V)\widehat{\otimes}K$, and the second one coincides with $(N\otimes_{\mathscr{A}}\mathscr{A}_V)\widehat{\otimes}K$. Hence the map $M\otimes_{\mathscr{A}}\mathscr{A}_V \to N\otimes_{\mathscr{A}}\mathscr{A}_V$ is injective, by Proposition 2.1.2(ii). ∎

By Propositions 2.2.3 and 2.2.4, we may refer to any affinoid domain V in X such that the algebra $\mathscr{A}_V$ is strictly k-affinoid as a *strictly affinoid domain.*

Let $\mathscr{V} = \{V_i\}_{i\in I}$ be a finite covering of X by affinoid domains and let M be an $\mathscr{A}$-module. These data determine the Čech complex $C^{\cdot}(\mathscr{V}, M)$ ($C^{\cdot}(\mathscr{V})$ if $M = \mathscr{A}$), which begins as follows: $0 \to M \to \prod_i M \otimes_{\mathscr{A}} \mathscr{A}_{V_i} \to \prod_{i,j} M \otimes_{\mathscr{A}} \mathscr{A}_{V_i\cap V_j} \to \cdots$ (see **[BGR]**, §8.2).

2.2.5. **Proposition** (Acyclicity Theorem). *For any $\mathscr{A}$-module (resp. finite Banach $\mathscr{A}$-module) M the complex $C^{\cdot}(\mathscr{V}, M)$ is exact (resp. exact and admissible).*

Proof. The case of an arbitrary $\mathscr{A}$-module M can be reduced to the case $M = \mathscr{A}$ by standard arguments (see **[BGR]**, 8.2.1/5), because, by Proposition 2.2.4(ii), the algebra $\mathscr{A}_V$ is flat over $\mathscr{A}$ for any affinoid domain V. If the algebra $\mathscr{A}$ is strictly k-affinoid and all V_i are strictly affinoid domains, the exactness assertion is Tate's Acyclicity Theorem (**[BGR]**, 8.2.1/1) and the admissibility assertion follows from Banach's Open Mapping Theorem. The general case is reduced to this case using Proposition 2.1.2(ii). ∎

By Remark 2.2.2(iv), the class of affinoid domains in X determines a Grothendieck topology (also called the G-topology, see **[BGR]**, 9.1.1/1). Namely, the admissible open subsets are the affinoid domains, and the admissible coverings are the finite coverings by affinoid domains. This G-topology is called *weak.* From Proposition 2.2.5 it follows that the correspondence $V \mapsto \mathscr{A}_V$ is a sheaf of k-affinoid algebras on the weak G-topology of X. Similarly, let $S(X)$ denote the class of closed subsets of X which are finite unions of affinoid domains (*special subsets* of X). The class $S(X)$ determines a G-topology on X which is called *special.*

2.2.6. **Corollary.** *For $V \in S(X)$, set $\mathscr{A}_V = \mathrm{Ker}(\prod_i \mathscr{A}_{V_i} \to \prod_{i,j}\mathscr{A}_{V_i\cap V_j})$, where $\mathscr{V} = \{V_i\}_{i\in I}$ is a finite covering of V by affinoid domains. Then:*

(i) *the commutative Banach k-algebra $\mathscr{A}_V$ does not depend on the covering $\mathscr{V}$;*

(ii) *the correspondence $V \mapsto \mathscr{A}_V$ is a sheaf on the special G-topology of X;*

(iii) *V is an affinoid domain if and only if the algebra $\mathscr{A}_V$ is k-affinoid.* ∎

2.2.7. **Corollary.** (i) *Let U and V be two closed subsets of X with empty intersection, and set $W = U \cup V$. Then W is an affinoid domain if and only if U and V are affinoid domains. If this is the case, then one has an isomorphism of k-affinoid algebras $\mathscr{A}_W \widetilde{\to} \mathscr{A}_U \times \mathscr{A}_V$.*

(ii) *Let V be an affinoid domain in X such that the homomorphism $\mathscr{A} \to \mathscr{A}_V$ is surjective and admissible. Then the complement $U = X\backslash V$ is an* *affinoid domain, and there is an isomorphism* $\mathscr{A} \widetilde{\to} \mathscr{A}_U \times \mathscr{A}_V$.

PROOF. (i) The converse implication follows immediately from Corollary 2.2.6. To verify the direct implication, we may assume that $W = X$. Since the sets U and V are also open, and since the Laurent neighborhoods of a point form a basis of closed neighborhoods of the point, it follows that U and V are special subsets of X. By Corollary 2.2.6, $\mathscr{A} \widetilde{\to} \mathscr{A}_U \times \mathscr{A}_V$. This implies that the algebras $\mathscr{A}_U$ and $\mathscr{A}_V$ are k-affinoid, and hence U and V are affinoid domains.

(ii) From **[BGR]**, 7.3.3/7, it follows that V is open in X. Hence (ii) follows from (i). ∎

2.2.8. COROLLARY. *A basis of the topology of* X *is formed by the open connected subsets which are countable at infinity.*

PROOF. Let $\mathscr{U}$ be an open neighborhood of a point $x \in X$. Choose a rational neighborhood $V = X(p^{-1}\frac{f}{g})$ of x in $\mathscr{U}$ and a sequence $0 < \varepsilon_1 < \varepsilon_2 < \cdots < 1$, and set $V_n = X((\varepsilon_n p)^{-1}\frac{f}{g})$. Since $\varepsilon_n < \varepsilon_{n+1}$, V_n lies in the topological interior of V_{n+1} in X. We now choose a connected component W_n of V_n for all $n \geq 1$ so that $W_1 \subset W_2 \subset \ldots$. Then W_n is a rational neighborhood of x lying in the topological interior of W_{n+1}. The set $\cup_{n=1}^{\infty} W_n$ is a connected open neighborhood of X which is countable at infinity and lies in $\mathscr{U}$. ∎

Given a closed subset $\Sigma \subset X$, let $\mathscr{A}_{(\Sigma)}$ denote the localization of $\mathscr{A}$ with respect to the elements which do not vanish on Σ. Given a bounded homomorphism $\varphi: \mathscr{A} \to \mathscr{B}$ to a commutative Banach k-algebra $\mathscr{B}$, let Σ_φ denote the image of $\mathscr{M}(\mathscr{B})$ in X (the *spectrum* of φ). It is clear that φ extends to a homomorphism $\mathscr{A}_{(\Sigma_\varphi)} \to \mathscr{B}$.

2.2.9. PROPOSITION. *The following properties of a bounded homomorphism of k-affinoid algebras* $\varphi: \mathscr{A} \to \mathscr{B}$ *are equivalent*:

(a) *the image of* $\mathscr{A}$ (*resp.* $\mathscr{A}_{(\Sigma_\varphi)}$) *in* $\mathscr{B}$ *is everywhere dense*;

(b) *there exists a Weierstrass* (*resp. rational*) *domain* $V \subset X$ *containing* Σ_φ *such that* φ *extends to an admissible epimorphism* $\mathscr{A}_V \to \mathscr{B}$.

PROOF. The implication (b) ⇒ (a) is trivial. Suppose that the image of $\mathscr{A}_{(\Sigma_\varphi)}$ in $\mathscr{B}$ is everywhere dense. Because $\mathscr{B}$ is k-affinoid, there exists an admissible epimorphism $\varphi': \mathscr{A}\{r_1^{-1}T_1, \ldots, r_n^{-1}T_n\} \to \mathscr{B}$ extending φ. By Proposition 2.1.7, we may assume that $\varphi'(T_i) = \varphi(f_i)/\varphi(g)$, where $f_i, g \in \mathscr{A}$ and $f_i/g \in \mathscr{A}_{(\Sigma_\varphi)}$, $1 \leq i \leq n$. Because g does not vanish on Σ_φ, there exists $r_{n+1} > 0$ with $|g(x)| \geq r_{n+1}^{-1}$ for all $x \in \Sigma_\varphi$. Let V denote the rational domain $X(r^{-1}\frac{f}{g})$, where $f = (f_1, \ldots, f_n, 1)$ and $r = (r_1, \ldots, r_n, r_{n+1})$. It is clear that $\Sigma_\varphi \subset V$ and that φ' extends to an admissible epimorphism $\psi: \mathscr{A}\{r_1^{-1}T_1, \ldots, r_{n+1}^{-1}T_{n+1}\} \to \mathscr{B}$ sending T_{n+1} to $\varphi(g)^{-1}$. Because $\psi(gT_i - f_i) = \psi(gT_{n+1} - 1) = 0$, $1 \leq i \leq n$, the homomorphism ψ

factors through $\mathcal{A}_V$. The first case of the implication (a) $\Rightarrow$ (b) is treated similarly. ∎

2.2.10. COROLLARY. *The following properties of an affinoid domain $V \subset X$ are equivalent:*

(a) *the image of $\mathcal{A}$ (resp. $\mathcal{A}_{(V)}$) in $\mathcal{A}_V$ is everywhere dense;*

(b) *V is a Weierstrass (resp. rational) domain.*

PROOF. (a) $\Rightarrow$ (b). By Proposition 2.2.9, there exists a rational domain $U \subset X$ containing V such that the homomorphism $\mathcal{A}_U \to \mathcal{A}_V$ is surjective and admissible. From Corollary 2.2.7(ii) it follows that V is open in U. Hence, $V = U(r^{-1}h)$ for some $r_i > 0$, $1 \le i \le n$, where $h_1, \dots, h_n$ are generators of the kernel of the homomorphism $\mathcal{A}_U \to \mathcal{A}_V$. As in the proof of Proposition 2.2.9, we choose elements $f_1, \dots, f_n, g \in \mathcal{A}$ such that $1/g \in \mathcal{A}_{(U)}$ and $\|f_i/g - h_i\| < r_i$ in $\mathcal{A}_U$, $1 \le i \le n$. If $r_{n+1} = \rho(g^{-1})$ in $\mathcal{A}_U$, then $V = X(r^{-1}\frac{f}{g}) \cap U$, where $f = (f_1, \dots, f_n, 1)$ and $r = (r_1, \dots, r_{n+1})$, i.e., V is a rational domain in X. ∎

2.2.11. COROLLARY (Transitivity Property). *Suppose that U is a Weierstrass (resp. rational) domain in V and V is a Weierstrass (resp. rational) domain in X. Then U is a Weierstrass (resp. rational) domain in X.* ∎

2.3. Affinoid spaces. Let X be the spectrum of a k-affinoid algebra $\mathcal{A}$. Using the Acyclicity Theorem, the space X can be provided with a sheaf of rings as follows. For an open set $\mathcal{U} \subset X$, we set

$$\Gamma(\mathcal{U}, \mathcal{O}_X) = \varprojlim \mathcal{A}_V,$$

where the limit is taken over all special subsets $V \subset \mathcal{U}$. It is easy to see that the stalk $\mathcal{O}_{X,x}$ of $\mathcal{O}_X$ at a point $x \in X$ is a local ring with maximal ideal $\mathbf{m}_x = \{f \in \mathcal{O}_{X,x} \mid |f(x)| = 0\}$. If $\mathcal{A}$ is strictly k-affinoid and $x \in \mathrm{Max}(\mathcal{A})$, then $\mathcal{O}_{X,x}$ coincides with the algebra of germs of affinoid functions on $\mathrm{Max}(\mathcal{A})$ considered in **[BGR]**, 7.3.2. From **[BGR]**, 7.3.2/3 one can deduce that $\mathcal{O}_{X,x}$ is a Noetherian ring for any $x \in X$ and any $\mathcal{A}$.

The locally ringed space X obtained in this way is called a *k-affinoid space*. If the algebra $\mathcal{A}$ is strictly k-affinoid, X is called a *strictly k-affinoid space*. The correspondence $\mathcal{A} \mapsto X = \mathcal{M}(\mathcal{A})$ is a faithful contravariant functor from the category of k-affinoid algebras to the category of locally ringed spaces. But this functor is not fully faithful (see Remark 2.1.13). We therefore define a *morphism of k-affinoid spaces* $X = \mathcal{M}(\mathcal{A}) \to Y = \mathcal{M}(\mathcal{B})$ to be a morphism of locally ringed spaces which comes from a bounded homomorphism $\mathcal{B} \to \mathcal{A}$. With this definition, the category of k-affinoid spaces is antiequivalent to the category of k-affinoid algebras. For any non-Archimedean field K over k, we have a *ground field extension functor* $X \mapsto X\widehat{\otimes}K = \mathcal{M}(\mathcal{A}\widehat{\otimes}K)$. An *affinoid space over k* is a K-affinoid space for some non-Archimedean field K over k. A morphism between two such spaces is defined to be a morphism of locally ringed spaces which is induced by a bounded k-homomorphism of the corresponding affinoid algebras over k.

To every $\mathcal{A}$-module M one can associate an $\mathcal{O}_X$-module $\mathcal{O}_X(M)$, where for an open subset $\mathcal{U} \subset X$ we set

$$\Gamma(\mathcal{U}, \mathcal{O}_X(M)) = \varprojlim M \otimes_{\mathcal{A}} \mathcal{A}_V.$$

Since $\mathcal{O}_X(M)_x \widetilde{\to} M \otimes_{\mathcal{A}} \mathcal{O}_{X,x}$, the functor $M \mapsto \mathcal{O}_X(M)$ from the category of $\mathcal{A}$-modules $\mathrm{Mod}(\mathcal{A})$ to the category of $\mathcal{O}_X$-modules $\mathrm{Mod}(X)$ is exact, fully faithful, and commutes with tensor products. It is easy to show that for a finite $\mathcal{A}$-module M, $\mathcal{O}_X(M)$ is a coherent $\mathcal{O}_X$-module. In particular, $\mathcal{O}_X$ is a coherent sheaf of rings. The following proposition is easily deduced from a result of Kiehl ([**Ki2**]).

2.3.1. PROPOSITION. *Every coherent $\mathcal{O}_X$-module is isomorphic to $\mathcal{O}_X(M)$ for some finite $\mathcal{A}$-module M.* ∎

2.3.2. COROLLARY. *Let $\varphi: Y = \mathcal{M}(\mathcal{B}) \to X = \mathcal{M}(\mathcal{A})$ be a morphism of k-affinoid spaces.*

(i) *$\varphi_*(\mathcal{O}_Y)$ is a coherent $\mathcal{O}_X$-module if and only if $\mathcal{B}$ is a finite Banach $\mathcal{A}$-module.*

(ii) *$\varphi_*(\mathcal{O}_Y)$ is a coherent $\mathcal{O}_X$-module and the homomorphism $\mathcal{O}_X \to \varphi_*(\mathcal{O}_Y)$ is surjective if and only if $\mathcal{A} \to \mathcal{B}$ is an admissible epimorphism.*

PROOF. Suppose that $\varphi_*(\mathcal{O}_Y)$ is a coherent $\mathcal{O}_X$-module. By Proposition 2.3.1, the $\mathcal{A}$-module $\mathcal{B} = \Gamma(X, \varphi_*(\mathcal{O}_Y))$ is finite. Let $\mathcal{B}'$ denote the algebra $\mathcal{B}$ endowed with a finite Banach $\mathcal{A}$-module structure (see Proposition 2.1.12). We have to show that the canonical bounded homomorphism $\nu: \mathcal{B}' \to \mathcal{B}$ is admissible. By Proposition 2.2.9, it suffices to show that the map $\psi: Y \to Y' = \mathcal{M}(\mathcal{B}')$ is bijective. We have $\varphi'_*(\mathcal{O}_{Y'}) \widetilde{\to} \varphi_*(\mathcal{O}_Y)$, where φ' denotes the morphism $Y' \to X$. From this it follows that $\varphi^{-1}(x) = \varphi'^{-1}(x)$ for any point $x \in X$. Hence ψ is bijective. The statement (ii) follows from (i). ∎

A morphism $\varphi: Y \to X$ satisfying the properties of Corollary 2.3.2(i) (resp. (ii)) is said to be *finite* (resp. a *closed immersion*).

Now let $\mathcal{X}$ denote the affine scheme $\mathrm{Spec}(\mathcal{A})$. The canonical map $\pi: X \to \mathcal{X}$, which associates to a character of $\mathcal{A}$ its kernel, is continuous and surjective. It is easy to see that this is a strictly flat morphism of locally ringed spaces. In particular, the functor $\mathcal{F} \mapsto \pi^*(\mathcal{F})$ from the category of $\mathcal{O}_{\mathcal{X}}$-modules to the category of $\mathcal{O}_X$-modules is exact and faithful. If $\mathcal{O}_{\mathcal{X}}(M)$ is the quasicoherent $\mathcal{O}_{\mathcal{X}}$-module associated with an $\mathcal{A}$-module M, then $\pi^*(\mathcal{O}_{\mathcal{X}}(M)) = \mathcal{O}_X(M)$. Proposition 2.3.1 implies that π^* induces an equivalence of categories of coherent sheaves $\mathrm{Mod}^h(\mathcal{X}) \widetilde{\to} \mathrm{Mod}^h(X)$. From this it follows that if $\mathcal{M}_X$ (resp. $\mathcal{M}_{\mathcal{X}}$) denotes the *sheaf of germs of meromorphic* (resp. *rational*) *functions* on X (resp. $\mathcal{X}$), then the natural homomorphism of sheaves $\mathcal{M}_{\mathcal{X}} \to \pi_*(\mathcal{M}_X)$ induces an isomorphism $\Gamma(\mathcal{X}, \mathcal{M}_{\mathcal{X}}) \widetilde{\to} \Gamma(X, \mathcal{M}_X)$.

By Corollary 2.3.2(i) (resp. (ii)), the category of finite morphisms (resp. closed immersions) over X is equivalent to the corresponding category over $\mathscr{X}$. In particular, X is irreducible if and only if $\mathscr{X}$ is irreducible.

There are also connections between the local properties of X and $\mathscr{X}$. Recall that the *Krull dimension* $\dim(A)$ of a commutative ring A is the maximal length n of a chain $\wp_0 \subset \wp_1 \subset \cdots \subset \wp_n$ of prime ideals of A. Let A be local, and let $\mathbf{m}$ be its maximal ideal. The *depth* of A (depth (A)) is the maximal length n of a regular sequence $f_1, \dots, f_n \in \mathbf{m}$. A sequence $f_1, \dots, f_n \in \mathbf{m}$ is *regular* if, for any $1 \le i \le n$, f_i is not a zero divisor in $A/(f_1, \dots, f_{i-1})$. We also set $\mathrm{e}(A) = \dim_{A/\mathbf{m}}(\mathbf{m}/\mathbf{m}^2)$. The ring A is called *Cohen-Macaulay* (resp. *regular*) if $\dim(A) = \mathrm{depth}(A)$ (resp. $\dim(A) = e(A)$).

2.3.3. PROPOSITION. *Every point* $x \in X$ *has an affinoid neighborhood* V *such that if* $\mathscr{V} = \mathrm{Spec}(\mathscr{A}_V)$, *then* $\dim(\mathscr{O}_{X,x}) = \dim(\mathscr{O}_{\mathscr{V},\pi(x)})$, $\mathrm{depth}(\mathscr{O}_{X,x}) = \mathrm{depth}(\mathscr{O}_{\mathscr{V},\pi(x)})$, $\mathrm{e}(\mathscr{O}_{X,x}) = \mathrm{e}(\mathscr{O}_{\mathscr{V},\pi(x)})$. ∎

X is said to be *reduced* (*normal, Cohen-Macaulay, regular*) *at a point* x if the ring $\mathscr{O}_{X,x}$ possesses the corresponding property. X is said to be *smooth at a point* x if, for any non-Archimedean field K over k, $X\widehat{\otimes}K$ is regular at any point x' over x. From results of **[BKKN]** it is easy to deduce that X is smooth at x if and only if $X\widehat{\otimes}K$ is regular at some point x' over x for some non-Archimedean field K over k which contains the field $k^{p^{-1}}$, where $p = \mathrm{char}(k)$.

2.3.4. PROPOSITION. X *is reduced* (*normal, Cohen-Macaulay, regular, smooth*) *at a point* x *if and only if* $\mathscr{X}$ *possesses the same property at the point* $\pi(x)$. ∎

The *dimension* $\dim(X)$ of a k-affinoid space $X = \mathscr{M}(\mathscr{A})$ is the Krull dimension of the algebra $\mathscr{A}\widehat{\otimes}K$ for some field K of the form $K_{r_1,\dots,r_n}$ such that $\mathscr{A}\widehat{\otimes}K$ is strictly K-affinoid. It is clear that $\dim(X)$ does not depend on the field K. But for $X = \mathscr{M}(K_r)$, where $r \notin \sqrt{|k^*|}$, we have: $\dim(X) = 1$ if X is regarded as a k-affinoid space and $\dim(X) = 0$ if X is regarded as a K_r-affinoid space. Note that for any finite affinoid covering $\{V_i\}$ of X, one has $\dim(X) = \max \dim(V_i)$.

The *dimension* $\dim_x(X)$ *of* X *at a point* $x \in X$ is the infimum of $\dim(\mathscr{A}_V)$ over all affinoid neighborhoods V of x in X. By this remark, $\dim(X)$ is equal to the supremum of $\dim_x(X)$ over all $x \in X$. X is said to be *pure dimensional* if $\dim_x(X) = \dim(X)$ for all $x \in X$.

2.3.5. PROPOSITION. X *is pure* d*-dimensional if and only if all of the irreducible components of* X *are of dimension* d.

PROOF. By Proposition 2.1.4(iii), the statement is easily reduced to the case when the valuation on k is nontrivial and X is strictly k-affinoid.

In this case X is pure d-dimensional if and only if $\dim_x(X) = d$ for all $x \in \mathrm{Max}(\mathcal{A})$. Hence, we can use Theorem III.4.3 from **[GR1]**. ∎

2.3.6. COROLLARY. *For every closed k-analytic subset $Y \subset X$ there exists a decomposition $Y = \bigcup_{d=0}^{\dim(Y)(d)} Y$, where ${}^{(d)}Y$ is a pure d-dimensional closed k-analytic subset. Furthermore, if $d \neq d'$, then* $\dim({}^{(d)}Y \cap {}^{(d')}Y) \leq \min(d, d') - 1$. ∎

2.3.7. COROLLARY (Ritt's Lemma). *A closed k-analytic subset $Y \subset X$ is nowhere dense in X if and only if* $\dim_x(Y) < \dim_x(X)$ *for all $x \in X$.* ∎

The space X is said to be *irreducible at a point* $x \in X$ if the stalk $\mathcal{O}_{X,x}$ is an integral domain. For example, X is irreducible at all normal points. X is said to be *pure dimensional at a point* $x \in X$ if there exists an open neighborhood $x \in \mathcal{U}$ such that $\dim_y(X) = \dim_x(X)$ for all $y \in \mathcal{U}$. If X is irreducible at a point $x \in X$, then the irreducible affinoid neighborhoods of x form a basis of affinoid neighborhoods of x, and X is pure dimensional at x.

2.4. The reduction map. Given a commutative Banach algebra $\mathcal{A}$, the set $\mathcal{A}^\circ = \{f \in \mathcal{A} \mid \rho(f) \leq 1\}$ is a ring and $\mathcal{A}^{\circ\circ} = \{f \in \mathcal{A} \mid \rho(f) < 1\}$ is an ideal in it. The residue ring $\mathcal{A}^\circ/\mathcal{A}^{\circ\circ}$ is denoted by $\widetilde{\mathcal{A}}$. Every homomorphism of commutative Banach algebras $\varphi: \mathcal{A} \to \mathcal{B}$ induces ring homomorphisms $\varphi^\circ: \mathcal{A}^\circ \to \mathcal{B}^\circ$ and $\widetilde{\varphi}: \widetilde{\mathcal{A}} \to \widetilde{\mathcal{B}}$. In particular, for any point $x \in \mathcal{M}(\mathcal{A})$ there is a homomorphism $\widetilde{\chi}_x: \widetilde{\mathcal{A}} \to \widetilde{\mathcal{H}(x)}$. Because $\widetilde{\mathcal{H}(x)}$ is a field, $\mathrm{Ker}(\widetilde{\chi}_x)$ is a prime ideal of $\widetilde{\mathcal{A}}$. Letting $\widetilde{k}(\widetilde{x})$ denote the fraction field of the ring $\widetilde{\mathcal{A}}/\mathrm{Ker}(\widetilde{\chi}_x)$, we obtain an embedding of fields $\widetilde{k}(\widetilde{x}) \to \widetilde{\mathcal{H}(x)}$ and the following map:

$$\pi: \mathcal{M}(\mathcal{A}) \to \mathrm{Spec}(\widetilde{\mathcal{A}}): x \mapsto \mathrm{Ker}(\widetilde{\chi}_x).$$

If the norm on $\mathcal{A}$ is trivial, then $\mathcal{A}^\circ = \mathcal{A}$, and $\mathcal{A}^{\circ\circ}$ is the nilradical of $\mathcal{A}$. Note that in this case the map $\pi: \mathcal{M}(\mathcal{A}) \to \mathrm{Spec}(\widetilde{\mathcal{A}}) = \mathrm{Spec}(\mathcal{A})$ does not coincide with the map defined in Remark 1.2.5(i). If the valuation on k is nontrivial and $\mathcal{A}$ is a strictly k-affinoid algebra, then the restriction of π to $\mathrm{Max}(\mathcal{A})$ coincides with the Tate map $\mathrm{Max}(\mathcal{A}) \to \mathrm{Max}(\widetilde{\mathcal{A}})$.

Given a commutative ring R with identity, one sets $V(\mathbf{a}) = \{\wp \in \mathrm{Spec}(R) \mid \mathbf{a} \subset \wp\}$, $\Lambda(\mathbf{a}) = \{\wp \in \mathrm{Spec}(R) \mid \wp \subset \mathbf{a}\}$, and $D(f) = \{\wp \in \mathrm{Spec}(R) \mid f \notin \wp\}$, where $\mathbf{a}$ is an ideal of R and $f \in R$. The sets of the form $V(\mathbf{a})$ are closed in the Zariski topology of Spec $(\mathcal{A})$. The sets of the form $D(f)$ form a basis of this topology. It is clear that $\Lambda(\mathbf{a}) = \bigcap_{f \notin \mathbf{a}} D(f)$.

2.4.1. LEMMA. *Let $\mathcal{A}$ be a commutative Banach k-algebra with identity.*

(i) *If an ideal $\mathbf{a} \subset \widetilde{\mathcal{A}}$ is generated by elements $\widetilde{f}_i$, $i \in I$, where $f_i \in A^\circ$, then* $\pi^{-1}(V(\mathbf{a})) = \{x \in \mathcal{M}(\mathcal{A}) \mid |f_i(x)| < 1, i \in I\}$.

(ii) *If $f \in \mathcal{A}^\circ \backslash \mathcal{A}^{\circ\circ}$, then the set* $\pi^{-1}(D(f)) = \{x \in \mathcal{M}(\mathcal{A}) \mid |f(x)| = 1\}$ *is nonempty.* ∎

2.4.2. Corollary. *If the ring $\mathcal{A}$ is Noetherian, the inverse image of an open (resp. closed) subset of Spec $(\widetilde{\mathcal{A}})$ is a closed (resp. open) subset of $\mathcal{M}(\mathcal{A})$.* ∎

2.4.3. Corollary. *The inverse image of a minimal prime ideal $\wp \in$ Spec$(\widetilde{\mathcal{A}})$ is nonempty and closed.*

Proof. Because $\Lambda(\wp) = \{\wp\}$, it follows that $\pi^{-1}(\wp) = \cap \pi^{-1}(D(\widetilde{f}))$, where f runs through all elements in $\mathcal{A}^\circ \backslash \mathcal{A}^{\circ\circ}$ with $\widetilde{f} \notin \wp$. The intersection is nonempty because the space $\mathcal{M}(\mathcal{A})$ is compact and $D(\widetilde{f}_1) \cap D(\widetilde{f}_2) = D(\widetilde{f_1 f_2})$. ∎

2.4.4. Proposition. *Let $\mathcal{A}$ be a strictly k-affinoid algebra. Set $X = \mathcal{M}(\mathcal{A})$, $\widetilde{X} =$ Spec$(\widetilde{\mathcal{A}})$ and denote by $\widetilde{X}_{gen}$ the set of generic points of the irreducible components of $\widetilde{X}$.*

(i) *The reduction map $\pi: X \to \widetilde{X}$ is surjective.*

(ii) *For any $\widetilde{x} \in \widetilde{X}_{gen}$, there exists a unique point $x \in X$ with $\pi(x) = \widetilde{x}$. If $|\mathcal{A}|_{\text{sup}} = |k|$, then for these points there is an isomorphism $\widetilde{k}(\widetilde{x}) \xrightarrow{\sim} \widetilde{\mathcal{H}(x)}$.*

(iii) *The set $\pi^{-1}(\widetilde{X}_{gen})$ is the Shilov boundary of $\mathcal{A}$.*

A closed subset Γ of the spectrum of a commutative Banach k-algebra $\mathcal{A}$ is called a boundary if every element of $\mathcal{A}$ has its maximum in Γ. The set of all boundaries is partially ordered by inclusion, and it satisfies the conditions of Zorn's Lemma. Hence, there exist minimal boundaries. If there exists a unique minimal boundary, it is said to be the *Shilov boundary* of $\mathcal{A}$, and it is denoted by $\Gamma(\mathcal{A})$.

Proof. If the valuation on k is trivial, then the norm on $\mathcal{A}$ is equivalent to the trivial norm. In this case $|\mathcal{A}|_{\text{sup}} = |k|$, and all of the statements are easily verified (see 1.4.2). So we assume that the valuation on k is nontrivial. In this case (i) is proved in the same way as Tate's Theorem ([**BGR**], 7.1.5/4).

(ii) First we consider the case when $\widetilde{X}$ is irreducible (i.e., $\widetilde{\mathcal{A}}$ has no zero divisors). Let $\widetilde{x}$ be the generic point of $\widetilde{X}$, and let $x \in \pi^{-1}(\widetilde{x})$. We claim that $|f(x)| = |f|_{\text{sup}}$ for all $f \in \mathcal{A}$. Indeed, if $|f|_{\text{sup}} = 0$, then $f \in \text{rad}(\mathcal{A})$ and $|f(x)| = 0$. If $|f|_{\text{sup}} > 0$, then there exist $a \in k^*$ and $n \geq 1$ with $|f|^n_{\text{sup}} = |a|$ ([**BGR**],6.2.1/4). Setting $g = a^{-1} f^n$, we have $|g|_{\text{sup}} = 1$. Since $g \in \mathcal{A}^\circ \backslash \mathcal{A}^{\circ\circ}$, it follows that $|g(x)| = 1$. Therefore, $|f(x)| = |f|_{\text{sup}}$. Hence, $\pi^{-1}(\widetilde{x}) = \{x\}$. Suppose that $|\mathcal{A}|_{\text{sup}} = |k|$. To show that $\widetilde{k}(\widetilde{x}) \xrightarrow{\sim} \widetilde{\mathcal{H}(x)}$ it suffices to assume that $\text{rad}(\mathcal{A}) = 0$. In this case the fraction field of $\mathcal{A}$ is everywhere dense in $\mathcal{H}(x)$, and so every element of $\widetilde{\mathcal{H}(x)}$ comes from an element f/g where $f, g \in \mathcal{A}$ and $|f|_{\text{sup}} = |g|_{\text{sup}}$. By the hypothesis, there exists $a \in k$ with $|f|_{\text{sup}} = |g|_{\text{sup}} = |a|$. We see that the element of $\widetilde{\mathcal{H}(x)}$ comes from f'/g', where $f' = a^{-1} f$ and $g' = a^{-1} g$ are elements of $\mathcal{A}^\circ \backslash \mathcal{A}^{\circ\circ}$.

Now suppose that the algebra $\mathscr{A}$ is arbitrary, and let $\widetilde{x} \in \widetilde{X}_{gen}$. We choose an element $f \in \mathscr{A}^\circ$ with $\widetilde{f}(\widetilde{x}) \neq 0$ such that $\widetilde{f}$ vanishes on all irreducible components of $\widetilde{X}$ which do not contain the point $\widetilde{x}$. Setting $\mathscr{B} = \mathscr{A}\{f^{-1}\}$, we have $\mathscr{M}(\mathscr{B}) = X(f^{-1})(= \{x \in X \| |f(x)| = 1\})$, and hence $\pi^{-1}(\widetilde{x}) \subset X(f^{-1})$. But $\widetilde{B} = \widetilde{A}[\widetilde{f}^{-1}]$ and $|\mathscr{B}|_{\sup} = |\mathscr{A}|_{\sup}$ (**[BGR]**,7.2.6). Because the statement (ii) is valid for $\mathscr{B}$, it is also valid for $\mathscr{A}$.

(iii) If $f \in \mathscr{A}^\circ \backslash \mathscr{A}^{\circ\circ}$, then there exists a point $\widetilde{x} \in \widetilde{X}_{gen}$ with $\widetilde{f}(\widetilde{x}) \neq 0$. So for the point $x := \pi^{-1}(\widetilde{x})$, we have $|f(x)| = 1 = |f|_{\sup}$. From this it follows that any element $f \in \mathscr{A}$ takes its maximum on $\pi^{-1}(\widetilde{X}_{gen})$. Now let $\widetilde{x} \in \widetilde{X}_{gen}$, and let $\mathscr{U}$ be an open neighborhood of the point $x := \pi^{-1}(\widetilde{x})$. Because $x = \bigcap_{\widetilde{f}(\widetilde{x}) \neq 0} \pi^{-1}(D(\widetilde{f}))$, there exists an element $f \in \mathscr{A}^\circ$ with $\widetilde{f}(\widetilde{x}) \neq 0$ and $\Sigma := \{y \| |f(y)| = 1\} \subset \mathscr{U}$. Furthermore, because $\Sigma = \bigcap_{\varepsilon > 0} \Sigma_\varepsilon$, where $\Sigma_\varepsilon = \{y \| |f(y)| \geq 1 - \varepsilon\}$, there exists $\varepsilon > 0$ with $\Sigma_\varepsilon \subset \mathscr{U}$. Hence, the open subset $\mathscr{U}' = \{y \| |f(y)| > 1 - \varepsilon\}$ is contained in $\mathscr{U}$. ∎

2.4.5. COROLLARY. *The Shilov boundary $\Gamma(\mathscr{A})$ exists and is finite for any k-affinoid algebra $\mathscr{A}$. Furthermore, if $x_0 \in \Gamma(\mathscr{A})$, then for any open neighborhood $\mathscr{U}$ of x_0, there exist $f \in \mathscr{A}$ and $\varepsilon > 0$ such that $|f(x_0)| = \rho(f)$ and $\{x \in X \| |f(x)| > \rho(f) - \varepsilon\} \subset \mathscr{U}$.*

PROOF. By the proof of Proposition 2.4.4(iii), to verify the assertions it suffices to show that, if they are true for the algebra $\mathscr{A}' = \mathscr{A} \widehat{\otimes} K_r$, then they are also true for $\mathscr{A}$. Let π denote the canonical map $X' = \mathscr{M}(\mathscr{A}') \to X$. We claim that the set $\pi(\Gamma(\mathscr{A}'))$ is the Shilov boundary of $\mathscr{A}$. It clearly suffices to verify the last assertion of the corollary for every point $x_0 \in \pi(\Gamma(\mathscr{A}'))$. Let $x_0' \in \Gamma(\mathscr{A}') \cap \pi^{-1}(x_0)$. We choose $f' = \sum_{i=-\infty}^{+\infty} a_i T^i \in \mathscr{A}'$ and $\varepsilon > 0$ so that $|f'(x_0')| = \rho(f')$ and $\{x' \in X' \| |f'(x')| > \rho(f') - \varepsilon\} \subset \pi^{-1}(\mathscr{U})$. Because $\rho(f') = \max |\rho(a_i)| r^i$ and $|f'(x_0')| = \max |a_i(x_0)| r^i$, there exists j with $|a_j(x_0)| = \rho(a_j)$ and $\rho(f') = |a_j(x_0)| r^j$. To satisfy (i) it suffices to take a_j and εr^{-j}, respectively. ∎

2.5. The relative interior. Let $\mathscr{A}$ be a k-affinoid algebra.

2.5.1. DEFINITION. A bounded $\mathscr{A}$-homomorphism $\varphi: \mathscr{B} \to \mathscr{D}$ from an $\mathscr{A}$-affinoid algebra $\mathscr{B}$ to a Banach $\mathscr{A}$-algebra $\mathscr{D}$ is called *inner with respect to $\mathscr{A}$* (or an *inner $\mathscr{A}$-homomorphism*) if there is an admissible epimorphism $\pi: \mathscr{A}\{r_1^{-1}T_1, \dots, r_n^{-1}T_n\} \to \mathscr{B}$ such that $\rho(\varphi(\pi(T_i))) < r_i$ for all $1 \leq i \leq n$.

2.5.2. PROPOSITION. *The following properties of a bounded $\mathscr{A}$-homomorphism $\varphi: \mathscr{B} \to \mathscr{D}$ are equivalent:*

(a) *φ is inner with respect to $\mathscr{A}$;*

(b) *for any bounded $\mathscr{A}$-homomorphism $\psi: \mathscr{A}\{r^{-1}X\} \to \mathscr{B}$, there exists a polynomial $P = X^n + a_1 X^{n-1} + \cdots + a_n$ such that $\rho(a_i) \leq r^i$, $1 \leq i \leq n$, and $\rho(\varphi(\psi(P))) < r^n$;*

(c) *for any* $0 < \varepsilon < 1$, *there exists an admissible epimorphism* π: $\mathcal{A}\{r_1^{-1}T_1, \dots, r_n^{-1}T_n\} \to \mathcal{B}$ *such that* $||\varphi(\pi(T_i))|| < \varepsilon r_i$, $1 \le i \le n$.

If the algebras $\mathcal{A}$ *and* $\mathcal{B}$ *are strictly* k*-affinoid, these properties are equivalent to*:

(d) *the ring* $\widetilde{\varphi}(\widetilde{\mathcal{B}})$ *is integral over* $\widetilde{\varphi}(\widetilde{\mathcal{A}})$.

PROOF. The implication (c) $\Rightarrow$ (a) is trivial.

We verify the implications (a) $\Rightarrow$ (d) $\Rightarrow$ (b) for strictly k-affinoid $\mathcal{A}$ and $\mathcal{B}$ (we assume that the valuation on k is nontrivial). Let $\pi: \mathcal{A}\{r_1^{-1}T_1, \dots, r_n^{-1}T_n\} \to \mathcal{B}$ be an admissible epimorphism with $\rho(\varphi(\pi(T_i))) < r_i$ for all $1 \le i \le n$. As in the proof of Corollary 2.1.6, we can reduce the situation to the case $r_i \in \sqrt{|k^*|}$, $1 \le i \le n$. Let $r_i^{s_i} = |c_i^{-1}|$ for $c_i \in k^*$, and define a monomorphism $\nu: \mathcal{A}' = \mathcal{A}\{T_1, \dots, T_n\} \to \mathcal{A}\{r_1^{-1}T_1, \dots, r_n^{-1}T_n\}$ by $\nu(T_i) = c_i T_i^{s_i}$, $1 \le i \le n$. This monomorphism is finite (see the proof of [**BGR**], 6.1.5/4). It follows that $\nu' = \pi \circ \nu: \mathcal{A}' \to \mathcal{B}$ is finite. By [**BGR**], 6.3.5/1, the induced homomorphism $\widetilde{\nu}': \widetilde{\mathcal{A}'} = \widetilde{\mathcal{A}}[T_1, \dots, T_n] \to \widetilde{\mathcal{B}}$ is finite. Hence, $\widetilde{\varphi}(\widetilde{\mathcal{B}})$ is integral over $\widetilde{\varphi}(\widetilde{\nu}'(\widetilde{\mathcal{A}}[T_1, \dots, T_n]))$. The latter ring coincides with $\widetilde{\varphi}(\widetilde{\mathcal{A}})$, because $\rho(\nu'(T_i)) < 1$, $1 \le i \le n$. The implication (d) $\Rightarrow$ (b) is verified in the same way. The implication (a) $\Rightarrow$ (b) in the general case now follows from the following lemma.

2.5.3. LEMMA. *If the property* (b) *is true for the homomorphism* $\varphi\widehat{\otimes}K_p$: $\mathcal{B}\widehat{\otimes}K_p \to \mathcal{D}\widehat{\otimes}K_p$ *with* $p \notin \sqrt{|k^*|}$, *then it is also true for the homomorphism* φ.

PROOF. Let $Q(X) = X^n + a_1X^{n-1} + \dots + a_n$ be a polynomial in $(\mathcal{A}\widehat{\otimes}K_p)[X]$ with $\rho(a_i) \le r^i$, $1 \le i \le n$, and $\rho(\varphi(\psi(Q))) < r^n$. If $a_i = \sum_{j=-\infty}^{+\infty} a_{i,j}T^j$, $a_{i,j} \in \mathcal{A}$, then the property (b) is true for the polynomial $P = X^n + a_{1,0}X^{n-1} + \dots + a_{n,0}$ in $\mathcal{A}[X]$. ∎

(b) $\Rightarrow$ (c). We may suppose that $\mathcal{B} = \mathcal{A}\{r^{-1}X\}$. By raising P to a large enough power, we may assume that $||\varphi(P)|| < \varepsilon r^n/||\varphi||$. Let q be a number with $q > r\max(||\varphi||/\varepsilon, 1)$. We claim that the homomorphism $\pi: \mathcal{A}\{q^{-1}T_0, r^{-n}T_1, r^{-(n+1)}T_2, \dots, r^{-(2n-1)}T_n\} \to \mathcal{A}\{r^{-1}X\}$ sending T_0 to X and T_i to $X^{i-1}P$, $1 \le i \le n$, satisfies the requirements. The inequalities for the norms are easily verified. To show that π is surjective and admissible, we need the following simple fact.

2.5.4. LEMMA. *Consider the polynomial rings* $R = \mathbf{Z}[A_1, \dots, A_n]$, $R[T_0, \dots, T_n]$, $R[X]$, *graded as follows*: $\deg(A_i) = i$, $\deg(T_i) = n + i - 1$, $1 \le i \le n$, $\deg(T_0) = \deg(X) = 1$. *Consider the homogeneous homomorphism* Φ: $R[T_0, \dots, T_n] \to R[X]$ *sending* T_0 *to* X *and* T_i *to* $X^{i-1}(X^n + A_1X^{n-1} + \dots + A_n)$, $1 \le i \le n$. *Then for all* $l \ge 0$ *there exist polynomials* $G_l \in R[T_1, \dots, T_n]$ *and* $H_l \in R[T_0]$ *such that* $\deg(G_l) = \deg(H_l) = l$,

$\deg_{T_0}(H_l) \le n-1$ *and* $T_0^l - G_l - H_l \in \mathrm{Ker}(\Phi)$. ■

Using the notation in Lemma 2.5.4, we consider the homomorphism ν: $R[T_0, \dots, T_n] \to \mathscr{A}\{q^{-1}T_0, \ r^{-n}T_1, r^{-(n+1)}T_2, \dots, r^{-(2n-1)}T_n\}$ sending A_i to a_i, $1 \le i \le n$, and we set

$$g_l = \nu(G_l),\, h_l = \nu(H_l) = a_1^{(l)}T_0^{n-1} + \dots + a_n^{(l)},\, l \ge 0.$$

Because $\rho(a_i) \le r^i$, $1 \le i \le n$, from Proposition 2.1.4(i) it follows that $||g_l|| \le Cr^l$ and $||a_i^{(l)}|| \le Cr^l$, $1 \le i \le n$, for all $l \ge 0$. Now choose an arbitrary element $f = \sum_{l=0}^{\infty} b_l X^l \in \mathscr{A}\{r^{-1}X\}$. Because $||b_l||r^l \to 0$ as $l \to \infty$, we can define elements $g = \sum_{l=0}^{\infty} b_l g_l$, $d_i = \sum_{l=0}^{\infty} b_l a_i^{(l)}$ and $h = d_1 T_0^{n-1} + \dots + d_n$. We have $||g_l|| \le C \max ||b_l|| r^l = C||f||$, $||h|| = \max_{1\le i\le n} ||d_i|| q^i \le C(\max_{1\le i \le n} q^i)||f||$ and $\pi(g+h) = f$. Hence the homomorphism is surjective and admissible. ■

2.5.5. Corollary. *If $\varphi: \mathscr{B} \to \mathscr{D}$ is an inner $\mathscr{A}$-homomorphism, then for any $\mathscr{A}$-homomorphisms of $\mathscr{A}$-affinoid algebras $\mathscr{B}' \to \mathscr{B}$ and of Banach $\mathscr{A}$-algebras $\mathscr{D} \to \mathscr{D}'$, the composition $\varphi': \mathscr{B}' \to \mathscr{D}'$ is an inner $\mathscr{A}$-homomorphism.* ■

2.5.6. Corollary. *Let $r \notin \sqrt{|k^*|}$, and suppose that the homomorphism $\varphi\widehat{\otimes}K_r: \mathscr{B}\widehat{\otimes}K_r \to \mathscr{D}\widehat{\otimes}K_r$ is inner with respect to $\mathscr{A}\widehat{\otimes}K_r$. Then φ is inner with respect to $\mathscr{A}$.* ■

Now let φ be a morphism of k-affinoid spaces $Y = \mathscr{M}(\mathscr{B}) \to \mathscr{M}(\mathscr{A})$.

2.5.7. Definition. The *relative interior of* φ is the set $\mathrm{Int}(Y/X)$ of points $y \in Y$ corresponding to the characters $\mathscr{B} \to \mathscr{H}(y)$ which are inner with respect to $\mathscr{A}$. The complement of $\mathrm{Int}(Y/X)$ is called the *relative boundary of* φ, and is denoted by $\partial(Y/X)$. If $\mathscr{A} = k$, the set $\mathrm{Int}(Y/X)$ (resp. $\partial(Y/X)$) is denoted by $\mathrm{Int}(Y)$ (resp. $\partial(Y)$) and is called the *interior* (resp. *boundary*) of Y.

It is easy to see that the set $\mathrm{Int}(Y/X)$ is open. But it may be empty. For example, this is the case for $\mathscr{B} = K_r$, where $r \notin \sqrt{|k^*|}$. The following properties of the relative interior follow easily from Proposition 2.5.2.

2.5.8. Proposition. (i) *For a morphism $\psi: X' \to X$, one has $\psi'^{-1}(\mathrm{Int}(Y/X)) \subset \mathrm{Int}(Y'/X')$, where ψ' is the canonical morphism $Y' = Y \times_X X' \to Y$.*

(ii) *For a non-Archimedean field K over k, one has $\psi^{-1}(\mathrm{Int}(Y/X)) \subset \mathrm{Int}(Y\widehat{\otimes}K/X\widehat{\otimes}K)$, where ψ is the canonical map $Y\widehat{\otimes}K \to Y$.*

(iii) *For morphisms $Z \xrightarrow{\psi} Y \xrightarrow{\varphi} X$, one has $\mathrm{Int}(Z/X) = \mathrm{Int}(Z/Y) \cap \psi^{-1}(\mathrm{Int}(Y/X))$.* ■

2.5.9. Proposition. *A bounded $\mathscr{A}$-homomorphism $\psi: \mathscr{B} \to \mathscr{D}$ to a Banach $\mathscr{A}$-algebra is inner with respect to $\mathscr{A}$ if and only if the image of $\mathscr{M}(\mathscr{D})$ in Y lies in $\mathrm{Int}(Y/X)$.*

2.5.10. LEMMA. *Let U be an affinoid domain in Y, and let V be a special subset of U admitting a finite affinoid covering $V = \cup_{i=1}^n V_i$ such that all of the homomorphisms $\theta_i: \mathcal{B}_U \to \mathcal{B}_{V_i}$ are inner with respect to $\mathcal{A}$. Then the homomorphism $\theta: \mathcal{B}_U \to \mathcal{B}_V$ is inner with respect to $\mathcal{A}$.*

PROOF. Let $\psi: \mathcal{A}\{r^{-1}X\} \to \mathcal{B}_U$ be a bounded $\mathcal{A}$-homomorphism. By hypothesis, there exist polynomials $P_i = X^{m_i} + a_1^{(i)} X^{m_i - 1} + \cdots + a_{m_i}^{(i)}$, $1 \le i \le n$, such that $\rho(a_j^{(i)}) \le r^j$ and $\rho(\theta_i(\psi(P_i))) < \rho(P_i)$. If $P = \prod_{i=1}^n P_i = X^m + a_1 X^{m-1} + \cdots + a^m$, then $\rho(a_i) \le r^i$ and $\rho(\theta(\psi(P))) < \rho(P)$, because the homomorphism $\mathcal{B}_V \to \prod_{i=1}^n \mathcal{B}_{V_i}$ is injective and admissible. ∎

2.5.11. LEMMA. *The following properties of a closed subset $\Sigma \subset Y$ are equivalent:*

(a) $\Sigma \subset \mathrm{Int}(Y/X)$;

(b) *For any $0 < \varepsilon < 1$, there exists an admissible epimorphism π: $\mathcal{A}\{r_1^{-1}T_1, \ldots, r_n^{-1}T_n\} \to \mathcal{B}$ such that $\Sigma \subset Y((\varepsilon r)^{-1} f)$, where $f_i = \pi(T_i)$, $1 \le i \le n$.*

PROOF. The implication (b) ⇒ (a) is trivial. Suppose that $\Sigma \subset \mathrm{Int}(Y/X)$. Then for every point $y \in \Sigma$ there exists a Weierstrass neighborhood V_y of y in Y such that the homomorphism $\mathcal{B} \to \mathcal{B}_{V_y}$ is inner with respect to $\mathcal{A}$. Since Σ is compact, it lies in a finite union $V = \cup_{i=1}^n V_{y_i}$, and we can apply Lemma 2.5.10 and Proposition 2.5.2. ∎

PROOF OF PROPOSITION 2.5.9. Let Σ denote the image of $\mathcal{M}(\mathcal{D})$ in Y, and let $0 < \varepsilon < 1$. By Lemma 2.5.11, there exists an admissible epimorphism $\pi: \mathcal{A}\{r_1^{-1}T_1, \ldots, r_n^{-1}T_n\} \to \mathcal{B}$ such that $\Sigma \subset Y((\varepsilon r)^{-1} f)$, where $f_i = \pi(T_i)$, $1 \le i \le n$. It follows that $\rho(\psi(\pi(T_i))) \le \varepsilon r_i$, $1 \le i \le n$. That is, ψ is inner with respect to $\mathcal{A}$. ∎

2.5.12. COROLLARY. *Let $r \notin \sqrt{|k^*|}$, $Y' = Y \widehat{\otimes} K_r$, $X' = X \widehat{\otimes} K_r$, and let ψ denote the canonical map $Y' \to X'$. Then* $\mathrm{Int}(Y/X) = \{y \in Y \mid \psi^{-1}(y) \subset \mathrm{Int}(Y'/X')\}$.

PROOF. Suppose that $\psi^{-1}(y) \subset \mathrm{Int}(Y'/X')$. Because $\psi^{-1}(y)$ coincides with the image of $\mathcal{M}(\mathcal{H}(y) \widehat{\otimes} K_r)$ in Y, Proposition 2.5.9 implies that the homomorphism $\mathcal{B} \widehat{\otimes} K_r \to \mathcal{H}(y) \widehat{\otimes} K_r$ is inner with respect to $\mathcal{A} \widehat{\otimes} K_r$. From Corollary 2.5.6 it follows that the character $\mathcal{B} \to \mathcal{H}(y)$ is inner with respect to $\mathcal{A}$. ∎

2.5.13. COROLLARY. (i) $\mathrm{Int}(Y/X) = Y$ *if and only if $\varphi: Y \to X$ is a finite morphism.*

(ii) *If Y is an affinoid domain in X, then* $\mathrm{Int}(Y/X)$ *coincides with the topological interior of Y in X.*

PROOF. By Corollary 2.5.12, we may assume that the valuation on k is nontrivial and that $\mathcal{A}$ and $\mathcal{B}$ are strictly k-affinoid algebras.

(i) If $\mathrm{Int}(Y/X) = Y$, then, by Proposition 2.5.9, the identity homomorphism $\mathcal{B} \to \mathcal{B}$ is inner with respect to $\mathcal{A}$. Proposition 2.5.2 implies that $\widetilde{\mathcal{B}}$ is integral over $\widetilde{\mathcal{A}}$. Hence, $\mathcal{B}$ is finite over $\mathcal{A}$.

(ii) Let $y \in \mathrm{Int}(Y/X)$. By the Gerritzen-Grauert Theorem ([**BGR**], 7.3.5/3), there exists a finite rational covering $\{X_i\}$ of X such that $Y_i = Y \cap X_i$ is a Weierstrass domain in X_i. Since $y \in \mathrm{Int}(Y_i/X_i)$ for all i with $y \in Y_i$, we may assume that Y is a Weierstrass domain in X. By the hypothesis, there exists an affinoid generating system $f_1, \dots, f_n$ of $\mathcal{B}$ over $\mathcal{A}$ such that $y \in \mathcal{U} = \{x \in Y \mid |f_i(x)| < 1, 1 \le i \le n\}$. We claim that the set $\mathcal{U}$ is open in X. Because the image of $\mathcal{A}$ in $\mathcal{B}$ is everywhere dense, there exist elements $g_1, \dots, g_n \in \mathcal{A}$ whose images in $\mathcal{B}$ form an affinoid generating system of $\mathcal{B}$ over $\mathcal{A}$ and such that $\|g_i - f_i\| < 1$ on Y, $1 \le i \le n$. We set $\mathcal{V} = \{x \in X \| g_i(x)| < 1, 1 \le i \le n\}$ and $V = \{x \in X \| g_i(x)| \le 1, 1 \le i \le n\}$. Then $Y \subset V$, and the homomorphism $\mathcal{A}_V \to \mathcal{B}$ is surjective. Hence $V = Y \cup V'$, where V' is an affinoid subdomain of X and $Y \cap V' = \varnothing$. Because $\mathcal{V}$ is open in X and $\mathcal{U} = \mathcal{V} \cap Y$, it follows that $\mathcal{U}$ is open in X. ∎

2.5.14. COROLLARY. *Let $U \subset V$ be affinoid domains in X. Then the canonical homomorphism $\mathcal{A}_V \to \mathcal{A}_U$ is inner with respect to $\mathcal{A}$ if and only if U lies in the topological interior of V in X.* ∎

2.5.15. PROPOSITION. *Let V be a Weierstrass (resp. rational) domain in X. Then the following properties of a closed subset $\Sigma \subset X$ are equivalent:*

(a) *Σ lies in the topological interior of V in X;*

(b) *for any $0 < \varepsilon < 1$, there exists a representation $V = X(r^{-1}f)$ (resp. $X(r^{-1}\frac{f}{g})$) such that $\Sigma \subset X((\varepsilon r)^{-1}f)$ (resp. $X((\varepsilon r)^{-1}\frac{f}{g})$).*

PROOF. Suppose that V is rational and $\Sigma \subset \mathrm{Int}(V/X)$. By Lemma 2.5.11, there exists an admissible epimorphism $\pi\colon \mathcal{A}\{r_1^{-1}T_1, \dots, r_n^{-1}T_n\} \to \mathcal{A}_V$ such that $\Sigma \subset V((\varepsilon r)^{-1}h)$, where $h_i = \pi(T_i)$, $1 \le i \le n$. By Proposition 2.1.7, we may assume that $h_i = f_i'/g' \in \mathcal{A}_{(V)}$. Let $r_{n+1} = \rho(g'^{-1})$ in $\mathcal{A}_V$. Set $W = X(r'^{-1}\frac{f'}{g'})$, where $f' = (f_1', \dots, f_n', 1)$, $r' = (r_1, \dots, r_n, r_{n+1})$. Then $V \subset W$, and the canonical map $\mathcal{A}_W \to \mathcal{A}_V$ is an admissible epimorphism. By Corollary 2.2.7, the complement $W \backslash V$ is an affinoid domain in X, and $\mathcal{A}_W \widetilde{\to} \mathcal{A}_V \times \mathcal{A}_{W\backslash V}$. Let e be an element of $\mathcal{A}_W$ which goes to 0 in $\mathcal{A}_V$ and to 1 in $\mathcal{A}_{W\backslash V}$, and let $0 < r_{n+2} < 1$. Since W is rational, there exists an element $f''/g'' \in \mathcal{A}_{(W)}$ such that $\|e - f''/g''\| < \varepsilon r_{n+2}$ in $\mathcal{A}_W$. If we set $f = (f_1'g'', \dots, f_n'g'', g'', f''g')$, $g = g'g''$ and $r = (r_1, \dots, r_{n+2})$, then $V = X(r^{-1}\frac{f}{g})$ and $\Sigma \subset X((\varepsilon r)^{-1}\frac{f}{g})$. ∎

2.5.16. COROLLARY. *Let $\psi\colon \mathcal{A} \to \mathcal{D}$ be a bounded homomorphism to a* *commutative Banach k-algebra $\mathcal{D}$. Then for any rational neighborhood V of the spectrum Σ_ψ, ψ extends to a unique bounded homomorphism $\psi_V : \mathcal{A}_V \to \mathcal{D}$.*

PROOF. Let $0 < \varepsilon < 1$. By Proposition 2.5.15, there exists a representation $V = X(r^{-1}\frac{f}{g})$ such that $\Sigma_\psi \subset X((\varepsilon r)^{-1}\frac{f}{g})$. Then $\psi(g)$ is invertible in $\mathcal{D}$, and $\rho(\psi(f_i)) \le \varepsilon r_i$, $1 \le i \le n$. Therefore, one can construct a bounded homomorphism $\psi' : \mathcal{A}\{r^{-1}T\} \to \mathcal{D}$ which extends ψ and sends T_i to $\psi(f_i)/\psi(g)$, $1 \le i \le n$. Because $\psi'(gT_i - f_i) = 0$, it follows that ψ' factors through $\mathcal{A}\{r^{-1}T\}/(gT_i - f_i) = \mathcal{A}_V$. ∎

2.5.17. PROPOSITION. *Let $X = \bigcup_{i\in I} X_i$ be a finite affinoid covering of X, $Y_i = \varphi^{-1}(X_i)$. Suppose that a point $y \in Y$ belongs to $\mathrm{Int}(Y_i/X_i)$ for all $i \in I$ with $y \in Y_i$. Then $y \in \mathrm{Int}(Y/X)$.*

PROOF. We may assume that the valuation on k is nontrivial and that X and Y are strictly k-affinoid. Using Lemmas 8.2.2/2, 8.2.2/3, and 8.2.2/4 from **[BGR]**, one easily reduces the situation to the case of the covering $X = X(f) \cup X(f^{-1})$, where $f \in \mathcal{A}$. By Proposition 2.5.2, in this case the proposition follows from the following lemma.

2.5.18. LEMMA. *Let A be a subring of a commutative ring B (both are rings with identity), and let x be an invertible element of B. Then the intersection of the integral closures of $A[x]$ and of $A[x^{-1}]$ in B coincides with the integral closure of the ring A in B.*

2.5.19. REMARK. This lemma makes precise Exercise 5, ch.V, §1, from the book **[Bou1]**, which asserts that the intersection of the rings $A[x]$ and $A[x^{-1}]$ is integral over A.

PROOF. Let an element $y \in B$ be integral over the rings $A[x]$ and $A[x^{-1}]$. It suffices to construct a faithful $A[y]$-module which is finitely generated over A. By hypothesis, the element y satisfies the equations:

$$y^n + P_{n-1}(x)y^{n-1} + \cdots + P_0(x) = 0$$
$$y^m + Q_{m-1}(x^{-1})y^{m-1} + \cdots + Q_0(x^{-1}) = 0$$

where $P_i(T), Q_j(T) \in A[T]$. We set $k = \max_{0\le i\le n-1}(\deg(P_i))$ and $\ell = \max_{0\le j\le m-1}(\deg(Q_j))$, and we let M denote the A-submodule of B which is generated by the elements $x^i y^j$ with $0 \le i \le n\ell + mk - 1$ and $0 \le j \le nm - 1$. We claim that M is an $A[y]$-module. It suffices to verify that the elements $x^i y^{nm}$, $0 \le i \le n\ell + mk - 1$, lie in M.

If $0 \le i \le n\ell - 1$, then

$$x^i y^{nm} = x^i[-P_{n-1}(x)y^{n-1} - \cdots - P_0(x)]^m;$$

and because $i + mk \le n\ell + mk - 1$, it follows that $x^i y^{nm} \in M$. If $n\ell \le i \le n\ell + mk - 1$, then

$$x^i y^{nm} = x^i[-Q_{m-1}(x^{-1})y^{m-1} - \cdots - Q_0(x^{-1})]^n;$$

and because $i - n\ell \geq 0$, it follows that $x^i y^{nm} \in M$. Hence, M is an $A[y]$-module. It is a faithful $A[y]$-module because it contains the identity of the ring B. ∎

2.5.20. PROPOSITION (Maximum Modulus Principle). *For an affinoid domain V in a k-affinoid space $X = \mathcal{M}(\mathcal{A})$ one has $\Gamma(\mathcal{A}) \cap V \subset \Gamma(\mathcal{A}_V) \subset \partial(V/X) \cup (\Gamma(\mathcal{A}) \cap V)$.*

PROOF. The first inclusion follows from Corollary 2.4.5. Let K be a field of the form $K_{r_1, \dots, r_n}$ with nontrivial valuation for which the algebras $\mathcal{A}\widehat{\otimes}K$ and $\mathcal{A}_V\widehat{\otimes}K$ are strictly K-affinoid, and let π denote the canonical map $X' = X\widehat{\otimes}K \to X$. Then $\partial(V/X) = \pi(\partial(\pi^{-1}(V)/X'))$, and, by the proof of Corollary 2.4.5, $\Gamma(\mathcal{A}_V) = \pi(\Gamma(\mathcal{A}_{\pi^{-1}(V)}))$. Therefore, the second inclusion is reduced to the case when the valuation on k is nontrivial and X and V are strictly k-affinoid.

We may assume that X is pure dimensional. In particular, the dimension of any irreducible component of $\operatorname{Spec}(\widetilde{\mathcal{A}}_V)$ is equal to $\dim(\widetilde{\mathcal{A}})$. Let $x \in \Gamma(\mathcal{A}_V)$, and suppose that $x \notin \partial(V/X)$. By Proposition 2.5.2, $\widetilde{\chi_x}(\widetilde{\mathcal{A}}_V)$ is integral over $\widetilde{\chi_x}(\widetilde{\mathcal{A}})$. Hence, the dimensions of these rings are equal. But the kernel of $\widetilde{\chi_x}$ in $\widetilde{\mathcal{A}}_V$ is a minimal prime ideal of $\widetilde{\mathcal{A}}_V$. It follows that the kernel of $\widetilde{\chi_x}$ in $\widetilde{\mathcal{A}}$ is a minimal prime ideal of $\widetilde{\mathcal{A}}$. Using Proposition 2.4.4, we find that $x \in \Gamma(\mathcal{A})$. ∎

2.5.21. REMARK. One can describe $X = \mathcal{M}(\mathcal{A})$ in terms of the maximal spectrum $X_0 = \operatorname{Max}(\mathcal{A})$ if the valuation on k is nontrivial and $\mathcal{A}$ is strictly k-affinoid. Let $\varphi: Y \to X$ be a morphism of k-affinoid spaces, and let U and V be special subsets of Y with $U \subset V$. We say that U is *inner in V with respect to X* and write $U \Subset_X V$ if there exist a special subset $U \subset W \subset V$ and finite coverings by affinoid subdomains $U = \bigcup_{i=1}^n U_i$ and $W = \bigcup_{i=1}^n W_i$ such that, for any $1 \leq i \leq n$, $U_i \subset W_i$ and the canonical homomorphism $\mathcal{B}_{W_i} \to \mathcal{B}_{U_i}$ is inner with respect to $\mathcal{A}$. From Lemma 2.5.10 it follows that if U and V are affinoid subdomains of Y, then $U \Subset_X V$ if and only if the canonical homomorphism $\mathcal{B}_V \to \mathcal{B}_U$ is inner with respect to $\mathcal{A}$. Furthermore, a nonempty subset $\Omega \subset S(Y)$ is called an *X-filter in $S(Y)$* if it satisfies the following conditions: (a) if $U \in \Omega$ and $U \subset V$, then $V \in \Omega$; (b) if U, $V \in \Omega$, then $U \cap V \neq \emptyset$ and $U \cap V \in \Omega$; (c) for any $V \in \Omega$ there exists $U \in \Omega$ with $U \Subset_X V$. The set of X-filters in $S(Y)$ is partially ordered by inclusion. Its maximal elements are called *X-ultrafilters in $S(Y)$*. If Ω is an X-filter, then $\Omega^* = \bigcap_{V \in \Omega} V$ is a nonempty compact subset of $\operatorname{Int}(Y/X)$. Conversely, if Σ is a nonempty compact subset of $\operatorname{Int}(Y/X)$, then $\Omega_\Sigma = \{V \in S(Y) | \Sigma \subset \operatorname{Int}(V/X)\}$ is an X-filter. From the above results it follows that the maps $\Omega \mapsto \Omega^*$ and $\Sigma \mapsto \Omega_\Sigma$ are inverse to one another. Now suppose that the valuation of k is nontrivial and $\mathcal{A}$ is strictly k-affinoid. Then the correspondence $V \mapsto V_0 = V \cap X_0$ gives an isomorphism between $S(X)$ and the lattice $S(X_0)$ of special subsets in X_0. The above fact implies that there is a one-to-one correspondence between points of X and X-ultrafilters in $S(X_0)$.

2.6. Functions analytic in a neighborhood of a closed subset. Let Σ be a closed subset in a k-affinoid space $X = \mathcal{M}(\mathcal{A})$. We set $\Gamma(\Sigma, \mathcal{O}_X) = \varinjlim \Gamma(\mathcal{U}, \mathcal{O}_X)$, where $\mathcal{U}$ runs through the open neighborhoods of Σ; and for $f \in \Gamma(\Sigma, \mathcal{O}_X)$ we set $||f||_\Sigma = \max_{x\in\Sigma} |f(x)|$. Let $\mathcal{H}(\Sigma)$ denote the completion of $\Gamma(\Sigma, \mathcal{O}_X)$ with respect to the seminorm $|| \,||_\Sigma$, and let $\mathcal{P}(\Sigma)$ (resp. $\mathcal{R}(\Sigma)$) denote the closure of the image of $\mathcal{A}$ (resp. $\mathcal{A}_{(\Sigma)}$) in $\mathcal{H}(\Sigma)$.

Next, the set $\Sigma^h = \{x \in X \,|\, |f(x)| \le ||f||_\Sigma \text{ for all } f \in \mathcal{A}\}$ is called the *holomorphically convex envelope of* Σ *in* X: The set $\Sigma^m = \{x \in X \,|\, |f(x)| \le ||\frac{f}{g}||_\Sigma |g(x)| \text{ for all } \frac{f}{g} \in \mathcal{A}_{(\Sigma)}\}$ is called the *meromorphically convex envelope of* Σ *in* X. The set Σ is said to be *holomorphically* (resp. *meromorphically*) *convex in* X if $\Sigma = \Sigma^h$ (resp. $\Sigma = \Sigma^m$). The following proposition follows easily from the results of §2.5.

2.6.1. PROPOSITION. (i) *Every Weierstrass (resp. rational) neighborhood of* Σ *is also a neighborhood of* Σ^h *(resp.* Σ^m*).*

(ii) *The intersection of all Weierstrass (resp. rational) neighborhoods of* Σ *coincides with* Σ^h *(resp.* Σ^m*).*

(iii) $\mathcal{M}(\mathcal{P}(\Sigma)) \xrightarrow{\sim} \Sigma^h$ *and* $\mathcal{M}(\mathcal{R}(\Sigma)) \xrightarrow{\sim} \Sigma^m$. ∎

2.6.2. COROLLARY. *Suppose that* Σ *is holomorphically (resp. meromorphically) convex in* X. *Then for any* $f \in \Gamma(\Sigma, \mathcal{O}_X)$ *and* $\varepsilon > 0$, *there exist an open neighborhood* $\mathcal{U}$ *of* Σ *and an element* $g \in \mathcal{A}$ *(resp.* $\mathcal{A}_{(\Sigma)}$*) such that* $|f(x) - g(x)| < \varepsilon$ *for all* $x \in \mathcal{U}$. ∎

2.6.3. PROPOSITION. *Let* X *be a* k*-affinoid space of dimension* 1. *Then:*

(i) *a finite union of affinoid domains in* X *is an affinoid domain;*

(ii) $\mathcal{M}(\mathcal{H}(\Sigma)) \xrightarrow{\sim} \Sigma$ *for any closed subset* $\Sigma \subset X$.

PROOF. (i) If the valuation on k is nontrivial and X is strictly k-affinoid, this is a result of Fieseler (**[Fi]**). To prove (i) in the general case, it suffices to show that, if V is a special subset of X such that $\pi^{-1}(V)$ is an affinoid domain in $X' = X\widehat{\otimes}K_r$, where $r \notin \sqrt{|k^*|}$ (π denotes the canonical map $X' \to X$), then V is an affinoid domain in X. From the Acyclicity Theorem 2.2.5 and Proposition 2.1.2 it follows that $\mathcal{A}_{\pi^{-1}(V)} = \mathcal{A}_V\widehat{\otimes}K_r$. By Corollary 2.1.8 and Proposition 2.2.6(iii), V is an affinoid domain in X.

(ii) From (i) it follows that Σ has a basis of closed neighborhoods formed by affinoid domains. Hence, (ii) follows from Proposition 2.2.4(i). ∎

2.6.4. REMARK. In the general case, an affinoid domain in a k-affinoid space of dimension 1 is not a rational domain. Consider the space $X = \mathcal{A}(k\{r^{-1}T\})$. If the valuation on k is nontrivial, all affinoid domains in X are rational. In this case it follows from Propositions 2.6.1(ii) and 2.6.3(i) that all closed subsets of X are meromorphically convex. But if the valuation on k is trivial, this is not the case. For example, if $r < 1$, then X is naturally homeomorphic to the closed interval $[0, r]$, and a closed subset of X is a rational domain if and only if it is connected.

CHAPTER 3

Analytic Spaces

3.1. Definition and general properties. A *k-quasiaffinoid space* is a pair $(\mathscr{U}, \varphi)$ consisting of a locally ringed space $\mathscr{U}$ and an open immersion φ of $\mathscr{U}$ in a k-affinoid space X. A closed subset $V \subset \mathscr{U}$ is called an *affinoid domain* if $\varphi(V)$ is an affinoid domain in X. The corresponding k-affinoid algebra is denoted by $\mathscr{A}_V$ (if $X = \mathscr{M}(\mathscr{A})$). A *morphism of k-quasiaffinoid spaces* $(\mathscr{U}, \varphi) \to (\mathscr{V}, \psi)$ is defined as follows. This is a morphism of locally ringed spaces $\theta: \mathscr{U} \to \mathscr{V}$ such that, for any pair of affinoid domains $U \subset \mathscr{U}$ and $V \subset \mathscr{V}$ with $\theta(U) \subset \mathrm{Int}(V/\mathscr{V})$ (the topological interior of V in $\mathscr{V}$), the induced homomorphism $\mathscr{B}_V \to \mathscr{A}_U$ is bounded. In this situation the bounded homomorphism $\mathscr{B}_V \to \mathscr{A}_U$ is defined for all U and V with $\theta(U) \subset V$. A k-quasiaffinoid space which is isomorphic to a pair $(\mathscr{U}, \varphi: \mathscr{U} \hookrightarrow X)$ with X strictly k-affinoid is called a *strictly k-quasiaffinoid space.*

Now let X be a locally ringed space. A *k-analytic atlas* on X is a collection of pairs $(\mathscr{U}_i, \varphi_i)$, $i \in I$, called *charts* of the atlas, satisfying the following conditions:

AT 1. Each $\mathscr{U}_i$ is an open subset of X, and the $\mathscr{U}_i$ cover X.

AT 2. Each φ_i is an open immersion of $\mathscr{U}_i$ in a k-affinoid space.

AT 3. The induced morphism of locally ringed spaces $\varphi_j\varphi_i^{-1}: \varphi_i(\mathscr{U}_i \cap \mathscr{U}_j) \to \varphi_j(\mathscr{U}_i \cap \mathscr{U}_j)$ is an isomorphism of k-quasiaffinoid spaces for each pair $i, j \in I$.

Suppose that we are given an open subset $\mathscr{U} \subset X$ and an open immersion φ of $\mathscr{U}$ in a k-affinoid space. Then $(\mathscr{U}, \varphi)$ is *compatible* with the atlas $\{(\mathscr{U}_i, \varphi_i)\}_{i \in I}$ if each morphism $\varphi_i\varphi^{-1}$ (defined as in AT 3) is an isomorphism of k-quasiaffinoid spaces. Two atlases are said to be *compatible* if every chart of one atlas is compatible with the other atlas. An equivalence class of k-analytic atlases on X is said to define a *k-analytic space* structure on X. The underlying topological space of X will be denoted by $|X|$.

Let X, Y be two k-analytic spaces, and let $f: X \to Y$ be a morphism of locally ringed spaces. Then f is called a *morphism of k-analytic spaces* if there exists an atlas $\{(\mathscr{U}_i, \varphi_i)\}$ of X and an atlas $\{(\mathscr{V}_j, \psi_j)\}$ of Y such that for each i, j the morphism $\psi_j f \varphi_i^{-1}: \varphi_i(\mathscr{U}_i) \to \psi_j(\mathscr{V}_j)$ is a morphism of k-quasiaffinoid spaces. It is clear that the same condition holds for any choice of atlases on X and Y defining the same structure. The category of

k-analytic spaces is denoted by $k\text{-}\mathcal{A}n$. It admits fibre products, and, for any non-Archimedean field K over k, we have a ground field extension functor $k\text{-}\mathcal{A}n \to K\text{-}\mathcal{A}n: X \mapsto X\widehat{\otimes}K$. The category of k-affinoid spaces $k\text{-}\mathcal{A}ff$ is a full subcategory of $k\text{-}\mathcal{A}n$.

An *analytic space over* k is a K-analytic space for some non-Archimedean field K over k. Morphisms between such spaces are defined as above, so that the morphisms between affinoid spaces over k are induced by bounded homomorphisms between the corresponding affinoid k-algebras. The category of analytic spaces over k is denoted by $\mathcal{A}n_k$. It does not admit fibre products except in special cases. The most important such case is the following. Let $\varphi: Y \to X$ be a morphism of k-analytic spaces, and let $x \in X$. Then the fibre product $Y_x = Y \times_X \mathcal{M}(\mathcal{H}(x))$ exists, because there is an isomorphism $Y_x \widetilde{\to} (Y\widehat{\otimes}\mathcal{H}(x)) \times_{X\widehat{\otimes}\mathcal{H}(x)} \mathcal{M}(\mathcal{H}(x))$. This is an $\mathcal{H}(x)$-analytic space, which is called the *fibre of* φ *at the point* x.

A morphism of k-analytic spaces $\varphi: Y \to X$ is called an *analytic domain* if φ induces a homeomorphism of Y with its image in X, and if, for any morphism $\psi: Z \to X$ of analytic spaces over k with $\psi(Z) \subset \varphi(Y)$, there exists a unique morphism $\sigma: Z \to Y$ such that $\psi = \varphi \circ \sigma$. As a rule, we identify an analytic domain with its image in X. If Y is isomorphic to a (strictly) k-affinoid (resp. (strictly) k-quasiaffinoid) space, it is called a (*strictly*) *affinoid* (resp. (*strictly*) *quasiaffinoid*) *domain.* The affinoid neighborhoods of a point $x \in X$ form a basis of neighborhoods of x.

A k-analytic space X is said to be *strictly* k-*analytic* if every compact subset of X lies in a finite union of strictly quasiaffinoid domains. That this definition is consistent with the notion of a strictly k-affinoid space follows from the following lemma. (However, in the general case, a k-quasiaffinoid space which is strictly k-analytic is not strictly k-quasiaffinoid.)

3.1.1. LEMMA. *Let* X *be a* k-*affinoid space. Suppose that every point of* X *is contained in a strictly quasiaffinoid domain. Then* X *is strictly* k-*affinoid.*

PROOF. Let $X = \mathcal{M}(\mathcal{A})$. By Corollary 2.1.6, it suffices to show that $|\mathcal{H}(x)^*| \subset \sqrt{|k^*|}$ for every point x in the Shilov boundary $\Gamma(\mathcal{A})$ of $\mathcal{A}$. Let $\mathcal{U}$ be a strictly quasiaffinoid domain in X containing the point x, and let V be an affinoid neighborhood of x in $\mathcal{U}$. The domain $\mathcal{U}$ is isomorphic to an open subset of a strictly k-affinoid space $X' = \mathcal{M}(\mathcal{A}')$. By Corollary 2.4.5, we have $x \in \Gamma(\mathcal{A}_V)$, and, by the Maximum Modulus Principle 2.5.20, $x \in \Gamma(\mathcal{A}')$. Since $\mathcal{A}'$ is strictly k-affinoid, it follows that $|\mathcal{H}(x)^*| \subset \sqrt{|k^*|}$. ∎

3.1.2. PROPOSITION. *Suppose that the valuation on* k *is nontrivial.*

(i) *A* k-*analytic space is strictly* k-*analytic if and only if every compact subset lies in a finite union of strictly affinoid domains.*

(ii) *Let* $\varphi: Y \to X$ *be a closed morphism* (*the definition is given below, after the proof*), *and suppose that* X *is strictly* k-*analytic. Then* Y *is also strictly* k-*analytic.*

PROOF. (i) The direct implication follows from Proposition 2.2.3(iii). The converse implication is trivial.

(ii) We may assume that $X = \mathcal{M}(\mathcal{A})$ is strictly k-affinoid and $Y = \mathcal{M}(\mathcal{B})$ is k-affinoid. If $y \in \mathrm{Int}(Y/X)$, then there exists a closed immersion of k-affinoid spaces $Y \hookrightarrow X \times E(0, r)$ ($E(0, r) = \mathcal{M}(k\{r^{-1}T\})$) with $|T(y_i)| < r_i$. Let $p_i \in \sqrt{|k^*|}$ be numbers with $|T_i(y)| < p_i \le r_i$. Then $V = Y \cap (X \times E(0, p))$ is an affinoid neighborhood of the point y. Moreover, V is strictly k-affinoid, because it is a closed k-analytic subspace of the strictly k-affinoid space $X \times E(0, p)$. ∎

Let $\varphi: Y \to X$ be a morphism of k-analytic spaces. The *relative interior of* φ is the set $\mathrm{Int}(Y/X)$ of points $y \in Y$ for which there exist affinoid neighborhoods Y' of y and X' of $\varphi(y)$ with $\varphi(Y') \subset X'$ and $y \in \mathrm{Int}(Y'/X')$. The complement of $\mathrm{Int}(Y/X)$ is called the *relative boundary of* φ and is denoted by $\partial(Y/X)$. If $X = \mathcal{M}(k)$, these sets are denoted by $\mathrm{Int}(Y)$ and $\partial(Y)$ and are called the *interior* and the *boundary of* Y, respectively. For example, $Y_0 := \{y \in Y | [\mathcal{H}(y): k] < \infty\} \subset \mathrm{Int}(Y)$. A morphism $\varphi: Y \to X$ is said to be *closed* if $\partial(Y/X) = \emptyset$. A k-analytic space X is said to be *closed* if the canonical morphism $X \to \mathcal{M}(k)$ is closed. The following proposition lists properties of the relative interior, which are easily deduced from the results of §2.5.

3.1.3. PROPOSITION. (i) *If Y is an analytic domain in X, then* $\mathrm{Int}(Y/X)$ *coincides with the topological interior of Y in X.*

(ii) *For a sequence of morphisms $Z \xrightarrow{\psi} Y \xrightarrow{\varphi} X$, one has* $\mathrm{Int}(Z/X) = \mathrm{Int}(Z/Y) \cap \psi^{-1}(\mathrm{Int}(Y/X))$.

(iii) *For a morphism $\psi: X' \to X$, one has* $\psi'^{-1}(\mathrm{Int}(Y/X)) \subset \mathrm{Int}(Y'/X')$, *where ψ' is the canonical morphism $Y' = Y \times_X X' \to Y$.*

(iv) *For a non-Archimedean field K over k, one has* $\psi^{-1}(\mathrm{Int}(Y/X)) \subset \mathrm{Int}(Y\widehat{\otimes}K/X\widehat{\otimes}K)$, *where ψ is the canonical map $Y\widehat{\otimes}K \to Y$. If $K = K_r$ for some $r \notin \sqrt{|k^*|}$, then* $\mathrm{Int}(Y/X) = \{y \in Y | \psi^{-1}(y) \subset \mathrm{Int}(Y'/X')\}$. ∎

A morphism $\varphi: Y \to X$ is called an *open immersion* if it is an isomorphism between Y and an open subset of X (with the induced k-analytic structure). For example, if φ is an analytic domain, then the induced morphism $\mathrm{Int}(Y/X) \to X$ is an open immersion. Furthermore, φ is called a *closed immersion* if it induces a homeomorphism of Y with a closed subset of X, $\varphi_*(\mathcal{O}_Y)$ is a coherent $\mathcal{O}_X$-module, and the homomorphism of sheaves $\mathcal{O}_X \to \varphi_*(\mathcal{O}_Y)$ is surjective. Open and closed immersions are preserved under finite compositions, under any base change functor, and under extension of the ground field.

3.1.4. PROPOSITION. (i) *If $\varphi: Y \to X$ is a closed immersion, then the inverse image of an affinoid domain is an affinoid domain.*

(ii) *There is a one-to-one correspondence between closed immersions* $\varphi\colon Y \to X$ *and coherent subsheaves of ideals* $\mathcal{J} \subset \mathcal{O}_X$.

Proof. By standard arguments, the statements are reduced to the following assertion. If $\varphi\colon Y \to X$ is a morphism which is a homeomorphism and has the property that $\mathcal{O}_X \widetilde{\to} \varphi_*(\mathcal{O}_Y)$, then φ is an isomorphism. Note that this property is preserved under any base change functor. So we may suppose that X is k-affinoid and Y is k-quasiaffinoid. Since Y is compact, from Corollary 2.2.7 it follows that Y is also k-affinoid. The required fact now follows from Corollary 2.3.2. ∎

If $\varphi\colon Y \to X$ is a closed immersion, the set $\varphi(Y)$ is called a *closed k-analytic subset of* X. A k-analytic space X is said to be *irreducible* if it is not a union of two distinct closed k-analytic subsets.

A morphism of k-analytic spaces $\varphi\colon Y \to X$ is said to be *separated* if the diagonal morphism $Y \to Y \times_X Y$ is a closed immersion. A k-analytic space X is said to be *separated* if the canonical morphism $X \to \mathcal{M}(k)$ is separated. A continuous map of topological spaces $B \to A$ is said to be *separated* if the image of B under the diagonal map $B \to B \times_A B$ is closed. It is said to be *proper* if it is separated and the inverse image of a compact set is compact.

3.1.5. Proposition. *A morphism* $\varphi\colon Y \to X$ *is separated if and only if the map* $|Y| \to |X|$ *is separated. In particular, a k-analytic space X is separated if and only if* $|X|$ *is Hausdorff.*

Proof. Since the map $|Y \times_X Y| \to |Y| \times_{|X|} |Y|$ is proper, the image of Y in $Y \times_X Y$ is closed if and only if it is closed in $|Y| \times_{|X|} |Y|$. The direct implication follows immediately from this remark. The converse implication also follows from the remark and the fact that k-quasiaffinoid spaces are separated. ∎

3.1.6. Corollary. *If a k-analytic space X is separated, then the affinoid domains are closed in X, and the intersection of two affinoid domains is also an affinoid domain.* ∎

A morphism $\varphi\colon Y \to X$ is said to be *compact* if the map $|Y| \to |X|$ is proper. It is said to be *proper* if it is compact and closed. A k-analytic space X is said to be *compact* (resp. *proper*) if the canonical morphism $X \to \mathcal{M}(k)$ is compact (resp. proper). A special case of proper morphisms is finite morphisms, which are defined as follows. A morphism $\varphi\colon Y \to X$ is *finite* if, for any point $x \in X$, there exists an affinoid neighborhood V of x such that $\varphi^{-1}(V) \to V$ is a finite morphism of k-affinoid spaces. For example, closed immersions and normalizations (see below) are finite morphisms. Note that compact (proper, finite) morphisms are preserved under finite compositions, under any base change functor and under extension of the ground field. The following properties of finite morphisms are standard, and can be deduced from the above results.

3.1.7. PROPOSITION. (i) *If $\varphi: Y \to X$ is a finite morphism, then the inverse image of an affinoid domain is an affinoid domain.*

(ii) *The correspondence $\varphi \mapsto \varphi_*(\mathcal{O}_Y)$ gives an equivalence between the category of finite morphisms over X and the category of coherent $\mathcal{O}_X$-algebras.* ■

A morphism of k-affinoid spaces $\varphi: Y = \mathcal{M}(\mathcal{B}) \to X = \mathcal{M}(\mathcal{A})$ is called a *Runge immersion* if the image of $\mathcal{A}$ is everywhere dense in $\mathcal{B}$. By Proposition 2.2.9, this is equivalent to requiring that φ admit a factorization $Y \xrightarrow{\varphi'} X' \xrightarrow{\psi} X$, where φ' is a closed immersion and X' is a Weierstrass domain in X via ψ. Furthermore, a morphism of separated k-analytic spaces $\varphi: Y \to X$ is called an *immersion* if φ induces a homeomorphism of $|Y|$ with its image in $|X|$ and if, for any point $y \in Y$, there exists an affinoid neighborhood X' of $\varphi(y)$ such that $\varphi^{-1}(X') \to X'$ is a Runge immersion of k-affinoid spaces. For example, open and closed immersions are immersions. It is easy to show that a morphism of strictly k-affinoid spaces is an immersion if and only if it is a locally closed immersion in the sense of **[BGR]**, 7.3.3.

The local properties of k-affinoid spaces introduced in §2.3 extend naturally to k-analytic spaces. A k-analytic space is said to be *reduced* (*normal, Cohen-Macaulay, regular*) if it has the corresponding property at all points. X is said to be *smooth of dimension* n if it is pure n-dimensional and all of its points are smooth. The *reduction of* X is the closed immersion $X_{\mathrm{red}} \to X$ corresponding to the nilradical of the sheaf $\mathcal{O}_X$. It is clear that X is irreducible if and only if X_{red} is irreducible. Furthermore, by Proposition 2.1.14(i), for every reduced k-analytic space X one can construct the *normalization*, consisting of a normal k-analytic space X' and a morphism $\pi: X' \to X$. (If $X = \mathcal{M}(\mathcal{A})$ is k-affinoid, then $X' = \mathcal{M}(\mathcal{A}')$, where $\mathcal{A}'$ is the integral closure of $\mathcal{A}$ in the full ring of fractions, regarded as a finite Banach $\mathcal{A}$-algebra.)

3.1.8. PROPOSITION. *Let X be a reduced k-analytic space, and let X' be the normalization of X. Then the following are equivalent:*

(a) *X is irreducible;*

(b) *X' is connected;*

(c) *X' is irreducible.* ■

3.1.9. COROLLARY. *Every k-analytic space X can be uniquely decomposed into irreducible components $X = \bigcup_{i\in I} X_i$. The covering $\{X_i\}_{i\in I}$ is locally finite.* ■

A morphism $\varphi: Y \to X$ is said to be *smooth of relative dimension* n if it is flat and if, for any point $x \in X$ with nonempty fibre Y_x, Y_x is a smooth $\mathcal{H}(x)$-analytic space of dimension n. A morphism $\varphi: Y \to X$ is said to be *unramified* if, for any point $x \in X$ with nonempty fibre Y_x, Y_x is isomorphic to $\mathcal{M}(K)$, where K is a finite separable algebra over $\mathcal{H}(x)$.

The morphism is said to be *étale* if it is unramified and flat. Étale morphisms are the same as smooth morphisms of relative dimension 0.

3.2. Topological properties of analytic spaces. From Corollary 2.2.8 it follows that analytic spaces are locally connected. The theorem that follows shows that they are locally arcwise connected. We recall that a topological space X is said to be arcwise connected if, for any two distinct points $x, y \in X$, there exists a homeomorphic embedding $\varphi: [0, 1] \to X$ with $\varphi(0) = x$ and $\varphi(1) = y$.

3.2.1. THEOREM. *Every connected k-analytic space is arcwise connected.*

PROOF. It suffices to show that an irreducible k-affinoid space $X = \mathcal{M}(\mathcal{A})$ is arcwise connected. By Proposition 2.1.4(iii), we may assume that X is an irreducible strictly k-affinoid space. We set $Y = \mathcal{M}(\mathcal{A}\{T\})$, and let φ denote the canonical map $Y \to X$. The fibre $\varphi^{-1}(x)$ at a point $x \in X$ is isomorphic to $\mathcal{M}(\mathcal{H}(x)\{T\})$. Let σ denote the map $X \to Y$ which sends a point $x \in X$ to the point of $\varphi^{-1}(x)$ corresponding to the norm of the $\mathcal{H}(x)$-affinoid algebra $\mathcal{H}(x)\{T\}$. Given a point $y \in Y$, we set $\ell_y = \{p(E(y, \rho)) | r(y) \le \rho \le 1\}$ (see §4.2). This is a path connecting the points y and $\sigma(\varphi(y))$ in $\varphi^{-1}(\varphi(y))$.

3.2.2. LEMMA. (i) *The map* $\sigma: X \to Y$ *is continuous.*

(ii) *For any open neighborhood* $\mathcal{U}$ *of* ℓ_y, *there exists an open neighborhood* $\mathcal{V}$ *of* y *such that* $\ell_z \subset \mathcal{U}$ *for all* $z \in \mathcal{V}$.

PROOF. (i). It suffices to verify that for any $f \in \mathcal{A}\{T\}$ the real valued function $x \mapsto |f(\sigma(x))|$ is continuous on X. If $f(T) = \sum_{i=0}^{\infty} a_i T^i$, where $a_i \in \mathcal{A}$ and $||a_i|| \to 0$ as $i \to \infty$, then

$$|f(\sigma(x))| = \max_i |a_i(x)|.$$

(ii). Let $y \in Y$ and $x = \varphi(y)$. One has:

$$\ell_y = \{z \in \varphi^{-1}(x) |\ |f(z)| \ge |f(y)| \text{ for all } f \in \mathcal{A}\{T\}\}.$$

Using standard compactness arguments, we see that there exist $f_1, \dots, f_m \in \mathcal{A}\{T\}$, $\varepsilon > 0$ and an open neighborhood $\mathcal{W}$ of the point x such that the open set

$$\mathcal{V} := \{z \in Y |\ |f_i(z)| > |f_i(y)| - \varepsilon,\ 1 \le i \le m\} \cap \varphi^{-1}(\mathcal{W})$$

lies in $\mathcal{U}$. It is clear that $y \in \mathcal{V}$ and $\ell_z \subset \mathcal{V}$ for all $z \in \mathcal{V}$. ∎

3.2.3. COROLLARY. *Let* $X = \mathcal{M}(k\{T_1, \dots, T_n\})$, *and let* x_0 *be the point of* X *corresponding to the norm of* $k\{T_1, \dots, T_n\}$. *There exists a map which associates to each point* $x \in X$ *a closed subset* $\ell_x \subset X$ *such that:*

(a) $\ell_{x_0} = \{x_0\}$ *and* ℓ_x, *for* $x \ne x_0$, *is homeomorphic to the unit interval with endpoints* x *and* x_0;

(b) *for any open neighborhood* $\mathscr{U}$ *of* ℓ_x, *there exists an open neighborhood* $\mathscr{V}$ *of* x *such that* $\ell_{x'} \subset \mathscr{U}$ *for all* $x' \in \mathscr{V}$. ∎

3.2.4. LEMMA. *Let* $\varphi: Y \to X$ *be a finite morphism of pure dimensional* k*-analytic spaces, and suppose that* X *is locally irreducible and* $\dim(X) = \dim(Y)$. *Then* φ *is an open map.*

PROOF. We may assume that X and Y are separated. Let $y \in Y$, $x = f(y)$ and $\varphi^{-1}(x) = \{y = y_1, \dots, y_n\}$. If V is a sufficiently small irreducible affinoid neighborhood of the point x, one has $\varphi^{-1}(V) = \bigcup_{i=1}^{n} W_i$, where W_i is an affinoid neighborhood of y_i and $W_i \cap W_j = \varnothing$ for $i \neq j$. Moreover, the affinoid subdomains W_1 obtained in this way form a basis of affinoid neighborhoods of the point y. We obtain a finite morphism $\varphi_1: W_1 \to V$. The image $\varphi_1(W_1)$ is a closed k-analytic subset of V. Since $\dim(\varphi_1(W_1)) = \dim(W_1) = \dim(V)$ and V is irreducible, it follows that $\varphi(W_1) = V$. Since $\mathrm{Int}(Y/X) = Y$ and $\mathrm{Int}(W_1/V) = W_1$, we have $\mathrm{Int}(W_1/Y) = \varphi_1^{-1}(\mathrm{Int}(V/X))$, i.e., φ_1 induces a surjective map of open subsets $\mathrm{Int}(W_1/Y) \to \mathrm{Int}(V/X)$. ∎

Let Y be the character space of an irreducible n-dimensional strictly k-affinoid algebra. By the Noether Normalization Lemma, there exists a finite surjective morphism $\pi: Y \to X = \mathscr{M}\ (k\{T_1, \dots, T_n\})$. By Lemma 3.2.4, the map π is open. Hence, the theorem follows from the following lemma.

3.2.5. LEMMA. *Let* $\pi: Y \to X$ *be a proper and open surjective map with finite fibres between connected compact Hausdorff spaces. Suppose that* X *satisfies the properties in Corollary 3.2.3 (for some* $x_0 \in X$*). Then the space* Y *is arcwise connected.*

PROOF. We will show that the arcwise connected components of Y are open. This will give the arcwise connectedness of Y, because Y is connected.

Let $y \in Y$ and $x = \pi(y)$, and let Σ be the connected component of the point y in $\pi^{-1}(\ell_x)$. We claim that $\pi(\Sigma) = \ell_x$. Indeed, this follows from the fact that the connected component of a point in a compact Hausdorff space coincides with the intersection of its open-closed neighborhoods, and the map π is open. Furthermore, since Σ has a countable basis of open sets, it is metrizable ([**En**], 4.2.8). But connected metrizable compact spaces are arcwise connected ([**En**], 6.3.11). Hence, Σ is arcwise connected.

Let $\Sigma_1 = \Sigma, \Sigma_2, \dots, \Sigma_n$ be the connected components of the set $\pi^{-1}(\ell_x)$, and let $\mathscr{W} = \mathscr{W}_1, \mathscr{W}_2, \dots, \mathscr{W}_n$ be open neighborhoods of $\Sigma_1, \dots, \Sigma_n$, respectively, such that $\mathscr{W}_i \cap \mathscr{W}_j = \varnothing$ for $i \neq j$. There exists an open neighborhood $\mathscr{U}$ of ℓ_x such that $\pi^{-1}(\mathscr{U}) \subset \bigcup_{i=1}^{n} \mathscr{W}_i$. Let $\mathscr{V}$ be an open neighborhood of the point x such that $\ell_{x'} \subset \mathscr{U}$ for all $x' \in \mathscr{V}$. We claim that every point $y' \in \pi^{-1}(\mathscr{V}) \cap \mathscr{W}$ lies in the arcwise connected component of the point y.

If $x' = \pi(y')$, then $\ell_{x'} \subset \mathscr{U}$, and hence, $\pi^{-1}(\ell_{x'}) \subset \bigcup_{i=1}^{n} \mathscr{W}_i$. Let Σ' be the connected component of the point y' in the set $\pi^{-1}(\ell_{x'})$. We know that Σ' is arcwise connected and maps surjectively onto $\ell_{x'}$. Moreover, since $y' \in \mathscr{W} = \mathscr{W}_1$, we have $\Sigma' \subset \mathscr{W}$. Hence, $\Sigma \cap \Sigma' \neq \varnothing$, because these sets contain a common point in $\pi^{-1}(x_0)$. From this it follows that the points y and y' lie in the same arcwise connected component of Y. ∎

Recall the definition of the dimension $\dim(X)$ of a normal topological space X. First of all, the *order* of a family of subsets of X is the largest integer n such that the family contains $n+1$ sets with nonempty intersection. The *dimension* $\dim(X)$ is the least integer n such that every finite open covering of X has a finite refinement of order $\leq n$.

We also recall some properties of $\dim(X)$ which will be used later.

(i) If Σ is a closed subset of X, then $\dim(\Sigma) \leq \dim(X)$.

(ii) If X is a locally finite union of closed subsets $X = \bigcup_{i \in I} \Sigma_i$ and $\dim(\Sigma_i) \leq n$, $i \in I$, then $\dim(X) \leq n$ (**[En]**, 7.2.3).

(iii) If $\dim(X) \leq n$, then every locally finite open covering of X has a refinement of order $\leq n$ (**[En]**, 7.2.4).

(iv) Suppose that for any pair A, B of disjoint closed subsets of X there exists an open subset $\mathscr{U} \subset X$ such that $A \subset \mathscr{U} \subset \overline{\mathscr{U}} \subset X \backslash B$ and $\dim(\partial(\mathscr{U}/X)) \leq n-1$. Then $\dim(X) \leq n$ (**[En]**, 7.2.6).

3.2.6. THEOREM. *Let $X = \mathscr{M}(\mathscr{A})$ be a strictly k-affinoid space. Then* $\dim(|X|) = \dim(\mathscr{A})$ *and* $\dim(\partial(X)) = \dim(\mathscr{A}) - 1$.

PROOF. Let $\mathscr{A} = k\{T_1, \dots, T_n\}$ and $a \in \operatorname{Max}(\mathscr{A})$. Every polydisc $E = E(a, r)$ of radius $r = (r_1, \dots, r_n)$ with center at a determines a multiplicative seminorm $|\ |_E$ on $\mathscr{A}$: $|f|_E = \sup_{x \in E} |f(x)|$. Consider the map $[0, 1]^n \to X : r \mapsto |\ |_{E(0, r)}$. This map is continuous and injective, and hence is a homeomorphism of $[0, 1]^n$ with its image in X. In particular, $\dim(|X|) \geq n$; moreover, every open subset $\mathscr{U} \subset X$ contains a closed (in X) subset of dimension n. (This is also seen from the construction in §6.1.) Using this fact and following lemma, we find that the inequality $\dim(|X|) \geq \dim(\mathscr{A})$ is valid for any strictly affinoid algebra $\mathscr{A}$.

3.2.7. LEMMA. *Let $\varphi: Y \to X$ be a finite surjective morphism of pure dimensional k-analytic spaces. Then φ is a local homeomorphism outside of a nowhere dense closed subset $\Lambda_\varphi \subset Y$.*

PROOF. We may assume that X and Y are reduced k-affinoid spaces. Furthermore, we may assume that X and Y are normal. Finally, it suffices to show that $\Lambda_\varphi \neq Y$.

Let $n = \max\{c(\varphi^{-1}(x)) |\ x \in X\}$ and $M = \{x \in X |\ c(\varphi^{-1}(x)) = n\}$. We claim that M is open and that φ is a local homeomorphism at any point $y \in \varphi^{-1}(M)$.

Let $x \in M$ and $\varphi^{-1}(x) = \{y_1, \dots, y_n\}$. Since Y is compact, the open subsets of the form $\varphi^{-1}(\mathscr{U})$, where $\mathscr{U}$ runs through the open neighborhoods of x, form a basis of open neighborhoods of the set $\varphi^{-1}(x)$. Hence, there exists a connected affinoid neighborhood V of x such that $\varphi^{-1}(V) = \bigcup_{i=1}^{n} W_i$, where W_i is an affinoid neighborhood of y_i and $W_i \cap W_j = \varnothing$ for $i \neq j$. The affinoid subdomain V is irreducible because it is connected and normal, and the induced morphisms $\varphi_i: W_i \to V$ are finite. Hence, all of the maps φ_i are surjective. From this it follows that $c(\varphi^{-1}(x')) = n$ for all points $x' \in V$, i.e., the maps φ_i are bijective. Since W_i and V are compact, all of the φ_i are homeomorphisms. Finally, since $\mathrm{Int}(Y/X) = Y$ and $\mathrm{Int}(W_i/V) = W_i$, we have $\mathrm{Int}(W_i/Y) = \varphi^{-1}(\mathrm{Int}(V/X))$. That is, φ_i induces a homeomorphism of open sets $\mathrm{Int}(W_i/Y) \tilde{\to} \mathrm{Int}(V/X)$. ∎

We now verify the required equalities by induction. If $\dim(\mathscr{A}) = 0$, then X is finite and $\partial(X) = \varnothing$. That is, $\dim(|X|) = 0$ and $\dim(\partial(X)) = -1$. Suppose that the equalities are valid for all strictly affinoid algebras of dimension $< \dim(\mathscr{A})$. Consider a finite admissible monomorphism $k\{T_1, \dots, T_n\} \hookrightarrow \mathscr{A}$, where $n = \dim(\mathscr{A})$. It induces maps $\pi_i: X \to Y_i = \mathscr{M}(k\{T_i\})$. Let y_i be the point of Y_i corresponding to the norm of $k\{T_i\}$. Since $\partial(Y_i) = \{y_i\}$, it follows that $\partial(X) = \bigcup_{i=1}^{n} \pi_i^{-1}(y_i)$. But $\pi_i^{-1}(y_i)$ is the character space of the strictly $\mathscr{H}(y_i)$-affinoid algebra $\mathscr{A}\widehat{\otimes}_{k\{T_i\}}\mathscr{H}(y_i)$. All of these algebras are of Krull dimension $\leq n-1$, and since $\dim(\mathscr{A}) = n$, we have $\dim(\mathscr{A}\widehat{\otimes}_{k\{T_i\}}\mathscr{H}(y_i)) = n-1$ for some i. By induction, $\dim(\partial(X)) = \max\{\dim(\pi_i^{-1}(y_i)) \mid 1 \leq i \leq n\} = n-1$. We now use the property (iv) to obtain the inequality $\dim(X) \leq n$. Let A and B be disjoint closed subsets of X, and let V be a special neighborhood of A such that $V \cap B = \varnothing$. Then $\dim(\partial(V/X)) \leq n-1$. Setting $\mathscr{U} = \mathrm{Int}(V/X)$, we obtain an open subset such that $A \subset \mathscr{U} \subset \overline{\mathscr{U}} \subset X \backslash B$ and $\dim(\partial(\mathscr{U}/X)) \leq n-1$. Hence, $\dim(|X|) \leq n$. ∎

3.2.8. Corollary. *The topological dimension of a paracompact separated strictly k-analytic space is equal to its dimension.* ∎

We denote the cardinality of a set X by $c(X)$. Let X be a topological space. The infimum of $c(\mathscr{B})$ over all bases $\mathscr{B}$ of open subsets of X is called the *weight* of X and is denoted by $w(X)$. The *character* $\chi(x, X)$ *of* X *at a point* $x \in X$ is the infimum of $c(\mathscr{B}(x))$ over all bases $\mathscr{B}(x)$ of open neighborhoods of x in X. The *character* of X is the supremum of $\chi(x, X)$ over all points $x \in X$. Finally, the *density* $d(X)$ of X is the infimum of $c(M)$ over all everywhere dense subsets $M \subset X$.

For example, if the valuation on k is nontrivial, then $c(k) = 2^{\sup(\aleph_0, c(\widetilde{k}))}$, $\chi(k) = \aleph_0$ and $w(k) = d(k) = \sup(\aleph_0, c(\widetilde{k}))$. If the valuation on k is trivial, then $\chi(k) = 1$ and $w(k) = d(k) = c(k)$.

3.2.9. Proposition. *Let X be the character space of a strictly k-affinoid algebra $\mathscr{A}$ of dimension ≥ 1. Then $c(X) = \sup(2^{\aleph_0}, c(k))$ and $w(X) = \chi(X) = d(X) = w(k)$.*

PROOF. If $X \to Y$ is a proper surjective map with finite fibres, then all of the above invariants of X and Y coincide. So we may assume that $\mathcal{A} = k\{T_1, \dots, T_n\}$, $n \geq 1$. We also assume that the valuation on k is nontrivial.

It is clear that $c(k) \leq c(X)$. The reverse inequality is verified by induction. If $n = 1$, it follows for algebraically closed k from the description of X. Since $c(X) \leq c(X \widehat{\otimes} \widehat{k^a})$ and $c(\widehat{k^a}) = c(k)$, it is also true for any k. We now consider the projection $\pi: X = \mathcal{M}(k\{T_1, \dots, T_n\}) \to Y = \mathcal{M}(k\{T_1\})$. The fibre $\pi^{-1}(y)$ of a point $y \in Y$ is the space $\mathcal{M}(\mathcal{H}(y)\{T_2, \dots, T_n\})$. By induction, $c(\pi^{-1}(y)) \leq c(\mathcal{H}(y))$. Since $c(\mathcal{H}(y)) = c(k)$ (see the end of §1.4.4) and $c(Y) = c(k)$, we have $c(X) \leq c(k)$.

The inequality $d(X) \leq d(k)$ holds because Max $(\mathcal{A})$ is everywhere dense in X. To verify the reverse inequality, it suffices to suppose that $\mathcal{A} = k\{T\}$. But in this case $d(k) \leq d(X)$, because the complement of the point $p(E(0, 1))$ is a disjoint union of open subsets over all closed points of the affine line over $\widetilde{k}$.

The open sets of the form Int (V/X), where V is a Laurent subdomain of X, form a basis of the topology of X. So $w(X) \leq d(k\{T_1, \dots, T_n\}) = d(k)$. Since $d(k) = d(X) \leq w(X)$, we have $w(X) = d(k)$.

Finally, to verify that $w(X) \leq \chi(X)$ it suffices to suppose that $\mathcal{A} = k\{T\}$. In this case it follows from the above remark that $d(k) \leq \chi(X)$. Since $w(X) = d(k)$, we have $\chi(X) = d(k)$. ∎

3.3. Connection with rigid analytic geometry. Recall the definition of a rigid k-analytic space (when we consider rigid analytic spaces, we suppose that the valuation on k is nontrivial). Let $\mathcal{X} = \mathrm{Max}(\mathcal{A})$ be the maximal spectrum of a strictly k-affinoid algebra $\mathcal{A}$. (At the moment we consider only strictly k-affinoid algebras and affinoid domains.) As above, $\mathcal{X}$ can be endowed with the weak and special G-topologies. It can also be endowed with a *strong* G-topology as follows ([**BGR**], 9.1.4/2).

(a) A subset $U \subset \mathcal{X}$ is admissible open if and only if it admits a covering $\{U_i\}_{i \in I}$ by affinoid subdomains such that, for any affinoid morphism $\varphi: \mathcal{Y} \to \mathcal{X}$ with $\varphi(\mathcal{Y}) \subset U$, the covering $\{\varphi^{-1}(U_i)\}_{i \in I}$ of $\mathcal{Y}$ has a finite affinoid covering which refines it.

(b) A covering $\{U_i\}_{i \in I}$ of an admissible open set $U \subset \mathcal{X}$ by admissible open sets U_i is admissible if and only if, for any affinoid morphism $\varphi: \mathcal{Y} \to \mathcal{X}$ with $\varphi(\mathcal{Y}) \subset U$, the covering $\{\varphi^{-1}(U_i)\}_{i \in I}$ of $\mathcal{Y}$ has a finite affinoid covering which refines it.

The strong G-topology is slightly finer than the special one. (See [**BGR**], 9.1.2/1.) Thus, every sheaf in the weak G-topology extends to a sheaf in the strong G-topology which is uniquely determined up to isomorphism. Furthermore, any homomorphism of sheaves in the weak G-topology extends uniquely to a homomorphism between the corresponding extensions ([**BGR**], 9.2.3/1). Because of this, the sheaf $V \mapsto \mathcal{A}_V$ extends to a sheaf of rings $\mathcal{O}_{\mathcal{X}}$ in the strong G-topology. The space $\mathcal{X}$ endowed with the strong G-topology and with the sheaf of rings $\mathcal{O}_{\mathcal{X}}$ is a *rigid k-affinoid space.* A *rigid k-analytic space* is a locally G-ringed space $(\mathcal{X}, \mathcal{O}_{\mathcal{X}})$ having an admissible affinoid covering such that its G-topology fulfills certain natural conditions ([**BGR**], 9.3.1/4).

Now let X be a separated strictly k-analytic space. By Corollary 3.1.6, the class of strictly affinoid domains in X determines a G-topology on X (called the *weak* G-topology). From Proposition 2.2.5 it follows that the correspondence $V \mapsto \mathcal{A}_V$ is a sheaf of strictly k-affinoid algebras on X. Similarly, the class of special subsets of X (i.e., finite unions of strictly affinoid domains) determines the *special* G-topology on X; and, by Corollary 2.2.6, the correspondence $V \mapsto \mathcal{A}_V$ is a sheaf of commutative Banach k-algebras on X. The special G-topology is slightly finer than the weak one.

We set $X_0 = \{x \in X | [\mathcal{H}(x): k] < \infty\}$. From Proposition 2.1.15 it follows that X_0 is everywhere dense in X. The weak (resp. special) G-topology of X induces a G-topology on X_0, which is also called the weak (resp. special) G-topology. We endow X_0 with a slightly finer G-topology (which is the only topology we consider below) as follows. A subset $U \subset X_0$ is admissible open if, for any strictly affinoid subdomain $Y \subset X$, the intersection $U \cap Y_0$ is admissible open in Y_0. Furthermore, a covering $\{U_i\}_{i \in I}$ of an admissible open subset $U \subset X_0$ by admissible open subsets $U_i \subset X_0$ is admissible if $\{U_i \cap Y_0\}_{i \in I}$ is an admissible covering of $U \cap Y_0$ for any strictly affinoid subdomain $Y \subset X$. Note that a family of strictly affinoid subdomains of X gives rise to an admissible covering of X_0 if and only if any compact subset of X lies in a finite union of some of the subdomains. By Proposition 3.1.2(i), X_0 has an admissible affinoid covering. The sheaves of rings $\mathcal{O}_{Y_0}$, where Y runs through the strictly affinoid subdomains of X, are compatible on intersections. So they glue together to form a sheaf of rings $\mathcal{O}_{X_0}$. Thus, we obtain a rigid k-analytic space $(X_0, \mathcal{O}_{X_0})$. The following proposition follows easily from the previous results.

3.3.1. PROPOSITION. (i) *The correspondence $X \mapsto X_0$ is a fully faithful functor which commutes with fibre products and with extension of the ground field.*

(ii) $\mathcal{O}(X) = \mathcal{O}(X_0)$ *and* $\mathcal{M}(X) = \mathcal{M}(X_0)$.

(iii) *X is reduced, normal, Cohen-Macaulay, regular, smooth or of dimension n if and only if X_0 has the same property.* ∎

Let $\varphi: Y \to X$ be a morphism of separated strictly k-analytic spaces, and let $\varphi_0: Y_0 \to X_0$ be the corresponding morphism of rigid k-analytic spaces.

3.3.2. PROPOSITION. *φ is a closed immersion, finite, compact, or proper if and only if φ_0 has the same property.*

A morphism of rigid k-analytic spaces is said to be *compact* if the inverse image of an affinoid subdomain is a finite union of affinoid subdomains. (Such a morphism is usually called quasicompact.)

PROOF. If φ_0 is compact, then φ is obviously compact. If φ_0 is proper, then φ is proper, by Proposition 2.5.17. The assertion for finite morphisms and closed immersions follows from this. ■

3.3.3. REMARK. The functor $X \mapsto X_0$ does not send open immersions to open immersions. For example, let X be an arbitrary separated strictly k-analytic space, and let $Y = X\backslash\{x\}$, where $x \in X\backslash X_0$. Then $Y \hookrightarrow X$ is an open immersion, but $Y_0 \to X_0$ is not an open immersion, because $Y_0 = X_0$ (as sets) and the G-topologies of X_0 and Y_0 do not coincide. Nevertheless, if T is a closed k-analytic subset of a separated strictly k-analytic space X, then there is a canonical isomorphism of rigid k-analytic spaces $(X\backslash T)_0 \xrightarrow{\sim} X_0\backslash T_0$.

Given a sheaf of sets $\mathcal{F}$ on X, the correspondence $V_0 \mapsto \varinjlim \Gamma(\mathcal{U}, \mathcal{F})$, where $\mathcal{U}$ runs through the open neighborhoods of the closure of V_0 in X, is a sheaf of sets in the special G-topology of X_0, and hence it determines a sheaf ${}^{\tau}\mathcal{F}$ on the G-topology of X_0. For an $\mathcal{O}_X$-module $\mathcal{F}$, the sheaf $\mathcal{F}_0 = {}^{\tau}\mathcal{F} \otimes_{{}^{\tau}\mathcal{O}_X} \mathcal{O}_{X_0}$ is an $\mathcal{O}_{X_0}$-module. From Proposition 2.3.1 it follows that the correspondence $\mathcal{F} \mapsto \mathcal{F}_0$ gives an equivalence between the categories of coherent sheaves on X and X_0. The following proposition is easily verified using the well-known properties of paracompact topological spaces.

3.3.4. PROPOSITION. *Let X be a separated paracompact strictly k-analytic space. Then*:

(i) *for any sheaf of abelian groups $\mathcal{F}$ on X and $n \geq 0$, there are canonical isomorphisms*:

$$H^n(X, \mathcal{F}) \xleftarrow{\sim} \check{H}^n(X, \mathcal{F}) \xrightarrow{\sim} \check{H}^n(X_0, {}^{\tau}\mathcal{F}) \xrightarrow{\sim} H^n(X_0, {}^{\tau}\mathcal{F});$$

(ii) *for any coherent $\mathcal{O}_X$-module $\mathcal{F}$ and $n \geq 0$, there is a canonical isomorphism $H^n(X, \mathcal{F}) \xrightarrow{\sim} H^n(X_0, \mathcal{F}_0)$*;

(iii) *for any sheaf of groups $\mathcal{F}$, there is a canonical bijection $\check{H}^1(X, \mathcal{F}) \xrightarrow{\sim} \check{H}^1(X_0, {}^{\tau}\mathcal{F})$. In particular, the categories of locally constant sheaves on X and X_0 are equivalent.* ■

Proposition 3.3.4 means that the cohomological results from rigid analytic geometry are applicable to our analytic spaces. For convenience, we list them, giving necessary references. Various consequences of the central results are obtained in the same way as in complex analysis. Note that now we do not suppose that the valuation on k is nontrivial.

3.3.5. PROPOSITION (Kiehl's Theorem). *Let $\varphi: Y \to X$ be a proper morphism of k-analytic spaces, and let $\mathcal{F}$ be a coherent $\mathcal{O}_Y$-module. Then the $\mathcal{O}_X$-modules $R^n\varphi_*(\mathcal{F})$, $n \geq 0$, are coherent. Moreover, if X is a k-affinoid space, then $R^n\varphi_*(\mathcal{F}) = \mathcal{O}_X(H^n(Y, \mathcal{F}))$, $n \geq 0$.*

PROOF. It suffices to assume that $X = \mathcal{M}(\mathcal{A})$ is a k-affinoid space. In this case Y is compact and, by Proposition 3.3.4(ii), the cohomology groups of $\mathcal{F}$ can be computed from a finite affinoid covering $\{V_i\}$ of Y. In particular, the groups $H^n(Y, \mathcal{F})$ can be endowed with a Banach k-spaces structure. Let K be a field of the form $K_{r_1, \dots, r_n}$ with nontrivial valuation such that $X\widehat{\otimes}K$ and all of the $V_i\widehat{\otimes}K$ are strictly K-affinoid. By Kiehl's Theorem ([**Ki1**]), $H^n(Y\widehat{\otimes}K, \pi^*(\mathcal{F}))$, $n \geq 0$, are finite $\mathcal{A}\widehat{\otimes}K$-modules ($\pi$ denotes the canonical morphism $Y\widehat{\otimes}K \to Y$). By Proposition 2.1.2(ii), $H^n(Y\widehat{\otimes}K, \pi^*(\mathcal{F})) = H^n(Y, \mathcal{F})\widehat{\otimes}K$. By Proposition 2.1.11, $H^n(Y, \mathcal{F})$, $n \geq 0$, are finite Banach $\mathcal{A}$-modules. The canonical homomorphism of $\mathcal{O}_X$-modules $\mathcal{O}_X(H^n(Y, \mathcal{F})) \to R^n\varphi_*(\mathcal{F})$ is an isomorphism, because the stalks of these sheaves coincide at any point $x \in X$. ∎

3.3.6. PROPOSITION (Remmert's Mapping Theorem). *For any proper morphism of k-analytic spaces $\varphi: Y \to X$, the image $\varphi(Y)$ is a closed k-analytic set in X.*

PROOF. By Kiehl's Theorem, the $\mathcal{O}_X$-module $\varphi_*(\mathcal{O}_Y)$ is coherent. Hence, $\operatorname{Supp}(\varphi_*(\mathcal{O}_Y))$ is a closed k-analytic set in X. Since $\varphi(Y) = \operatorname{Supp}(\varphi_*(\mathcal{O}_Y))$, the required statement follows. ∎

3.3.7. PROPOSITION (Stein Factorization Theorem). *Let $\varphi: Y \to X$ be a proper morphism of k-analytic spaces. Then φ admits a unique factorization*

$$Y \xrightarrow{f} X' \xrightarrow{g} X$$

through a k-analytic space X' with the following properties:

(i) *f is a proper surjection, g is finite and $\varphi = g \circ f$;*

(ii) *$f_*(\mathcal{O}_Y) = \mathcal{O}_{X'}$ and all of the fibres of f are connected;*

(iii) *if $h: Y \to Z$ is an arbitrary morphism which is constant on the connected fibre components of φ, then there exists a unique morphism $h': X' \to Z$ such that $h = h' \circ f$.*

PROOF. By Kiehl's Theorem, $\varphi_*(\mathcal{O}_Y)$ is a coherent sheaf of $\mathcal{O}_X$-algebras. Therefore, one can associate with $\varphi_*(\mathcal{O}_Y)$ a k-analytic space X' together with a finite morphism $g: X' \to X$ such that $g_*(\mathcal{O}_{X'}) = \varphi_*(\mathcal{O}_Y)$. There is also a morphism $f: Y \to X'$ such that $\varphi = g \circ f$. It is clear that the morphism f is proper and $f_*(\mathcal{O}_Y) = \mathcal{O}_{X'}$. In particular, f is surjective. To show that all fibres of f are connected, it suffices to assume that X is k-quasiaffinoid and, in particular, separated. Suppose that there exists a point $x \in X'$ such that the fibre Y_x is not connected, i.e., $Y_x = Y' \cup Y''$, where Y' and Y'' are disjoint closed subsets of Y_x. Then there exists an open neighborhood $\mathcal{U}$ of x such that $f^{-1}(\mathcal{U}) = \mathcal{U}' \cup \mathcal{U}''$ with $Y' \subset \mathcal{U}'$, $Y'' \subset \mathcal{U}''$ and $\mathcal{U}' \cap \mathcal{U}'' = \varnothing$. But this contradicts the equality $f_*(\mathcal{O}_Y) = \mathcal{O}_{X'}$.

Let $h: Y \to Z$ be a morphism which is constant on the connected fibre components of φ. Then there is a unique map $h': |X'| \to |Z|$ between the underlying sets such that $h = h' \circ f$. This map is continuous because h is continuous and f is closed (i.e., it sends closed sets to closed sets). Since $\mathcal{O}_{X'} \xrightarrow{\sim} f_*(\mathcal{O}_Y)$, we have $h'_*(\mathcal{O}_{X'}) \xrightarrow{\sim} h_*(\mathcal{O}_Y)$. Hence, the homomorphism $\mathcal{O}_Z \to h_*(\mathcal{O}_Y)$ corresponding to h canonically determines a homomorphism $\mathcal{O}_Z \to h'_*(\mathcal{O}_{X'})$. This proves that h' is a morphism of k-analytic spaces. ∎

3.3.8. COROLLARY. *Let $\varphi: Y \to X$ be a morphism of k-analytic spaces. Then φ is finite if and only if φ is proper and has finite fibres.* ∎

The proof of the following facts is completely analogous to the proof of the corresponding facts in algebraic geometry (**[M1]**, Chapter 5).

3.3.9. PROPOSITION. *Let $\varphi: Y \to X$ be a proper morphism of k-analytic spaces with $X = \mathcal{M}(\mathcal{A})$ affinoid, and let $\mathcal{F}$ be a coherent sheaf on Y that is flat over X. There is a finite complex $K^{\cdot}: 0 \to K^0 \to K^1 \to \cdots \to K^n \to 0$ of finite projective $\mathcal{A}$-modules such that for any morphism $\psi: X' = \mathcal{M}(\mathcal{A}') \to X$ of k-affinoid spaces there is a canonical isomorphism*

$$H^p(Y', \psi'^*(\mathcal{F})) \xrightarrow{\sim} H^p(K^{\cdot} \underset{\mathcal{A}}{\otimes} \mathcal{A}'), \; p \geq 0,$$

where $Y' = Y \underset{X}{\times} X'$ and ψ' is the projection $Y' \to Y$. ∎

3.3.10. COROLLARY. *In the situation of Proposition 3.3.9, suppose that the $\mathcal{A}$-algebra $\mathcal{A}'$ is flat. Then*

$$H^p(Y', \psi'^*(\mathcal{F})) \xrightarrow{\sim} H^p(Y, \mathcal{F}) \underset{\mathcal{A}}{\otimes} \mathcal{A}'. \qquad ∎$$

Given a morphism $\varphi: Y \to X$ and a point $x \in X$, we let $\mathcal{F}_x$ denote the inverse image of a coherent $\mathcal{O}_Y$-module $\mathcal{F}$ on the $\mathcal{H}(x)$-analytic space $Y_x = \varphi^{-1}(x)$ (i.e., $\mathcal{F}_x = \mathcal{F} \otimes_{\mathcal{O}_x} \mathcal{H}(x)$). The following corollaries are analogs of Corollaries 1, 2, and 3 in **[M1]**, Chapter 5.

3.3.11. COROLLARY (Semicontinuity Theorem). *Let $\varphi: Y \to X$ be a proper morphism of k-analytic spaces, and let $\mathcal{F}$ be a coherent sheaf on Y which is flat over X. Then:*

(i) *for all $p \geq 0$ and $n \geq 0$, the set $\{x \in X \mid \dim_{\mathcal{H}(x)} H^p(Y_x, \mathcal{F}_x) \geq n\}$ is closed k-analytic in X;*

(ii) *the function $X \to \mathbf{Z}$ defined by*

$$x \mapsto \chi(\mathcal{F}_x) = \sum_{p=0}^{\infty} (-1)^p \dim_{\mathcal{H}(x)} H^p(Y_x, \mathcal{F}_x)$$

is locally constant on X. ∎

3.3.12. COROLLARY. *In the situation of Corollary 3.3.11, suppose that X is reduced and connected. Then for all p the following are equivalent:*

(a) *$x \mapsto \dim_{\mathcal{H}(x)} H^p(Y_x, \mathcal{F}_x)$ is a constant function;*

(b) $R^p\varphi_*(\mathscr{F})$ *is a locally free sheaf on* X, *and for all* $x \in X$ *the natural map*

$$R^p\varphi_*(\mathscr{F}) \underset{\mathscr{O}_X}{\otimes} \mathscr{H}(x) \to H^p(Y_x, \mathscr{F}_x)$$

is an isomorphism.

If these conditions are fulfilled, then the map

$$R^{p-1}\varphi_*(\mathscr{F}) \underset{\mathscr{O}_X}{\otimes} \mathscr{H}(x) \to H^{p-1}(Y_x, \mathscr{F}_x)$$

is an isomorphism for all $x \in X$. ∎

3.3.13. COROLLARY. *In the situation of Corollary* 3.3.11, *suppose that for some* p *one has* $H^p(Y_x, \mathscr{F}_x) = 0$ *for all* $x \in X$. *Then the natural map*

$$R^{p-1}\varphi_*(\mathscr{F}) \underset{\mathscr{O}_X}{\otimes} \mathscr{H}(x) \to H^{p-1}(Y_x, F_x)$$

is an isomorphism for all $x \in X$. ∎

A closed subset T of a k-analytic space X is said to be *thin of order* $n \geq 1$ (or simply *thin*, if $n = 1$) if every point $x \in X$ has an open neighborhood $\mathscr{U}$ such that $T \cap \mathscr{U}$ is contained in a nowhere dense closed k-analytic subset Y of $\mathscr{U}$ with $\dim_y(Y) \leq \dim_y(X) - n$ for all points $y \in Y$. A function $f \in \mathscr{O}(X\backslash T)$ is said to be *bounded near* T if every point $x \in T$ has a neighborhood $\mathscr{U}$ such that $f|_{\mathscr{U}\backslash T}$ is bounded.

3.3.14. PROPOSITION (Riemann Extension Theorem). *Let* X *be a normal* k*-analytic space. Let* T *be a thin set in* X, *and let* f *be a holomorphic function on* $X\backslash T$. *Assume that* f *is bounded near* T, *or that* T *is thin of order* 2. *Then* f *has a unique holomorphic extension to* X.

PROOF. The assertion is easily reduced to the case when the valuation on k is nontrivial, X is strictly k-affinoid, and T is a closed k-analytic set in X. But in this case it follows from Lütkebohmert's Theorem 1.6 in **[Lu]**, because, by Proposition 3.3.2(ii) and Remark 3.3.3, $\mathscr{O}(X) = \mathscr{O}(X_0)$ and $\mathscr{O}(X\backslash T) = \mathscr{O}(X_0\backslash T_0)$. ∎

3.3.15. COROLLARY. *If* X *is a connected normal* k*-analytic space, then for every thin set* T *in* X *the space* $X\backslash T$ *is connected.* ∎

3.3.16. COROLLARY (Normality Criterion). *Let* X *be a reduced* k*-analytic space. Then the following properties of a point* $x \in X$ *are equivalent*:

(a) x *is a normal point of* X;

(b) *there exist arbitrarily small open neighborhoods* $\mathscr{U}$ *of* x *such that, for any thin subset* $T \subset \mathscr{U}$, *every bounded holomorphic function on* $\mathscr{U}\backslash T$ *has a holomorphic extension to* $\mathscr{U}$. ∎

3.3.17. COROLLARY. *Let* $\varphi: Y \to X$ *be a proper morphism of* k*-analytic spaces, and let* $Y \xrightarrow{f} X' \xrightarrow{g} X$ *be the Stein factorization of* φ. *If* Y *is normal, then* X' *is also normal.* ∎

3.3.18. Corollary. *Let* $\varphi: Y \to X$ *be a morphism of reduced* k*-analytic spaces, and let* $\pi: X' \to X$ *and* $\pi': Y' \to Y$ *be normalizations of* X *and* Y. *If the inverse image* $\varphi^{-1}(N(X))$ *of the set of non-normal points of* X *is nowhere dense in* Y, *then there exists a unique morphism* $\varphi': Y' \to X'$ *such that the following diagram is commutative*:

$$\begin{array}{ccc} Y & \xrightarrow{\varphi} & X \\ \uparrow{\scriptstyle \pi'} & & \uparrow{\scriptstyle \pi} \\ Y' & \xrightarrow{\varphi'} & X' \end{array}$$

■

3.3.19. Proposition (Remmert-Stein Theorem). *Let* X *be a* k*-analytic space, and let* T *be a closed* k*-analytic set in* X *of dimension* $< d$. *Let* Y *be a closed* k*-analytic set in* $X\backslash T$ *which is of dimension* $\geq d$ *at all of its points. Then the topological closure* $\overline{Y}$ *of* Y *in* X *is a closed* k*-analytic set in* X.

Proof. The assertion is easily reduced to the case when the valuation on k is nontrivial and X is a strictly k-affinoid space. Let $\mathcal{Z}$ denote the closure of Y_0 in X_0. By Lütkebohmert's Theorem 3.1 in **[Lu]**, $\mathcal{Z}$ is a closed k-analytic set in the rigid k-affinoid space X_0. It follows that $\mathcal{Z} = Z_0$ for some closed k-analytic set Z in X, and that $Z = \overline{Y}$. ■

3.3.20. Corollary. *Let* X *be an irreducible* k*-analytic space, and let* T *be a closed* k*-analytic set in* X. *Then the space* $X\backslash T$ *is connected.* ■

3.3.21. Corollary. *Let* X *be a reduced* k*-analytic space, and let* X' *be its normalization. The following are equivalent*:

(a) X *is irreducible*;
(b) *every proper closed* k*-analytic subset of* X *is nowhere dense in* X;
(c) $\mathcal{M}(X)$ *is a field*;
(d) $\mathcal{O}(X')$ *is an integral domain*;
(e) X' *is connected*;
(f) X' *is irreducible.* ■

As in complex analysis, the Remmert-Stein Theorem implies the following two results (see also **[Lu]**). The formulation of the second result uses the notions introduced in §3.4.

3.3.22. Proposition (Levi Extension Theorem). *Let* X *be a reduced* k*-analytic space, and let* T *be a thin set in* X *of order* 2. *Then the restriction map* $\mathcal{M}(X) \to \mathcal{M}(X\backslash T)$ *is an isomorphism.* ■

3.3.23. Proposition (Chow's Theorem). *Every closed* k*-analytic subset of the projective space* $\mathbf{P}^n$ *is algebraic.* ■

3.3.24. Corollary. *Let* X *be a projective scheme over* k. *Then* X *is irreducible* (*resp. connected*) *if and only if the corresponding* k*-analytic space* X^{an} *is irreducible* (*resp. connected*). ■

Hartog's Theorem in complex analysis asserts that every function which is holomorphic in a neighborhood of the boundary of the n-dimensional disc, $n \geq 2$, has a unique holomorphic extension to the disc. The following statement is an analog of Hartog's Theorem.

3.3.25. PROPOSITION. *Let X be a reduced k-affinoid space of dimension ≥ 2 which has no irreducible components of dimension ≤ 1. Let $\mathscr{U}$ be a connected open neighborhood of the boundary $\partial(X)$. Then $\mathscr{M}(X) \tilde{\to} \mathscr{M}(\mathscr{U})$. If X is normal, then $\mathscr{O}(X) \tilde{\to} \mathscr{O}(\mathscr{U})$.*

PROOF. We may assume that the valuation on k is nontrivial and that $X = \mathscr{M}(\mathscr{A})$ is strictly k-affinoid. Let π denote the reduction map $X \to \widetilde{X} = \mathrm{Spec}(\widetilde{\mathscr{A}})$. We know that $\partial(X) = \pi^{-1}(\widetilde{X} \backslash \widetilde{X}_0)$, where $\widetilde{X}_0 = \mathrm{Max}(\widetilde{\mathscr{A}})$. For a point $p \in \widetilde{X}_0$ the inverse image $\pi^{-1}(\widetilde{X} \backslash \{p\})$ is closed in X. Since $\partial(X) = \bigcap_{p \in \widetilde{X}_0} \pi^{-1}(\widetilde{X} \backslash \{p\})$ and X is compact, we have $R := \bigcap_{i=1}^{n} \pi^{-1}(\widetilde{X} \backslash \{p_i\}) \subset \mathscr{U}$ for a finite number of points $p_1, \dots, p_n \in \widetilde{X}_0$. If $f_1, \dots, f_s$ are elements of $\mathscr{A}^{\circ}$ such that the set of all common zeros of $\widetilde{f}_1, \dots, \widetilde{f}_s$ on $\widetilde{X}$ coincides with $\{p_1, \dots, p_n\}$, then $R = \{x \in X \mid |f_i(x)| \geq 1 \text{ for } 1 \leq i \leq n\}$. That is, R_0 is a "Kugelfigur" of codimension ≥ 2 (see [**Ba**]). If f is a meromorphic (resp. holomorphic, if X is normal) function on $\mathscr{U}$, then f induces a meromorphic (resp. holomorphic) function on the rigid k-analytic space R_0. By Bartenwerfer's Theorem ([**Ba**], 3.5), $\mathscr{M}(X_0) \tilde{\to} \mathscr{M}(R_0)$ (resp. $\mathscr{O}(X_0) \tilde{\to} \mathscr{O}(R_0)$). Hence, f has a unique meromorphic (resp. holomorphic) extension to X. ∎

3.4. GAGA: the case of nontrivial valuation. Let X be a scheme of locally finite type over k. Let Φ be the functor from the category $\mathscr{A}n_k$ of analytic spaces over k to the category of sets which associates to every analytic space $\mathscr{X}$ the set of morphisms of k-ringed spaces $\mathrm{Hom}_k(\mathscr{X}, X)$.

3.4.1. THEOREM. *The functor Φ is represented by a closed k-analytic space X^{an} and a morphism $\pi: X^{an} \to X$. They have the following properties:*

(i) *For any non-Archimedean field K over k, there is a bijection $X^{an}(K) \tilde{\to} X(K)$. Furthermore, the map π is surjective and induces a bijection $X_0^{an} \tilde{\to} X_0$.*

(ii) *For any point $x \in X^{an}$, the homomorphism $\pi_x: \mathscr{O}_{X, \pi(x)} \to \mathscr{O}_{X^{an}, x}$ is local and strictly flat. Furthermore, if $x \in X_0^{an}$, then π_x induces an isomorphism of completions $\hat{\pi}_x: \hat{\mathscr{O}}_{X, \pi(x)} \tilde{\to} \hat{\mathscr{O}}_{X^{an}, x}$.*

PROOF. We first note that if X^{an} exists, the property (i) follows easily from the universal property of X^{an}. The construction of X^{an} will be made in several steps.

(1). Let $X = \mathrm{Spec}(k[T_1, \dots, T_n])$. We claim that in this case X^{an} is the n-dimensional affine space $\mathbf{A}^n$ defined in §1.5. One has

$$\mathbf{A}^n = \bigcup_{r>0} E(0, r) = \bigcup_{r>0} D(0, r)$$

(equality of sets), where $D(0, r)$ is the open subset $\{x \in E(0, r) \mid |T_i(x)| < r_i, 1 \le i \le n\}$ of the k-affinoid space $E(0, r) = \mathcal{M}(k\{r^{-1}T\})$. From Corollary 2.6.2 it follows that $D(0, r) \to \mathbf{A}^n$ is an open immersion of locally ringed spaces. In particular, $\mathbf{A}^n$ is a k-analytic space (obviously closed). The functor $\mathscr{X} \mapsto \mathrm{Hom}_k(\mathscr{X}, X)$ is represented by $\mathbf{A}^n$, because $\mathrm{Hom}_k(\mathscr{X}, \mathbf{A}^n) \xrightarrow{\sim} \Gamma(\mathscr{X}, \mathscr{O}_{\mathscr{X}})^n \xrightarrow{\sim} \mathrm{Hom}_k(\mathscr{X}, X)$.

(2). If the theorem is valid for a scheme Y, then it is also valid for any subscheme $X \subset Y$. Suppose that X is an open subscheme of Y. If $\varphi: Y^{an} \to Y$ is the canonical morphism, then it is easy to verify that the open subset $\varphi^{-1}(X)$ endowed with the induced ringed structure is the k-analytic space X^{an} associated with X and satisfying (ii).

Now let X be a closed subscheme of Y. If $\mathscr{J}$ is the coherent $\mathscr{O}_Y$-ideal which defines X, then $\mathscr{J}\mathscr{O}_{Y^{an}}$ is a coherent $\mathscr{O}_{Y^{an}}$-ideal which defines a closed k-analytic subspace X^{an} of Y^{an}. It is clear that X^{an} is the k-analytic space associated with X. Let $\pi: X^{an} \to X$ be the canonical morphism. For a point $x \in X^{an}$, the homomorphism π_x coincides with the homomorphism $\mathscr{O}_{Y,\varphi(x)}/\mathscr{J}_{\varphi(x)} \to \mathscr{O}_{Y^{an},x}/\mathscr{J}_{\varphi(x)}\mathscr{O}_{Y^{an},x}$ induced by φ_x. If $x \in X_0$, then $\hat{\varphi}_x$ is an isomorphism, and therefore $\hat{\pi}_x$ is an isomorphism.

(3). From (1) and (2) it follows that the theorem is valid for any affine scheme $X = \mathrm{Spec}(A)$, where A is a finitely generated ring over k. If X is arbitrary and $\{X_i\}_{i\in I}$ is a covering of X by open affine subschemes of the above type, then it follows from the universal property and from (2) that one can glue together all of the X_i^{an}'s and obtain the k-analytic space X^{an} associated with X.

It remains to show that π_x is a flat homomorphism for $x \in X^{an}$. We may suppose that $X = \mathrm{Spec}(A)$, where A is a finitely generated ring over k. It suffices to show that the ring $\mathscr{O}_{X^{an},x}$ is flat over A. This is true for all points $x \in X_0^{an} = X_0$. Hence, the ring $\mathscr{A}_V$ is flat over A for any affinoid subdomain $V \subset X^{an}$. But $\mathscr{O}_{X^{an},x} = \varinjlim \mathscr{A}_V$, where the limit is taken over all affinoid neighborhoods of x in X^{an}. Therefore, $\mathscr{O}_{X^{an},x}$ is flat over A. ∎

3.4.2. REMARK. If $X = \mathrm{Spec}(A)$, where A is a finitely generated ring over k, then X^{an} coincides with the set of all multiplicative seminorms on A whose restriction to k is the norm on k. If X is arbitrary, X^{an} (as a set) can be described as follows. Let $\mathscr{X}$ denote the union $\cup X(K)$ over all non-Archimedean fields K over k. The set $\mathscr{X}$ is endowed with the following equivalence relation. If $x' \in X(K')$ and $x'' \in X(K'')$, then $x' \sim x''$ if there exist a non-Archimedean field K and embeddings $K \to K'$ and $K \to K''$ such that the points x' and x'' come from the same point of $X(K)$. Then X^{an} coincides with the set of equivalence classes in $\mathscr{X}$.

The next proposition follows easily from Proposition 2.3.4.

3.4.3. PROPOSITION. *A scheme X of locally finite type over k is reduced, normal, Cohen-Macaulay, regular, smooth or of dimension n if and only if X^{an} has the same property.* ∎

3.4.4. PROPOSITION. *Let T be a locally constructible subset of X. Then $\pi^{-1}(\overline{T}) = \overline{\pi^{-1}(T)}$.*

PROOF. We may suppose that T is an everywhere dense open subset of X. Let Y be the reduced closed subscheme of X which is the complement of T. Then Y^{an} is a closed k-analytic subspace of X^{an} and $|Y^{an}| = |X^{an}|\backslash\pi^{-1}(T)$. We have to show that Y^{an} is nowhere dense in X^{an}. But this follows from Ritt's Lemma (Corollary 2.3.7), because $\dim_x(Y^{an}) = \dim(\mathscr{O}_{Y^{an},x}) = \dim(\mathscr{O}_{Y,\pi(x)}) < \dim(\mathscr{O}_{X,\pi(x)}) \doteq \dim_x(X^{an})$ for all $x \in Y_0^{an}$. ∎

3.4.5. COROLLARY. *In the situation of Proposition 3.4.4, the set T is open (resp. closed, resp. everywhere dense) in X if and only if $\pi^{-1}(T)$ is open (resp. closed, resp. everywhere dense) in X^{an}.* ∎

3.4.6. PROPOSITION. *Let $\varphi: Y \to X$ be a morphism of schemes of locally finite type over k, and let $\varphi^{an}: Y^{an} \to X^{an}$ be the corresponding morphism of k-analytic spaces. Then φ is (1) flat, (2) unramified, (3) étale, (4) smooth, (5) separated, (6) injective, (7) surjective, (8) an open immersion, (9) an isomorphism, and (10) a monomorphism if and only if φ^{an} has the same property.*

PROOF. Let $\pi: X^{an} \to X$ be the canonical morphism. For a point $x \in X^{an}$, the field $k(\pi(x)) = \mathscr{O}_{X,\pi(x)}/\mathbf{m}_{\pi(x)}$ is naturally embedded in $\mathscr{H}(x)$, and there is a canonical isomorphism of $\mathscr{H}(x)$-analytic spaces:

$$(*) \qquad (Y^{an})_x \xrightarrow{\sim} \left(Y_{\pi(x)} \underset{k(\pi(x))}{\otimes} \mathscr{H}(x)\right)^{an}.$$

For example, the morphism φ (resp. φ^{an}) is surjective if and only if all of its fibres are nonempty. Hence, (7) follows from $(*)$.

(1) The morphism φ (resp. φ^{an}) is flat if and only if it is flat at all points $y \in Y_0 = Y_0^{an}$. Therefore, (1) follows from Theorem 3.4.1(ii) and **[Bou1]**, Ch. III, 5.4.4.

(2) Suppose that φ is unramified. Then for any point $x \in X^{an}$, the fibre $Y_{\pi(x)}$ is the spectrum of a finite separable $k(\pi(x))$-algebra, and φ^{an} is unramified because of $(*)$. Conversely, if φ^{an} is unramified, then φ is unramified at all points of Y_0. Since the set of points of Y where φ is unramified is open in Y, it follows that φ is unramified everywhere.

(3) follows from (1) and (2); (4) follows from (1), the isomorphism $(*)$ and Proposition 3.4.3; (5) follows from Corollary 3.4.5.

(6) If φ^{an} is injective, then φ is injective by $(*)$. Conversely, suppose that φ is injective, i.e., $|\varphi^{-1}(\pi(x))| \le 1$ for any point $x \in X^{an}$. From this we deduce that $|{\varphi^{an}}^{-1}(x)| \le 1$ for any $x \in X^{an}$. This is clear for $x \in X_0^{an}$,

because for such a point $k(\pi(x)) = \mathscr{H}(x)$. Let x be an arbitrary point of X^{an} with $|\varphi^{-1}(\pi(x))| = 1$. We may assume that φ is a dominant morphism of finite type, that X is irreducible and that $\pi(x)$ is the generic point of X. Since there exists an open neighborhood $\mathscr{U}$ of $\pi(x)$ in X such that the induced morphism $\varphi^{-1}(\mathscr{U}) \to \mathscr{U}$ is finite ([**H**], ch. II, ex. 3.7), we may assume that φ is finite. Then φ^{an} is finite. In particular, the inverse image of an affinoid domain is an affinoid domain. So it suffices to show that a morphism of strictly k-affinoid spaces $\mathscr{M}(\mathscr{B}) \to \mathscr{M}(\mathscr{A})$ is injective if the induced map $\mathrm{Max}(\mathscr{B}) \to \mathrm{Max}(\mathscr{A})$ is injective. This is verified in the same way as Proposition 2.2.4(i).

(8) If φ is an open immersion, then φ^{an} is also an open immersion by the proof of Theorem 3.4.1. Conversely, suppose that φ^{an} is an open immersion. To show that φ is an open immersion, we use the fact that an étale and radical morphism is an open immersion ([**SGA**], I, §5.1). The morphism φ is étale by (3). The fibres of φ over the points of X_0 are obviously radical. Since the set of points $x \in X$ for which the fibre Y_x is radical forms a locally constructible set ([**EGA**], IV, 9.5.1.), it follows that all of the fibres of φ are radical.

(9) follows from (7) and (8). Finally, φ (resp. φ^{an}) is a monomorphism if the diagonal morphism $\Delta\colon Y \to Y \times_X Y$ (resp. $\Delta^{an}\colon Y^{an} \to Y^{an} \times_{X^{an}} Y^{an}$) is an isomorphism. Hence, (10) follows from (9). ∎

3.4.7. Proposition. *Suppose that the morphism φ is of finite type. Then φ is* (1) *dominant,* (2) *a closed immersion,* (3) *proper,* (4) *finite if and only if φ^{an} has the same property.*

Proof. (1) Suppose that φ is dominant, i.e., $\overline{\varphi(Y)} = X$. Since $\varphi(Y)$ is a constructible subset of X, the fact that $\overline{\varphi^{an}(Y^{an})} = X^{an}$ follows from Proposition 3.4.4. The converse implication is obvious.

(2) If φ is a closed immersion, then, by the proof of Theorem 3.4.1, φ^{an} is also a closed immersion. Conversely, suppose that φ^{an} is a closed immersion. Then φ^{an} is a proper monomorphism. By Proposition 3.4.6(10), φ is a monomorphism. Hence, the conclusion that φ is a closed immersion follows from (3).

(3) Suppose that φ is proper. Since the analytic spaces arising from schemes are closed, it suffices to show that the map $|\varphi^{an}|\colon |Y^{an}| \to |X^{an}|$ is proper. We may assume that X is an affine scheme. If Y is a projective space over X, then it is obvious that the map $|\varphi^{an}|$ is proper. From this and from the direct implication in (2) it follows that $|\varphi^{an}|$ is proper in the case when Y is a projective X-scheme. If Y is arbitrary, then by Chow's Lemma, there exist a projective X-scheme Y' and a projective surjective morphism $\psi\colon Y' \to Y$. Since the map $|\psi^{an}\varphi^{an}|\colon |Y'^{an}| \to |X^{an}|$ is proper and, by Proposition 3.4.6(7), $|\psi^{an}|\colon |Y'^{an}| \to |Y^{an}|$ is surjective, it follows that the map $|\varphi^{an}|\colon |Y^{an}| \to |X^{an}|$ is proper.

Conversely, suppose that φ^{an} is proper. By Proposition 3.4.6(5), φ is separated. We have to show that φ is universally closed. It suffices to show that φ is closed. Let T be a closed subset of Y. Since $\varphi(T)$ is a constructible subset of X, it follows from Proposition 3.4.4 that $\pi^{-1}(\overline{\varphi(T)}) = \overline{\pi^{-1}(\varphi(T))}$. But $\pi^{-1}(\varphi(T)) = \varphi^{an}(\pi^{-1}(T))$. Since φ^{an} is proper, $\varphi^{an}(\pi^{-1}(T))$ is closed in X^{an}, and hence $\pi^{-1}(\overline{\varphi(T)}) = \pi^{-1}(\varphi(T))$. It follows that $\overline{\varphi(T)} = \varphi(T)$, because π is surjective.

(4) The morphism φ (resp. φ^{an}) is finite if and only if it is proper and has finite fibres. Therefore, (4) follows from (3) and from the isomorphism $(*)$ in the proof of Proposition 3.4.6. ∎

3.4.8. THEOREM. (i) *X is separated $\Leftrightarrow$ $|X^{\mathrm{an}}|$ is Hausdorff;*

(ii) *X is proper $\Leftrightarrow$ $|X^{\mathrm{an}}|$ is Hausdorff and compact;*

(iii) *X is connected $\Leftrightarrow$ $|X^{\mathrm{an}}|$ is arcwise connected;*

(iv) *the dimension of X is equal to the topological dimension of $|X^{\mathrm{an}}|$.*

PROOF. (i) follows from Corollary 3.4.5 and Proposition 3.1.5. (ii) follows from (i), Proposition 3.4.7(3) and the fact that the k-analytic space X^{an} is closed.

(iii) By Theorem 3.2.1, it suffices to show that if X is connected, then X^{an} is connected. We may assume that X is affine and irreducible. Moreover, by Proposition 3.1.8, we may assume that X is normal. Consider a compactification $i: X \to P$, where P is a projective scheme and i is a dominant open immersion. Since X is irreducible, P is irreducible. From Corollary 3.3.24 it follows that P^{an} is connected. So it suffices to show that the complement of a nowhere dense closed k-analytic subset of a normal connected strictly k-affinoid space is connected. This follows from the Riemann Extension Theorem 3.3.14.

(iv) follows from Proposition 3.4.3 and Theorem 3.2.6. ∎

For completeness, we list the cohomological comparison theorems and some of their corollaries, all of which are proved in the same way as in the complex case. Note that if $\mathscr{F}$ is an $\mathscr{O}_X$-module, the inverse image $\pi^*(\mathscr{F}) = \mathscr{F}^{an}$ is a sheaf of $\mathscr{O}_{X^{an}}$-modules on X^{an}. From Theorem 3.4.1(ii) it follows that the functor $\mathscr{F} \mapsto \mathscr{F}^{an}$ is exact, faithful, and takes coherent sheaves to coherent sheaves.

3.4.9. PROPOSITION. *Let $\varphi: Y \to X$ be a proper morphism of schemes of locally finite type over k. Then for any coherent $\mathscr{O}_Y$-module F, there is a canonical isomorphism $(R^p\varphi_* F)^{an} \xrightarrow{\sim} R^p\varphi_*^{an}(F^{an})$, $p \geq 0$.* ∎

3.4.10. COROLLARY. *Let X be a proper k-scheme, and let F be a coherent $\mathscr{O}_X$-module. Then, for any integer $p \geq 0$, the canonical homomorphism $H^p(X, F) \to H^p(X^{an}, F^{an})$ is an isomorphism.* ∎

3.4.11. PROPOSITION. *Let X be a proper k-scheme. The functor that associates to a coherent $\mathscr{O}_X$-module F its inverse image F^{an} in X^{an} is an equivalence of categories.* ∎

3.4.12. COROLLARY. *The functor which associates to a proper k-scheme X the k-analytic space X^{an} is fully faithful.* ∎

3.4.13. COROLLARY. *Let X be a proper k-scheme. The functor $X' \mapsto X'^{an}$ induces an equivalence between the category of finite (resp. finite étale) schemes over X and the category of finite (resp. finite étale) k-analytic spaces over X^{an}.* ∎

3.4.14. COROLLARY. *Every reduced proper k-analytic space X of dimension 1 is algebraic. That is, there exists a projective algebraic curve Y over k such that $X \xrightarrow{\sim} Y^{an}$.* ∎

3.5. GAGA: the case of trivial valuation. Given a k-analytic space X, we let X_t denote the set of point $x \in X$ for which the canonical valuation on $\mathscr{H}(x)$ is trivial. The set X_t is closed in X, and $X_0 \subset X_t$. It is clear that every morphism $Y \to X$ induces a continuous map $Y_t \to X_t$.

Let X be a scheme of locally finite type over k. Let Φ be the functor defined at the beginning of §3.4.

3.5.1. THEOREM. *The functor Φ is represented by a closed strictly k-analytic space X^{an} and a morphism $\pi: X^{an} \to X$. They have the following properties:*

(i) *For any non-Archimedean field K over k, there is a bijection $X^{an}(K) \xrightarrow{\sim} X(K)$. Furthermore, $X_t^{an} \xrightarrow{\sim} X$, $X_0^{an} \xrightarrow{\sim} X_0$, and $\overline{X_0^{an}} = X_t^{an}$.*

(ii) *For any point $x \in X$, the homomorphism $\pi_x: \mathscr{O}_{X,\pi(x)} \to \mathscr{O}_{X^{an},x}$ is local and strictly flat. Furthermore, if $x \in X_t^{an}$, then π_x induces an isomorphism of completions $\hat{\pi}_x: \hat{\mathscr{O}}_{X,\pi(x)} \xrightarrow{\sim} \hat{\mathscr{O}}_{X^{an},x}$.*

(iii) *The functor $\mathscr{F} \mapsto \mathscr{F}^{an} = \pi^*(\mathscr{F})$ gives an equivalence between the categories of coherent sheaves on X and X^{an}. Furthermore, for a coherent $\mathscr{O}_X$-module $\mathscr{F}$ there are canonical isomorphisms $H^n(X, \mathscr{F}) \xrightarrow{\sim} H^n(X^{an}, \mathscr{F}^{an})$, $n \geq 0$.*

(iv) *The canonical map $\mathscr{M}(X) \to \mathscr{M}(X^{an})$ is an isomorphism; every meromorphic function on X^{an} is rational.*

(v) *The correspondence $X \mapsto X^{an}$ is a fully faithful functor that commutes with fibre products and extension of the ground field.*

PROOF. The k-analytic space is constructed in the same way as in the case of nontrivial valuation. It suffices to verify that if $X = \mathrm{Spec}(A)$, where $A = k[T_1, \dots, T_n]/\mathbf{a}$, then X^{an} has all of the required properties.

It is a simple matter to show that for $x \in X^{an}$, the homomorphism π_x is local and strictly flat. Suppose that $x \in X_t^{an}$, and let $\wp$ be the prime ideal of A corresponding to the point $\pi(x)$. In order to prove that $\hat{\pi}_x$ is an isomorphism, it suffices to assume that $\wp^d = 0$ for some $d \geq 1$. Note that for any $r \geq 1$ we have $k\{r_1^{-1}T_1, \dots, r_n^{-1}T_n\} = k[T_1, \dots, T_n]$, and hence, $\Gamma(X, \mathscr{O}_X) = A$. A basis of affinoid neighborhoods of the point x is formed by the Laurent neighborhoods of the form $V = X(r^{-1}f, g^{-1})$, where $f_1, \dots, f_m \in \wp$, $r_1, \dots, r_m < 1$ and $g_1, \dots, g_l \notin \wp$. We claim that $\mathscr{A}_V$ coincides with the localization of A with respect to the multiplicative system generated by $g_1, \dots, g_l$ (this will also give the equality $\overline{X_0^{\mathrm{an}}} = X_t^{\mathrm{an}}$). Indeed, if $f \in \wp$, then $f^d = 0$; and for $V = X(r^{-1}f)$, $r \geq 1$, one has $\mathscr{A}_V = A[[T]]/(T - f) = A$. If $g \notin \wp$, then for $V = X(g^{-1})$ one has $\mathscr{A}_V = A[T]/(gT - 1) \tilde{\to} A_{(g)}$.

Since the morphism π is strictly flat, the functor $\mathscr{F} \mapsto \mathscr{F}^{\mathrm{an}}$ is exact, faithful and takes coherent sheaves to coherent sheaves. Let F be a coherent sheaf on X. We denote $X(r) = \nu^{-1}(E(0, r))$, where ν is the closed immersion $X \to \mathbf{A}^n$. This is an affinoid domain in X^{an}, and for $r \geq 1$ one has $A \tilde{\to} \mathscr{A}_{X(r)}$. Hence, the inverse image of F on $X(r)$ is associated with a finite A-module M. It is clear that M does not depend on r and that $F = \mathscr{O}_X(M)^{\mathrm{an}}$. The properties (iii), (iv) and (v) are easily deduced from this.

It remains to show that X^{an} is a strictly k-analytic space. First of all, the projective space $\mathbf{P}^n = (\mathrm{Proj}(k[T_0, \dots, T_n]))^{\mathrm{an}}$ is a union of the strictly k-affinoid spaces $\mathscr{M}(k\{\frac{T_0}{T_i}, \dots, \frac{T_n}{T_i}\})$, $0 \leq i \leq n$. It follows that the analitification of any projective scheme over k is a finite union of strictly k-affinoid spaces and, in particular, it is strictly k-analytic. Hence, since X^{an} is an open subset of a projective scheme, it is strictly k-analytic. ∎

3.5.2. COROLLARY. *The functor $X' \to X'^{\mathrm{an}}$ induces an equivalence between the category of closed immersions (resp. finite morphisms) over X and the category of closed immersions (resp. finite morphisms) over X^{an}.* ∎

The other GAGA results mentioned in §3.5 are also true in this situation. Here we limit ourselves to the following fact.

3.5.3. THEOREM. (i) *X is separated $\Leftrightarrow$ $|X^{\mathrm{an}}|$ is Hausdorff;*
(ii) *X is proper $\Leftrightarrow$ $|X^{\mathrm{an}}|$ is Hausdorff and compact;*
(iii) *X is connected $\Leftrightarrow$ $|X^{\mathrm{an}}|$ is arcwise connected;*
(iv) *the dimension of X is equal to the topological dimension of $|X^{\mathrm{an}}|$.*

PROOF. The statements (i), (ii), and (iii) follow from Theorem 3.4.8; (iv) follows from Theorem 3.2.6. ∎

3.6. Meromorphic functions. In this subsection the field k is arbitrary.

Let X be a reduced irreducible k-analytic space. A reduced k-analytic space Y together with a proper morphism $\mu\colon Y \to X$ is called a *proper modification* of X if there is a thin subset $T \subset X$ such that $\mu^{-1}(T)$ is thin in Y and the induced morphism $Y \backslash \mu^{-1}(T) \to X \backslash T$ is an isomorphism; then clearly $\mu(Y) = X$.

3.6.1. PROPOSITION. *Let $\mu\colon Y \to X$ be a proper modification of an irreducible normal k-analytic space X. Then:*
(i) *the space Y is irreducible, and all of the fibres of μ are connected;*
(ii) *the canonical map $\mathscr{M}(X) \to \mathscr{M}(Y)$ is an isomorphism.*

PROOF. Since $X\backslash T$ is connected, the normal space $Y\backslash\mu^{-1}(T)$ is connected, and hence it is irreducible. Therefore, Y itself is irreducible, because $\mu^{-1}(T)$ is thin in Y.

Let $h \in \mathscr{M}(Y)$. We have to show that the function $h \circ \mu^{-1} \in \mathscr{M}(X\backslash T)$ extends meromorphically to every point of T. Since the sheaf $\mathscr{O}_Y + h\mathscr{O}_Y$ is coherent on Y, the sheaf $\mathscr{F} = \mu_*(\mathscr{O}_Y + h\mathscr{O}_Y)$ is coherent on X, by Kiehl's Theorem. If P denotes the polar locus of h, the support of the coherent $\mathscr{O}_X$-module $\mathscr{F}/\mathscr{O}_X$ is contained in $T \cup \mu(P)$, because $\mathscr{O}_Y + h\mathscr{O}_Y$ and $\mathscr{O}_Y$ coincide outside of P and because $\mu\colon Y\backslash\mu^{-1}(T) \to X\backslash T$ is an isomorphism. The set $\mu(P)$ is closed k-analytic in X, and hence the set $T \cup \mu(P)$ is thin in X. Therefore, every point $x \in T \cup \mu(P)$ has an open neighborhood $\mathscr{U}$ such that $g \cdot (\mathscr{F}/\mathscr{O}_X)|_{\mathscr{U}} = 0$ for some nonzero function $g \in \mathscr{O}(\mathscr{U})$. It follows that $(g \circ \mu) \cdot h|_{\mu^{-1}(\mathscr{U})} = f \circ \mu$ with $f \in \mathscr{O}(\mathscr{U})$. Since $f/g \in \mathscr{M}(\mathscr{U})$, we conclude that f/g is a meromorphic extension of $h \circ \mu^{-1}|_{\mathscr{U}\backslash T}$ onto $\mathscr{U}$.

It remains to show that all of the fibres of μ are connected. The direct image $\mu_*(\mathscr{O}_Y)$ is a coherent $\mathscr{O}_X$-module and lies in $\mu_*(\mathscr{M}_Y)$. But $\mu_*(\mathscr{M}_Y) = \mathscr{M}_X$. Since X is normal, we have $\mu_*(\mathscr{O}_Y) = \mathscr{O}_X$. It follows that all of the fibres of μ are connected. ∎

3.6.2. PROPOSITION (Hurwitz-Weierstrass Theorem). *Let X be a reduced proper scheme over k. Then the canonical map $\mathscr{M}(X) \to \mathscr{M}(X^{an})$ is an isomorphism.*

PROOF. It suffices to assume that the valuation on k is nontrivial (see Theorem 3.5.1(iv)) and that X is irreducible. If X is projective, the statement is deduced from the theorems of Remmert-Stein 3.3.19 and of Levi 3.3.22 in the same way as in the complex case (see **[GR2]**, §9.5.3). Now suppose that X is arbitrary. We may suppose that X is normal. By Chow's Lemma, there exist a projective k-scheme X' and a projective surjective morphism $\varphi\colon X' \to X$, which is an isomorphism outside of a proper closed subset. It is clear that $\mathscr{M}(X) \xrightarrow{\sim} \mathscr{M}(X')$. Since the induced morphism $\varphi^{an}\colon X'^{an} \to X^{an}$ is a proper modification of X^{an}, we have $\mathscr{M}(X^{an}) \xrightarrow{\sim} \mathscr{M}(X'^{an})$, by Proposition 3.6.1. Therefore, the required statement is reduced to the projective case. ∎

Let X be an irreducible normal k-analytic space, and let $f_1, \dots, f_n$ be meromorphic functions on X. The union T of the polar sets of $f_1, \dots, f_n$ is a thin closed k-analytic set in X. Each of the functions f_i induces a morphism $X\backslash T \to \mathbf{A}^1 \hookrightarrow \mathbf{P}^1$, and so we have the canonical morphism $\varphi\colon X\backslash T \to \mathbf{A}^n \hookrightarrow (\mathbf{P}^1)^n$ of $X\backslash T$ to the n-fold direct product $(\mathbf{P}^1)^n = \mathbf{P}^1 \times \cdots \times \mathbf{P}^1$ of the projective line (the Osgood space), regarded as a k-analytic space. Let Γ denote the graph of φ. This is an irreducible normal closed k-analytic subspace of $(X\backslash T) \times (\mathbf{P}^1)^n$ for which the projection $p\colon \Gamma \to X\backslash T$ is an isomorphism. The following proposition is proved in the same way as the corresponding fact in **[GR2]**, §9.6.3.

3.6.3. PROPOSITION. (i) *The topological closure* $\overline{\Gamma}$ *of* Γ *in* $X \times (\mathbf{P}^1)^n$ *is a reduced closed* k*-analytic subspace.*

(ii) $\overline{\Gamma}$ *together with the induced projection* $\overline{p}: \overline{\Gamma} \to X$ *is a proper modification of* X.

(iii) *If* X *is compact, then the closure of the set* $\varphi(X\backslash T)$ *in* $(\mathbf{P}^1)^n$ *is the image of* $\overline{\Gamma}$ *under the induced projection* $\overline{\varphi}: \overline{\Gamma} \to (\mathbf{P}^1)^n$. ∎

The subspace $\overline{\Gamma} \subset X \times (\mathbf{P}^1)^n$ together with the morphisms $\overline{p}: \overline{\Gamma} \to X$ and $\overline{\varphi}: \overline{\Gamma} \to (\mathbf{P}^1)^n$ is called the *graph of the meromorphic functions* $f_1, \dots, f_n$. From Propositions 3.6.1 and 3.6.3 one obtains:

3.6.4. COROLLARY. *The graph* $\overline{\Gamma}$ *is irreducible, all of the fibres of* $\overline{p}: \overline{\Gamma} \to X$ *are connected, and the canonical map* $\mathcal{M}(X) \to \mathcal{M}(\overline{\Gamma})$ *is an isomorphism.* ∎

The functions $f_1, \dots, f_n$ are said to be *analytically dependent* if $\overline{\varphi}(\overline{\Gamma}) \neq (\mathbf{P}^1)^n$. This property is equivalent to the requirement that there exist a nonempty open subset $\mathcal{U} \subset X\backslash T$ and a function $g \not\equiv 0$ holomorphic in a connected open subset of $\mathbf{A}^n$ containing $\varphi(\mathcal{U})$ such that $g \circ \varphi \equiv 0$ on $\mathcal{U}$. The functions $f_1, \dots, f_n$ are *algebraically dependent* if there exists a polynomial $P(z_1, \dots, z_n) \not\equiv 0$ such that $P(f_1(x), \dots, f_n(x)) \equiv 0$ on $X\backslash T$.

3.6.5. PROPOSITION. *Let* X *be an irreducible normal proper* k*-analytic space. Meromorphic functions* $f_1, \dots, f_n$ *on* X *are analytically dependent if and only if they are algebraically dependent.*

PROOF. The converse implication is obvious. Suppose that $f_1, \dots, f_n$ are analytically dependent. Let $\overline{\Gamma}$ be the graph of $f_1, \dots, f_n$. Since X is proper and the projection $\overline{p}: \overline{\Gamma} \to X$ is proper, it follows that $\overline{\Gamma}$ is also proper. By Remmert's Mapping Theorem 3.3.6, the image $\overline{\varphi}(\overline{\Gamma})$ is a closed k-analytic subset of $(\mathbf{P}^1)^n$. Since $f_1, \dots, f_n$ are analytically dependent, we have $\overline{\varphi}(\overline{\Gamma}) \neq (\mathbf{P}^1)^n$. By Chow's Theorem 3.3.23, there exists a polynomial $P(z_1, \dots, z_n) \not\equiv 0$ which vanishes on $\overline{\varphi}(\overline{\Gamma}) \cap \mathbf{A}^n$. This implies that $P(f_1, \dots, f_n) = 0$. ∎

3.6.6. THEOREM. *Let* X *be an irreducible normal proper* k*-analytic space. Then there exist a proper modification* $X' \to X$, *where* X' *is a normal* k*-analytic space, and a proper surjective morphism* $X' \to Y^{an}$, *where* Y *is a normal projective scheme over* k *of dimension* $\leq \dim(X)$, *such that* $\mathcal{M}(Y) \xrightarrow{\sim} \mathcal{M}(X') \xleftarrow{\sim} \mathcal{M}(X)$.

3.6.7. COROLLARY. *If* X *is a reduced irreducible proper* k*-analytic space, then* $\mathcal{M}(X)$ *is a finitely generated field over* k *of transcendence degree* $\leq \dim(X)$. ∎

PROOF. Since any $\dim(X) + 1$ meromorphic functions on X are analytically dependent, they are also algebraically dependent. Hence, the transcendence degree n of $\mathcal{M}(X)$ over k is at most $\dim(X)$. Let $f_1, \dots, f_n$ be a transcendence basis of $\mathcal{M}(X)$ over k. Let X' be the normalization of the graph $\overline{\Gamma}$ of $f_1, \dots, f_n$ (the *normal graph* of $f_1, \dots, f_n$), and consider the induced morphisms $p\colon X' \to \overline{\Gamma} \to X$ and $\varphi\colon X' \to \overline{\Gamma} \to (\mathbf{P}^1)^n$. Clearly, the morphism $p\colon X' \to X$ is a proper modification of X, and so $\mathcal{M}(X) \xrightarrow{\sim} \mathcal{M}(X')$. Since X is proper, the morphism φ is proper and has a Stein factorization:

$$X' \xrightarrow{\eta} Y \xrightarrow{\pi} (\mathbf{P}^1)^n .$$

Since X' is normal, it follows from Corollary 3.3.17 that Y is also normal. The morphism π is finite and surjective. By Corollaries 3.4.13 and 3.5.2, the k-analytic space Y is associated with a proper scheme over k which is obviously normal and projective. By the Hurwitz-Weierstrass Theorem, their fields of meromorphic and of rational functions coincide. We claim that $\mathcal{M}(Y) \xrightarrow{\sim} \mathcal{M}(X')$.

Let $f \in \mathcal{M}(X') = \mathcal{M}(X)$. We let X', p, φ (resp. X'_f, p_f, φ_f) denote the normal graph of $f_1, \dots, f_n$ (resp. $f_1, \dots, f_n, f$), and we let

$$X' \xrightarrow{\eta} Y \xrightarrow{\pi} (\mathbf{P}^1)^n \quad \left(\text{resp. } X'_f \xrightarrow{\eta_f} Y_f \xrightarrow{\pi_f} (\mathbf{P}^1)^{n+1}\right)$$

denote the Stein factorization of φ (resp. φ_f).

3.6.8. LEMMA. *The projection $(\mathbf{P}^1)^{n+1} \to (\mathbf{P}^1)^n$ onto the first n factors of $(\mathbf{P}^1)^{n+1}$ induces proper modifications $\xi\colon X'_f \to X'$ and $v\colon Y_f \to Y$ such that the diagram*

$$\begin{array}{ccccccc} & & X'_f & \xrightarrow{\eta_f} & Y_f & \xrightarrow{\pi_f} & (\mathbf{P}^1)^{n+1} \\ & \swarrow p_f & & & & & \\ X & & \downarrow\xi & & \downarrow v & & \downarrow \\ & \nwarrow p & & & & & \\ & & X' & \xrightarrow{\eta} & Y & \xrightarrow{\pi} & (\mathbf{P}^1)^n \end{array}$$

is commutative.

PROOF. (1) Construction of ξ. Let Q denote the joint polar locus of $f_1, \dots, f_n, f$. The projection $X \times (\mathbf{P}^1)^{n+1} \to X \times (\mathbf{P}^1)^n$ induces a surjective morphism $\chi\colon \overline{\Gamma}_f \to \overline{\Gamma}$ of the graph spaces $\overline{\Gamma}_f \subset X \times (\mathbf{P}^1)^{n+1}$ and $\overline{\Gamma} \subset X \times (\mathbf{P}^1)^n$ of the systems $f_1, \dots, f_n, f$ and $f_1, \dots, f_n$, respectively, which maps $\overline{\Gamma}_f \backslash (Q \times (\mathbf{P}^1)^{n+1}) \approx X \backslash Q$ isomorphically onto $\overline{\Gamma} \backslash (Q \times (\mathbf{P}^1)^n) \approx X \backslash Q$. Thus, the inverse image $\chi^{-1}(N(\overline{\Gamma}))$ of the set of non-normal points of $\overline{\Gamma}$ is nowhere dense in $\overline{\Gamma}_f$. By Corollary 3.3.18, χ uniquely induces a morphism $\xi\colon X'_f \to X'$ which is obviously a proper modification.

(2) Construction of v. The map $\eta \circ \xi$ is constant on the connected fibre components of φ_f. Hence, by the universal property of Stein factorizations (Proposition 3.3.7(iii)), we obtain a unique morphism $v\colon Y_f \to Y$ such that $v \circ \eta_f = \eta \circ \xi$. Since all of the fibres of η, ξ are connected, the same is

true of all of the fibres of v. We have to show that $v\colon Y_f \to Y$ is a proper modification.

Since $f_1, \ldots, f_n$ form a transcendence basis of $\mathcal{M}(X)$, it follows that $\pi(Y) = (\mathbf{P}^1)^n$ and $Z := \pi_f(Y_f) \neq (\mathbf{P}^1)^{n+1}$. Hence, the projection $(\mathbf{P}^1)^{n+1} \to (\mathbf{P}^1)^n$ induces a morphism $\nu\colon Z \to (\mathbf{P}^1)^n$ of the irreducible closed k-analytic subspace Z of $(\mathbf{P}^1)^{n+1}$ onto $(\mathbf{P}^1)^n$. We can choose a proper closed k-analytic subset $M \subset (\mathbf{P}^1)^n$ such that the induced morphism $Z \backslash \nu^{-1}(M) \to (\mathbf{P}^1)^n \backslash M$ is finite. If $T = \pi^{-1}(M)$, then the induced morphism $Y_f \backslash v^{-1}(T) \to Y \backslash T$ is finite, because the morphisms π, π_f and $Z \backslash \nu^{-1}(M) \to (\mathbf{P}^1)^n \backslash M$ are finite. We set $T' = T \cup \eta(Q)$, where Q is the joint polar locus of $f_1, \ldots, f_n, f$. As we saw in (1), ξ induces an isomorphism $X'_f \backslash \xi^{-1}(Q) \xrightarrow{\sim} X' \backslash Q$. Since $\eta_*(\mathcal{O}_{X'}) = \mathcal{O}_Y$ and $\eta_{f*}(\mathcal{O}_{X'_f}) = \mathcal{O}_{Y_f}$, it follows that for the induced finite morphism $v'\colon Y'_f = Y_f \backslash v^{-1}(T') \to Y' = Y \backslash T'$ we have $v'_*(\mathcal{O}_{Y'_f}) = \mathcal{O}_{Y'}$. The latter equality, together with the fact that the morphism v' is finite, imply that v' is an isomorphism. Hence, v is a proper modification. ∎

From Proposition 3.6.1 it follows that $\mathcal{M}(Y) \xrightarrow{\sim} \mathcal{M}(Y_f)$. Since f lies in the image of $\mathcal{M}(Y_f)$ in $\mathcal{M}(X')$, we have $f \in \mathcal{M}(Y)$. Hence, $\mathcal{M}(Y) \xrightarrow{\sim} \mathcal{M}(X')$. ∎

CHAPTER 4

Analytic Curves

4.1. One-dimensional quasipolyhedra. At this point we introduce a class of topological spaces that is useful for the description of analytic curves. This class contains one-dimensional polyhedra which admit a countable, locally finite triangulation. For brevity we call them simply *polyhedra*.

4.1.1. DEFINITION. (i) A connected locally compact Hausdorff space X is called a (one-dimensional) *quasipolyhedron* if:

(a) a basis of the topology of X is formed by the open subsets $\mathcal{U}$ that have the following properties:

(a_1) the set $\overline{\mathcal{U}}\backslash\mathcal{U}$ is finite;

(a_2) for any pair of distinct points $x, y \in \mathcal{U}$, there exists a unique closed subset $\ell_{x,y} \subset \mathcal{U}$ which is homeomorphic to the unit interval with endpoints x and y;

(b) all connected open subsets of X are countable at infinity.

(ii) A quasipolyhedron X is said to be *simply connected* if (a_2) is also valid for $\mathcal{U} = X$.

4.1.2. REMARK. We list some properties of a simply connected quasipolyhedron X.

(i) Any connected subset of X which is locally compact in the induced topology is a simply connected quasipolyhedron.

(ii) For any pair of connected closed subsets $\Sigma', \Sigma'' \subset X$ with empty intersection, there exists a unique closed subset $\ell_{\Sigma',\Sigma''} \subset X$ which is homeomorphic to the unit interval with endpoints $x \in \Sigma'$ and $y \in \Sigma''$ and has the property that $\ell_{\Sigma',\Sigma''} \cap \Sigma' = \{x\}$ and $\ell_{\Sigma',\Sigma''} \cap \Sigma'' = \{y\}$. The set $\ell_{\Sigma',\Sigma''}\backslash\{x, y\}$ consists of the points $z \in X$ such that Σ' and Σ'' lie in different connected components of $X\backslash\{z\}$.

(iii) For a connected closed subset $\Sigma \subset X$, the map $\tau_\Sigma : X \to \Sigma$ that is the identity on Σ and sends a point $x \in X\backslash\Sigma$ to the second endpoint of $\ell_{x,\Sigma}$, is continuous.

(iv) Let Σ be a closed subset of X. Then there is a minimal connected subset $L(\Sigma) \subset X$ containing Σ. Namely, $L(\Sigma) = \bigcup_{x,y\in\Sigma} \ell_{x,y}$. Moreover, $L(\Sigma)$ is closed in X, and $L(\Sigma)\backslash\Sigma$ is a polyhedron.

For a simply connected quasipolyhedron X, let $\widehat{X}$ denote the set of all partial orderings $\leq$ on X with the following properties:

(a) for any two points $x, y \in X$, there exists a unique minimal point $\sup(x, y) \in X$ which is greater than (or equal to) x and y;

(b) if $x \leq y$, then $x \leq z \leq y$ for all points $z \in \ell_{x,y}$.

The set X is naturally embedded in $\widehat{X}$ as follows. A point $x \in X$ determines the partial ordering $\leq$ for which $y \leq z$ if and only if $z \in \ell_{x,y}$. Let $\leq$ be an element of $\widehat{X}$. For $x \in X$ we set $\ell_x = \{y \in X | \; x \leq y\}$. If the partial ordering $\leq$ is associated with a point of X, then the set ℓ_x is a point or else is homeomorphic to $[0, 1]$. Otherwise, ℓ_x is homeomorphic to $[0, 1[$. We endow the set $\widehat{X}$ with a topology as follows. The basis of the topology consists of open subsets $\mathcal{U} \subset X$ with compact closure in X and of sets $\widetilde{\mathcal{U}} \subset \widehat{X}$ of the following type. $\mathcal{U}$ is a connected component of the complement of a point $x \in X$ with non-compact closure in X, and $\widetilde{\mathcal{U}}$ is the union of $\mathcal{U}$ and the partial orderings in $\partial(X) = \widehat{X} \backslash X$ for which $\ell_x \cap \mathcal{U} \neq \varnothing$.

4.1.3. PROPOSITION. (i) *The space $\widehat{X}$ is the unique compactification of X which is a simply connected quasipolyhedron. The complement $\partial(X)$ (the set of endpoints of X) is compact and totally disconnected, and it has a countable basis of open subsets.*

(ii) *The set $\Delta(X) = X \cap L(\partial(X))$ is a contractible polyhedron (it is empty if X has at most one endpoint), and $\partial(\Delta(X)) \xrightarrow{\sim} \partial(X)$.*

(iii) *$\Delta(X)$ consists of the points of X which have no connected open neighborhoods with a single endpoint.* ∎

Now let X be an arbitrary quasipolyhedron. The *skeleton* $\Delta(X)$ of X is the set of points of X having no open neighborhoods which are simply connected quasipolyhedra with a single endpoint. This is a closed subset of X.

4.1.4. PROPOSITION. (i) *$\Delta(X)$ is empty if and only if X is a simply connected quasipolyhedron having at most one endpoint.*

Suppose that $\Delta(X)$ is nonempty.

(ii) *$\Delta(X)$ is a connected polyhedron which does not contain points having open neighborhoods homeomorphic to $[0, 1[$. If X is compact, $\Delta(X)$ is also compact.*

(iii) *The connected components of the set $X \backslash \Delta(X)$ are simply connected quasipolyhedra with one endpoint in $\Delta(X)$.* ∎

4.1.5. DEFINITION. (i) A simply connected quasipolyhedron X is said to be *special* if, for any partial ordering $\leq$ in $\widehat{X}$, there exists a function $\theta: X \to [0, 1]$ that for any $x \in X$ induces a homeomorphic embedding of ℓ_x in $[0, 1]$ that preserves the ordering of these sets.

(ii) A quasipolyhedron X is said to be *special* if all of the simply connected subquasipolyhedra of X are special.

Note that we do not demand continuity of the function θ. It is easy to see that if the function θ exists for one partial ordering, then it exists for every partial ordering. Moreover, if X is special, then $\widehat{X}$ is also special. Finally, any connected polyhedron is a special quasipolyhedron.

4.1.6. PROPOSITION. *Let X be a special quasipolyhedron. Then every nonempty closed connected subset $\Sigma \subset X$ containing $\Delta(X)$ is a strong deformation retract of X. In particular, if X is simply connected, then it is contractible.*

PROOF. Let $\tau: X \to \Sigma$ be the retraction map in 4.1.2(iii). For any point $x \in \Sigma$ we choose a function θ_x on $\tau^{-1}(x)$ associated with the partial ordering induced by x. We may assume that $\theta_x(x) = 1$ for all $x \in X$. We now define a function $\theta: X \to [0, 1]$ by $\theta(x) = \theta_{\tau(x)}(x)$. The deformation retraction $\Phi: X \times [0, 1] \to X$ is constructed as follows. For a pair $(x, t) \in X \times [0, 1]$, $\Phi(x, t)$ is the point $y \in \ell_{x, \tau(x)}$ for which $\theta(y) = \max(\theta(x), t)$. It is easy to verify that the map Φ is continuous. ∎

4.1.7. PROPOSITION. *Let a pro-finite group G act continuously on a (special) quasipolyhedron X. Then the quotient space X/G is also a (special) quasipolyhedron, and the skeleton $\Delta(X/G)$ coincides with the maximal subpolyhedron of $\Delta(X)/G$ with no endpoints. Moreover, if X is simply connected, then $\partial(X)/G \tilde{\to} \partial(X/G)$.*

PROOF. Suppose that X is simply connected. For any point $x \in X \backslash X^G$, the orbit Gx is compact and the set $\Sigma_x = L(Gx) \backslash Gx$ is a polyhedron invariant under G. From this it follows that the stabilizer of a point in Σ_x has finite index in G. In particular, the set $X^G \cap \Sigma_x$ is nonempty. It is clear that X^G is connected. If τ denotes the retraction map $X \to X^G$, then $X^G \cap \Sigma_x = \{\tau(x)\}$. It follows that Σ_x is connected, and the canonical map $\pi: X \to X/G$ is injective on ℓ_{x, x_0}, where $x_0 = \tau(x)$. If $y \in \tau^{-1}(x_0)$, there exists $y' \in Gy \cap \tau^{-1}(x_0)$ such that $\ell_{x, x_0} \backslash \ell_{y', x_0} \subset \ell_{x, x_0} \backslash \ell_{y'', x_0}$ for all $y'' \in Gy \cap \tau^{-1}(x_0)$. It follows that π is injective on $\ell_{x, y'}$, and its image (for $y \notin Gx$) is a unique closed subset of X/G homeomorphic to the unit interval with endpoints $\pi(x)$ and $\pi(y)$. Hence, X/G is a simply connected quasipolyhedron, and $\partial(X)/G \tilde{\to} \partial(X/G)$. Suppose that X is special. If $x_0 \in X^G$ and $\theta: X \to [0, 1]$ is the corresponding function, then the function defined by $\theta'(x) = \max_{\sigma \in G} \theta({}^{\sigma}x)$ induces a function on X/G which satisfies the condition 4.1.5(i). The case of arbitrary X follows easily from this. ∎

4.1.8. PROPOSITION. *Any special quasipolyhedron X has a universal covering Ω. Ω is a simply connected special quasipolyhedron, and $\Delta(\Omega)/\pi_1(X) \tilde{\to} \Delta(X)$.* ∎

4.2. The projective line. Throughout this section we suppose that the valuation on k is nontrivial.

Let $\mathbf{A}^1$ and $\mathbf{P}^1$ denote the k-analytic spaces associated with the algebra-geometric affine and projective lines, respectively. Recall that $\mathbf{A}^1$ is the set of all multiplicative seminorms on the ring of polynomials $k[T]$ whose restriction to k coincides with the valuation on k. The topology of $\mathbf{A}^1$ is the weakest one with respect to which all functions $\mathbf{A}^1 \to \mathbf{R}_+$ of the form $|\ | \mapsto |f|$, where $f \in k[T]$, are continuous. The affine line $\mathbf{A}^1$ is locally compact, and $\mathbf{P}^1 = \mathbf{A}^1 \bigcup \{\infty\}$ is the minimal compactification of $\mathbf{A}^1$.

In §1.4.4 the points of the character space $\mathcal{M}(k\{r^{-1}T\})$ were classified for algebraically closed k. This also gives a classification of the points of $\mathbf{P}^1$ for arbitrary k. Indeed, $\mathcal{M}(\widehat{k^a}\{r^{-1}T\})/G \xrightarrow{\sim} \mathcal{M}(k\{r^{-1}T\})$, where $G = \mathrm{Gal}(k^a/k)$. The action of G does not change the type of the points, and so we have the same classification of the points of $\mathcal{M}(k\{r^{-1}T\})$ for arbitrary k. Furthermore, the space $\mathcal{M}(k\{r^{-1}T\})$ is naturally embedded in $\mathbf{A}^1$ as the closed disc $E(0, r)$ with center at 0, and $\mathbf{A}^1 = \bigcup_{r>0} E(0, r)$. Finally, $\mathbf{P}^1 \backslash E(0, 1)$ is isomorphic to the open unit disc $D(0, 1)$.

Let π denote the canonical map $\mathbf{P}^1_{\widehat{k^a}} \to \mathbf{P}^1$. The map π is surjective, proper, and open. The group G sends closed (resp. open) discs, of $\mathbf{A}^1_{\widehat{k^a}}$ to closed (resp. open) discs, and does not change their radii. The image under π of a closed (resp. open) disc of $\mathbf{A}^1_{\widehat{k^a}}$ is called a *closed* (resp. *open*) *disc* in $\mathbf{A}^1$. A *closed* (resp. *open*) *disc* in $\mathbf{P}^1$ is a closed (resp. open) disc in $\mathbf{A}^1$ or a set of the type $\mathbf{P}^1 \backslash D$ (resp. $\mathbf{P}^1 \backslash E$), where D is an open (resp. E is a closed) disc in $\mathbf{A}^1$. The points of $\mathbf{P}^1$ correspond bijectively to families of embedded closed discs. For a point $x = x(\mathcal{E}) \in \mathbf{A}^1$ we let $r(x)$ denote the infimum of the radii of the discs $E^{(\rho)} \in \mathcal{E}$ (the *radius* of x). For example, $r(x) = 0$ if and only if x is of type (1); and if x is associated with a closed disc E, then $r(x) = r(E)$.

A closed (resp. open) subset of $\mathbf{P}^1$ of the type $E \backslash \bigcup_{i=1}^n D_i$ (resp. $D \backslash \bigcup_{i=1}^n E_i$) is called a *standard* subset. Here E, E_i are closed discs and D, D_i are open discs in $\mathbf{P}^1$. A basis of the topology of $\mathbf{P}^1$ is formed by the standard open subsets. Note that if U and V are standard closed subsets with nonempty intersection, then the union $U \bigcup V$ either coincides with $\mathbf{P}^1$ or else is a standard closed subset. The class of affinoid subdomains of $\mathbf{P}^1$ coincides with the class of disjoint finite unions of standard closed sets. Every affinoid subdomain of $\mathbf{P}^1$ is rational. That is, the set of rational functions defined on it is everywhere dense in the corresponding k-affinoid algebra. An affinoid subdomain of $\mathbf{A}^1$ is a Weierstrass domain (i.e., of the form $\{x \mid |f_i(x)| \le p_i,\ 1 \le i \le n\}$, $f_i \in k[T]$) if and only if it is a finite union of closed discs. In particular, from Proposition 2.6.1 it follows that any compact subset of $\mathbf{A}^1$ is rationally convex. Moreover, it is polynomially convex if and only if its complement is connected.

Note that if x is of type (1) or (4), then $\mathbf{P}^1 \backslash \{x\}$ is connected. If x is of type (3), then $\mathbf{P}^1 \backslash \{x\}$ is a union of two open discs in $\mathbf{P}^1$. If x is of type

(2), then $\mathbf{P}^1\backslash\{x\}$ is a disjoint union of open discs in $\mathbf{P}^1$ which correspond to closed points of the nonsingular projective curve over $\widetilde{k}$ with field of rational functions $\widetilde{\mathscr{H}(x)}$.

4.2.1. THEOREM. *$\mathbf{P}^1$ is a simply connected special quasipolyhedron. In particular, all of the connected locally compact subsets of $\mathbf{P}^1$ are contractible.*

PROOF. For every $\rho \geq r(x) \in \mathbf{A}^1$, we define a closed disc $E(x, \rho)$ of radius ρ with center at x as follows. If $\rho > r(x)$, there exists a unique closed disc E of radius ρ containing x, and we set $E(x, \rho) = E$. If x is associated with a closed disc E, we set $E(x, r(x)) = E$. If x is of type (1) or (4), we set $E(x, r(x)) = \{x\}$. In the latter case we also set $p(E(x, r(x))) = x$. For points $x, y \in \mathbf{A}^1$ we let $r(x, y)$ denote the radius of the minimal closed disc containing x and y (this is the disc $E(x, r(x, y)) = E(y, r(x, y))$). We set $\ell'_{x,y} = \{p(E(x, t))| \; r(x) \leq t \leq r(x, y)\}$, $\ell''_{x,y} = \{p(E(y, t))| \; r(y) \leq t \leq r(x, y)\}$, and $\ell_{x,y} = \ell'_{x,y} \cup \ell''_{x,y}$. We also set $\ell_{x,\infty} = \{p(E(x, t))| \; r(x) \leq t < \infty\} \cup \{\infty\}$ and $\ell_{\infty,\infty} = \{\infty\}$. It is easy to verify that for $x \neq y$ the set $\ell_{x,y}$ is closed and homeomorphic to the unit interval with endpoints x and y. Moreover, $\ell_{x,y}\backslash\{x, y\}$ consists of all points $z \in \mathbf{P}^1$ such that x and y lie in different connected components of the set $\mathbf{P}^1\backslash\{z\}$. From this it follows that $\ell_{x,y}$ is the unique closed subset of $\mathbf{P}^1$ which is homeomorphic to the unit interval with endpoints x and y.

Let $\mathscr{U} = D\backslash\bigcup_{i=1}^n E_i$ be a standard open subset of $\mathbf{A}^1$. It is easy to see that if $x, y \in \mathscr{U}$, then $\ell_{x,y} \subset \mathscr{U}$. Moreover, if $D = D(a, r)$, then $\overline{\mathscr{U}}\backslash\mathscr{U} = \{p(E(a, r)), p(E_1), \ldots, p(E_n)\}$.

Furthermore, let $\leq$ be the partial ordering on $\mathbf{P}^1$ corresponding to the point $p(E(0, 1))$. Then the function θ in 4.1.5(i) can be defined as follows. If $x \in E(0, 1)$, then $\theta(x) = r(x)$. If $x \notin E(0, 1)$, then $\theta(x) = r(\sigma(x))$, where σ is the automorphism $T \mapsto 1/T$.

It remains to show that any connected open subset $\mathscr{U} \subset \mathbf{P}^1$, where $\mathscr{U} \neq \varnothing, \mathbf{P}^1$, is countable at infinity. We construct a decreasing sequence $V_1 \supset V_2 \supset \ldots$ of affinoid neighborhoods of $\Sigma = \mathbf{P}^1\backslash\mathscr{U}$ such that $\bigcap_{n=1}^\infty V_n = \Sigma$. It suffices to assume that $\Sigma \subset \mathbf{A}^1$.

Let N denote the set of affinoid neighborhoods of Σ of the form $V = \bigcup_{i=1}^n E_i$, where the E_i are discs in $\mathbf{A}^1$ with $E_i \cap \Sigma \neq \varnothing$. Since $\mathscr{U}$ is connected, N is a basis of neighborhoods of Σ. Let M denote the set of discs which occur in the decompositions of $V \in N$. For a point $x \in \Sigma$ and an integer $n \geq 0$, we denote by $M_n(x)$ the set of discs E with $x \in E \in M$ such that if $x \in E' \subset E$ and $E' \in M$, then $r(E) - r(E') \leq 1/n$. It is clear that $M_n(x)$ is nonempty. The required sets V_n are constructed inductively. V_1 is an arbitrary element of N which is a union of discs in $M_1(x)$. Suppose that V_{n-1}, $n \geq 1$, has already been constructed, and let $V_{n-1} = \bigcup_{i=1}^l E_i$.

For every point $x \in \Sigma$ we choose $E_x \in M_n(x)$ which is contained in some E_i. Then Σ lies in a finite union V_n of the E_x's. By construction, $V_n \in N$. It is easy to verify that $\bigcap_{n=1}^{\infty} V_n = \Sigma$. ∎

4.2.2. Corollary. *Let X be a smooth geometrically irreducible projective curve of genus zero. Then the space $|X^{\mathrm{an}}|$ is a simply connected special quasipolyhedron. In particular, $|X^{\mathrm{an}}|$ is contractible.* ∎

4.2.3. Remark. The partial ordering on $\mathbf{A}^1$ corresponding to the point ∞ is described as follows: $x \leq y$ if and only if $|f(x)| \leq |f(y)|$ for all $f \in k[T]$.

4.2.4. Remark. Let $\mathbf{\Omega}$ be a connected open subset of $\mathbf{P}^1$ whose complement lies in $\mathbf{P}^1(k)$ and contains at least two points, and let $\tau : \mathbf{\Omega} \to \Delta(\Omega)$ be the retraction map in 4.1.2(iii).

(i) Let $\leq$ be the partial ordering on $\mathbf{\Omega}$ defined as follows: $x \leq y$ if and only if $|f(x)| \leq |f(y)|$ for all $f \in \mathcal{O}(\Omega)$. Then the set of maximal points with respect to this ordering coincides with $\Delta(\Omega)$. Moreover, the ordering $\leq$ induces the partial ordering corresponding to x, $x \in \Delta(\Omega)$, on each fibre $\tau^{-1}(x)$.

(ii) For any compact connected subset $\Sigma \subset \Delta(\Omega)$, the set $\tau^{-1}(\Sigma)$ is a standard affinoid domain.

(iii) For any invertible holomorphic function $f \in \mathcal{O}(\Omega)^*$, one has $|f(x)| = |f(\tau(x))|$, $x \in \mathbf{\Omega}$. If the function $x \mapsto |f(x)|$ on $\Delta(\Omega)$ is constant, then f is constant.

(iv) The class of functions $x \mapsto |f(x)|$ on $\Delta(\Omega)$ determined by $f \in \mathcal{O}(\Omega)^*$ is described as follows. First of all, the points of $\Delta(\Omega)$ having no open neighborhoods homeomorphic to $]0, 1[$ are called vertices. The closure of a connected component of the complement of the set of vertices is called an edge (each edge is homeomorphic to $[0, 1]$ or to $[0, 1[$). Assume that $\infty \notin \mathbf{\Omega}$, and let $f \in \mathcal{O}(\Omega)^*$. Then for any edge ℓ there exist $\alpha \in |k^*|$ and $n \in \mathbf{Z}$ such that $|f(x)| = \alpha r(x)^n$ for $x \in \ell$. Furthermore, the function f determines a $\mathbf{Z}$-valued function c_f on the set of oriented edges as follows. If the orientation of ℓ is in the direction of increase of the function $x \mapsto r(x)$, then $c_f(x) = n$; otherwise, $c_f(x) = -n$. It is easy to see that the sum of the values of c_f on the edges emanating from a vertex is equal to zero, i.e., c_f is a harmonic 1-cochain. Moreover, for any harmonic $\mathbf{Z}$-valued 1-cochain c on $\Delta(\Omega)$, there exists $f \in \mathcal{O}(\Omega)^*$ with $c_f = c$ (f can be constructed using infinite products of rational functions).

(v) Suppose that the field k is locally compact, and set $\mathbf{\Omega} = \mathbf{P}^1 \backslash \mathbf{P}^1(k)$. This is a connected open subset of $\mathbf{P}^1$. The skeleton $\Delta(\Omega)$ coincides with the Bruhat-Tits building of the group $\mathrm{SL}_2(k)$. (See §5.5 for a more general construction.) One can show that $\mathbf{\Omega}$ admits a flat $\mathrm{SL}_2(k)$-invariant metric whose restriction to $\Delta(\Omega)$ coincides with the metric of the building (a metric m on a simply connected quasipolyhedron is said to be flat if $m(x, z) =$

$m(x, y) + m(y, z)$ for all points x, y, z with $y \in \ell_{x,z}$).

4.3. Curves of genus ≥ 1. All of the analytic spaces and affinoid subdomains considered until the end of this section are assumed to be strict. We recall the definitions and results on reduction of *k-analytic curves* (separated irreducible one dimensional k-analytic spaces) which will be used below.

An affinoid domain V in a k-affinoid space X is said to be *formal* if the induced morphism of the reductions $\widetilde{V} \to \widetilde{X}$ is an open immersion. Let X be a separated k-analytic space. An admissible (see §3.3) affinoid covering $\mathscr{U} = \{U_i\}_{i\in I}$ of X is said to be *formal* if the intersection $U_i \cap U_j$ is a formal subdomain of U_i for every $i, j \in I$. Two such coverings $\mathscr{U} = \{U_i\}_{i\in I}$ and $\mathscr{V} = \{V_j\}_{j\in J}$ are said to be *equivalent* if $U_i \cap V_j$ is a formal subdomain of both U_i and V_j for all $i \in I$ and $j \in J$. A formal covering $\mathscr{U} = \{U_i\}_{i\in I}$ gives rise to a reduced scheme $\widetilde{X}$ of locally finite type over $\widetilde{k}$ and to a map $\pi: X \to \widetilde{X}$ that extends all of the reduction maps $\pi_i: U_i \to \widetilde{U}_i$. The scheme $\widetilde{X}$ is called the *reduction of X with respect to the formal covering $\mathscr{U}$*, and it is denoted by $\widetilde{X}_{\mathscr{U}}$ if $\mathscr{U}$ needs to be specified. Equivalent formal coverings of X give rise to the same reduction.

A k-affinoid algebra $\mathscr{A}$ is said to be *distinguished* if the spectral norm on $\mathscr{A}$ is equal to the residue norm for some epimorphism $k\{T_1, \dots, T_n\} \to \mathscr{A}$. In the case when k is stable ([**BGR**], 3.6; discrete valuation fields and algebraically closed fields are stable), $\mathscr{A}$ is distinguished if and only if $\mathscr{A}$ is reduced and the values of the spectral norm on $\mathscr{A}$ lie in $|k|$ ([**BGR**], 6.4.3/1). For example, the algebra $\mathscr{A}_V$ of a standard affinoid subdomain $V = E\backslash\bigcup_{i=1}^n D_i$ of $\mathbf{A}^1$ is distinguished if $r(E), r(D_i) \in |k^*|$. A formal covering $\mathscr{U} = \{U_i\}_{i\in I}$ of a separated k-analytic space X is said to be *distinguished* if all of the k-affinoid spaces U_i are distinguished. This property only depends on the equivalence class of $\mathscr{U}$. If $\mathscr{U}$ is a distinguished formal covering of X such that $\widetilde{X}_{\mathscr{U}}$ is geometrically reduced, then for any non-Archimedean field K over k, $\mathscr{U}_K = \{U_i\widehat{\otimes}K\}_{i\in I}$ is a distinguished formal covering of $X\widehat{\otimes}K$, and its reduction is canonically isomorphic to $\widetilde{X}_{\mathscr{U}} \otimes_{\widetilde{k}} \widetilde{K}$ ([**Bo1**], 6.4).

Next, a pure one-dimensional algebraic curve $\mathscr{X}$ over a field K is said to be *semistable* if $\mathscr{X} \otimes K^a$ is reduced and all of its singularities are ordinary double points. If, in addition, every nonsingular rational component of $\mathscr{X} \otimes K^a$ (if there is more than one) meets the other components in more than two points, then $\mathscr{X}$ is said to be *stable*. Suppose that $\mathscr{X}$ is semistable. Let x be a double point of $\mathscr{X}$ belonging to only one component of $\mathscr{X}$, and let L denote the algebraic closure of K in the field of rational functions of the component. There are two possibilities: $K(x)$ coincides with L, or $K(x)$ is a separable quadratic extension of L. In the first case we say that the point x is *split* or, if necessary, *L-split*, and in the second case we say that x is *nonsplit*. The *incidence graph* $\Delta(\mathscr{X})$ of $\mathscr{X}$ is defined as follows. The vertices v of $\Delta(\mathscr{X})$ are the irreducible components $\mathscr{X}_v$ of $\mathscr{X}$. Two different vertices

$\mu, \nu \in \Delta(\mathscr{X})$ are joined by as many edges as there are intersection points in $\mathscr{X}_\mu \cap \mathscr{X}_\nu$. Furthermore, for each double point x belonging to only one component $\mathscr{X}_\nu$, one attaches a loop to the vertex ν if x is split, and a closed interval if x is nonsplit. (The second endpoint of the interval is not considered a vertex.) We say that the curve $\mathscr{X}$ is *K-split* if: (a) all double points are K-points; (b) all double points belonging to only one component are K-split; (c) $H^0(\mathscr{X}, \mathscr{O}_{\mathscr{X}}) = K$.

The Stable Reduction Theorem (**[BL1]**) asserts that, for every smooth geometrically connected projective curve X over k of genus ≥ 1, there exists a finite separable extension K over k such that the K-analytic space $(X \otimes K)^{\mathrm{an}}$ admits a distinguished formal covering with $\widetilde{K}$-split stable reduction. Furthermore, let X be a smooth geometrically connected projective curve over k of genus $g \geq 1$ admitting a distinguished formal covering $\mathscr{U}$ of X^{an} with $\widetilde{k}$-split semistable reduction $\widetilde{X}_{\mathscr{U}}$. Then one can construct a distinguished formal covering $\mathscr{V}$ of X^{an} with $\widetilde{k}$-split stable reduction $\widetilde{X}_{\mathscr{V}}$ such that $\mathscr{U}$ refines $\mathscr{V}$. Let $\mathscr{V}'$ be a second distinguished formal covering of X^{an} with stable reduction $\widetilde{X}_{\mathscr{V}'}$. If $g \geq 2$, or if $g = 1$ and $\widetilde{X}_{\mathscr{V}}$ is smooth, then $\mathscr{V}'$ is equivalent to $\mathscr{V}$ (in particular, $\widetilde{X}_{\mathscr{V}} = \widetilde{X}_{\mathscr{V}'}$). If $g = 1$ and $\widetilde{X}_{\mathscr{V}}$ is not smooth (a curve with this property is called a *Tate elliptic curve*), then this is not the case. If $\{\mathscr{X}_i\}$ are the irreducible components of $\widetilde{X}$ and b is the Betti number of $\Delta(\widetilde{X})$, then $g = b + \sum g(\mathscr{X}_i)$.

4.3.1. THEOREM. *Let X be a smooth k-analytic curve admitting a semistable reduction $\widetilde{X}$; let $\pi: X \to \widetilde{X}$ be the reduction map; let $\mathscr{X}_i$, $i \in I$, be the irreducible components of $\widetilde{X}$; and let $\widetilde{x} \in \widetilde{X}$.*

(i) *If $\widetilde{x}$ is the generic point of $\mathscr{X}_i$, then there exists a unique point $x_i \in X$ with $\pi(x_i) = \widetilde{x}$.*

(ii) *If $\widetilde{x}$ is a smooth closed point belonging to $\mathscr{X}_i$, then $\pi^{-1}(\widetilde{x})$ is a simply connected special quasipolyhedron with one endpoint at x_i.*

(iii) *Suppose that $\widetilde{x}$ is a double point of $\widetilde{X}$. Then $\pi^{-1}(\widetilde{x})$ is a simply-connected special quasipolyhedron. Furthermore:*

(a) *if $\widetilde{x}$ belongs to two different components $\mathscr{X}_i$ and $\mathscr{X}_j$, then $\pi^{-1}(\widetilde{x})$ has two endpoints, and $\overline{\pi^{-1}(\widetilde{x})} = \pi^{-1}(\widetilde{x}) \cup \{x_i, x_j\} = \widehat{\pi^{-1}(\widetilde{x})}$;*

(b) *if $\widetilde{x}$ belongs to only one component $\mathscr{X}_i$ and is split, then $\pi^{-1}(\widetilde{x})$ has two endpoints, and $\overline{\pi^{-1}(\widetilde{x})} = \pi^{-1}(\widetilde{x}) \cup \{x_i\}$ is obtained from $\widehat{\pi^{-1}(\widetilde{x})}$ by identifying the endpoints of $\pi^{-1}(\widetilde{x})$;*

(c) *if $\widetilde{x}$ belongs to only one component $\mathscr{X}_i$ and is nonsplit, then $\pi^{-1}(\widetilde{x})$ has one endpoint and $\overline{\pi^{-1}(\widetilde{x})} = \pi^{-1}(\widetilde{x}) \cup \{x_i\} = \widehat{\pi^{-1}(\widetilde{x})}$.*

PROOF. (i) follows from Proposition 2.4.4(ii). Note that if $\widetilde{x}$ is a closed point of $\widetilde{X}$, then $\overline{\pi^{-1}(\widetilde{x})}$ is contained in $\pi^{-1}(\widetilde{x}) \cup \{x_i | \widetilde{x} \in \mathscr{X}_i\}$ because the latter set is closed. Moreover, from Theorem 3.2.1 it follows that the set $\overline{\pi^{-1}(\widetilde{x})}$ is arcwise connected.

(ii). If $\widetilde{x}$ is a $\widetilde{k}$-point, then the fibre $\pi^{-1}(\widetilde{x})$ is isomorphic to the open unit disc $D \subset \mathbf{A}^1$ (**[Bo5]**, 6.3, **[BL1]**, 2.2). From Theorem 4.2.1 it follows that $\overline{\pi^{-1}(\widetilde{x})}$ is a simply connected special quasipolyhedron with one endpoint and $\overline{\pi^{-1}(\widetilde{x})} = \widehat{\pi^{-1}(\widetilde{x})} = \pi^{-1}(\widetilde{x}) \cup \{x_i\}$. Suppose that $\widetilde{x}$ is arbitrary, and let K be a finite normal extension of k such that all of the inverse images $\widetilde{y}_1, \dots, \widetilde{y}_n$ of $\widetilde{x}$ in $\widetilde{X \otimes K} = \widetilde{X} \otimes \widetilde{K}$ are $\widetilde{K}$-points. Let L be the maximal subfield of K which is purely inseparable over k, and let $\widetilde{y}$ be the inverse image of $\widetilde{x}$ in $\widetilde{X} \otimes \widetilde{L}$. Then there are homeomorphisms $|\pi_L^{-1}(\widetilde{y})| \widetilde{\rightarrow} |\pi^{-1}(\widetilde{x})|$ and $|\pi_K^{-1}(\widetilde{y}_1)|/G \widetilde{\rightarrow} |\pi_L^{-1}(\widetilde{y})|$, where π_L and π_K are the reduction maps for $X \otimes L$ and $X \otimes K$, respectively, and G is the stabilizer of the point $\widetilde{y}_1$ in $\mathrm{Gal}(K/L)$. Hence, the required property of $\pi^{-1}(\widetilde{x})$ follows from the previous case and Proposition 4.1.7.

(iii). First suppose that $\widetilde{x}$ is a $\widetilde{k}$-split $\widetilde{k}$-point. According to a result of Bosch and Lütkebohmert (**[BL1]**, 2.3), the fibre $\pi^{-1}(\widetilde{x})$ is isomorphic to an open annulus $R = \{x \in \mathbf{A}^1 \mid \alpha < |T(x)| < 1\}$ with $\alpha \in |k^*|$. This is a simply connected special quasipolyhedron with two endpoints. (a) and (b) follow from this.

Now suppose that $\widetilde{x}$ is arbitrary. Let K be a finite normal extension of k such that all of the inverse images $\widetilde{y}_1, \dots, \widetilde{y}_n$ of the point $\widetilde{x}$ in $\widetilde{Y}$, where $Y = X \otimes K$, are $\widetilde{K}$-split $\widetilde{K}$-points. We may assume that K is separable over k. If G is the stabilizer of the point $\widetilde{y} = \widetilde{y}_1$, then $|\pi_K^{-1}(\widetilde{y})|/G \widetilde{\rightarrow} |\pi^{-1}(\widetilde{x})|$. By Proposition 4.1.7, $\pi^{-1}(\widetilde{x})$ is a simply connected special quasipolyhedron with at most two endpoints. We identify $\pi_K^{-1}(\widetilde{y})$ with $R = \{x \in \mathbf{A}_K^1 \mid \alpha < |T(x)| < 1\}$, $\alpha \in |K|$. Setting $\mathscr{O}^\circ(R) = \{f \in \mathscr{O}(R) \mid |f(y)| \le 1 \text{ for all } y \in R\}$, $\mathscr{O}^{\circ\circ}(R) = \{f \in \mathscr{O}(R) \mid |f(y)| < 1 \text{ for all } y \in R\}$, and $\widetilde{\mathscr{O}}(R) = \mathscr{O}^\circ(R)/\mathscr{O}^{\circ\circ}(R)$, we have:

$$\mathscr{O}^\circ(R) = \{\sum_{i=0}^{\infty} a_i T^i + \sum_{j=1}^{\infty} b_j (\frac{a}{T})^j \mid a_i, b_j \in K^\circ\};$$

$$\widetilde{\mathscr{O}}(R) = \{\sum_{i=0}^{\infty} \widetilde{a}_i T^i + \sum_{j=1}^{\infty} \widetilde{b}_j S^j \mid \widetilde{a}_i, \widetilde{b}_j \in \widetilde{K},\ TS = 0\},$$

where $a \in K$ with $|a| = \alpha$ and S is the image of $\frac{a}{T}$ in $\widetilde{\mathscr{O}}(R)$. The ring $\widetilde{\mathscr{O}}(R)$ is local, and its maximal ideal is invariant under the action of G. The set of zero-divisors in $\widetilde{\mathscr{O}}(R)$ coincides with $\mathbf{a} \cup \mathbf{b}$ where $\mathbf{a} = \{\sum_{i=1}^{\infty} \widetilde{a}_i T^i \mid \widetilde{a}_i \in \widetilde{K}\}$ and $\mathbf{b} = \{\sum_{j=1}^{\infty} \widetilde{b}_j S^j \mid \widetilde{b}_j \in \widetilde{K}\}$. Hence, there are two possibilities:

(1) the ideals $\mathbf{a}$ and $\mathbf{b}$ are invariant under the action of G;

(2) there exists $\sigma \in G$ with ${}^\sigma\mathbf{a} = \mathbf{b}$ (and ${}^\sigma\mathbf{b} = \mathbf{a}$).

Since $\widetilde{\mathscr{O}}(R) \widetilde{\rightarrow} \widehat{\mathscr{O}}_{\widetilde{Y}, \widetilde{y}}$, the case (1) (resp. (2)) takes place if and only if the point $\widetilde{x}$ is split (resp. nonsplit). On the other hand, the skeleton $\Delta(R) =$

$\{p(E(0, t)) \mid \alpha < t < 1\}$ is invariant under the action of G, and hence there are two possibilities:

(1′) G acts as the identity on $\Delta(R)$;

(2′) G acts nontrivially through a quotient group of order two.

We claim that the cases (1) and (1′) (resp. (2) and (2′)) correspond to one another. Indeed, let $y = p(E(0, t))$, $\sqrt{\alpha} < t < 1$, and $f \in \mathcal{O}^{\circ\circ}(R)$. Then $|f(y)| = t$ for $\widetilde{f} \in \mathbf{a}\backslash\mathbf{a}^2$, and $|f(y)| < t$ for $\widetilde{f} \in \mathbf{b}$. Hence, the ideal $\mathbf{a}$ is invariant under the action of G if and only if the point y is fixed under the same action. ∎

Let X be a smooth k-analytic curve which admits a distinguished formal covering $\mathcal{U}$ with semistable reduction $\widetilde{X}$ having at most countably many components. From Theorem 4.3.1 it follows that $|X|$ is a connected quasipolyhedron. The covering $\mathcal{U}$ determines a closed subset $\Delta_{\mathcal{U}}(X)$ of X which has the structure of a locally finite graph with at most countably many vertices. Namely, the vertices of $\Delta_{\mathcal{U}}(X)$ are the inverse images of the generic points of $\widetilde{X}$, and the edges of $\Delta_{\mathcal{U}}(X)$ correspond to the double points $\widetilde{x} \in \widetilde{X}$ as follows. If $\widetilde{x}$ belongs to two different components of $\widetilde{X}$ or is split, then the corresponding edge is the closure of the skeleton $\Delta(\pi^{-1}(\widetilde{x}))$ in $\overline{\pi^{-1}(\widetilde{x})}$. If $\widetilde{x}$ belongs to only one component of $\widetilde{X}$ and is nonsplit, then the corresponding edge is the closure of the image of $\Delta(\pi_K^{-1}(\widetilde{y}))$ in $\pi^{-1}(\widetilde{x})$, where K is a finite normal extension of k which is large enough so that some inverse image $\widetilde{y}$ of $\widetilde{x}$ in $\widetilde{X} \otimes \widetilde{K}$ is $\widetilde{K}$-split. In this case the edge is homeomorphic to the unit interval (but its second endpoint is not considered a vertex of $\Delta_{\mathcal{U}}(X)$). The graph $\Delta_{\mathcal{U}}(X)$ is isomorphic to the incidence graph of the reduction $\widetilde{X}$ and is called the *analytic skeleton of X with respect to the covering $\mathcal{U}$*.

Now let X be a smooth geometrically connected projective curve over k of genus ≥ 1. Let K be a finite Galois extension of k such that the K-analytic space $X^{\mathrm{an}} \otimes K$ admits a distinguished formal covering $\mathcal{U}$ with $\widetilde{K}$-split stable reduction $\widetilde{X \otimes K}$. Since $X^{\mathrm{an}} \otimes K$ is a special quasipolyhedron, X^{an} is also a special quasipolyhedron, by Proposition 4.1.7. Further let $G = \mathrm{Gal}(K/k)$. If $X \otimes K$ is not a Tate elliptic curve, the covering ${}^{\sigma}\mathcal{U}$ of $X^{\mathrm{an}} \otimes K$ is equivalent to $\mathcal{U}$ for any $\sigma \in G$. In particular, the graph $\Delta_{\mathcal{U}}(X^{\mathrm{an}} \otimes K)$ is invariant under the action of G. If $X \otimes K$ is a Tate elliptic curve, then ${}^{\sigma}\mathcal{U}$ might not be equivalent to $\mathcal{U}$. But, nevertheless, the sets $\Delta_{\mathcal{U}}(X^{\mathrm{an}} \otimes K)$ and $\Delta_{{}^{\sigma}\mathcal{U}}(X^{\mathrm{an}} \otimes K)$ coincide. Indeed, they are homeomorphic to the unit circle and coincide with the skeleton of the quasipolyhedron $|X^{\mathrm{an}} \otimes K|$. In both cases, we set $\Delta^{\mathrm{an}}(X) = \Delta_{\mathcal{U}}(X^{\mathrm{an}} \otimes K)/G$. This set is a graph (which may have no vertices), which is called the *analytic skeleton* of X.

4.3.2. THEOREM. *Let X be a smooth geometrically connected projective curve over k of genus $g \geq 1$. Then X^{an} is a special quasipolyhedron whose Betti number is at most g. Its skeleton coincides with the maximal subpoly-*

hedron of $\Delta^{\mathrm{an}}(X)$ *which has no endpoints.* ∎

The skeleton of X^{an} is called the *topological skeleton* of X and is denoted by $\Delta^{\mathrm{top}}(X)$.

4.3.3. COROLLARY. *Any one-dimensional k-analytic space is locally contractible. In particular, if it is separated and connected, it has a universal covering.*

PROOF. It suffices to show that a connected one-dimensional k-affinoid space X is a special quasipolyhedron. By Proposition 4.1.7, we may assume that the ground field is algebraically closed. If X' is the normalization of X (we may assume that X is reduced), then X is obtained from X' by gluing together a finite number of points. Therefore, we may assume that X is smooth. By a result of van der Put ([**P2**]), X is isomorphic to a k-affinoid domain in the analytification of a smooth projective curve. The required fact now follows from Theorem 4.3.2. ∎

4.3.4. COROLLARY. *For any connected finite graph* Σ, *there exists a commutative Banach field* $\mathcal{K}$ *over* k *with* $\mathcal{M}(\mathcal{K})\widetilde{\rightarrow}\Sigma$.

PROOF. We may assume that the field k is algebraically closed. If Σ is simply connected, it can be embedded in the affine line $\mathbf{A}^1$ in such a way that its image does not contain points of $\mathbf{A}^1(k)$. Then $\mathcal{K} = \mathcal{H}(\Sigma)$ is a field. Since Σ is rationally convex in $\mathbf{A}^1$, we have $\mathcal{M}(\mathcal{K})\widetilde{\rightarrow}\Sigma$. Suppose that Σ is not simply connected, and let Σ' be the maximal subgraph of Σ with no endpoints. One can construct a smooth geometrically irreducible projective curve X whose analytic skeleton is homeomorphic to Σ' (for example, using Mumford curves (see §4.4)). As above, Σ can be embedded in X^{an} in such a way that its image does not contain points of $X(k)$; we set $\mathcal{K} = \mathcal{H}(\Sigma)$. From Proposition 2.6.3(ii) it follows that $\mathcal{M}(\mathcal{K})\widetilde{\rightarrow}\Sigma$. ∎

Suppose that X^{an} has stable reduction. Then the Betti number $b(X^{\mathrm{an}})$ is equal to the Betti number of the graph of the reduction. In particular, $b(X^{\mathrm{an}}) = g(X)$ if and only if the reduction is $\widetilde{k}$-split and all of its components are rational. In this case one says that X^{an} has *$\widetilde{k}$-split totally degenerate reduction.* For such a curve $\Delta^{\mathrm{top}}(X) = \Delta^{\mathrm{an}}(X)$.

4.3.5. REMARK. If the valuation on the ground field k is trivial, then any connected k-analytic curve is a special quasipolyhedron (contractible if the curve is smooth), for trivial reasons (see §1.4.2). Corollary 4.3.4 also holds. We remark that the definitions of formal coverings and distinguished k-affinoid algebras are applicable in this case (we shall need them in §6.4). For example, a strictly k-affinoid algebra $\mathcal{A}$ is distinguished if and only if it is reduced.

4.4. Uniformization. We first recall the definition of Mumford curves. Let Γ be a subgroup of $\mathrm{PGL}_2(k)$. A point $x \in \mathbf{P}^1(\widehat{k^a})$ is called a *limit point of* Γ if there exist $y \in \mathbf{P}^1(\widehat{k^a})$ and an infinite sequence $\{\gamma_n\}_{n\geq 0} \subset \Gamma$ with

$\lim_{n\to\infty} \gamma_n(y) = x$. For example, if Γ is not discrete in $\mathrm{PGL}_2(k)$, every point of $\mathbf{P}^1(\widehat{k}^a)$ is a limit point of Γ. Let Σ_Γ denote the set of limit points of Γ. The group Γ is said to be *discontinuous* if $\Sigma_\Gamma \neq \mathbf{P}^1(\widehat{k}^a)$ and if, for any point $x \in \mathbf{P}^1(\widehat{k}^a)$, the closure $\overline{\Gamma x}$ of the orbit of x in $\mathbf{P}^1(\widehat{k}^a)$ is compact. The group Γ is called a *Schottky group* if: (a) Γ is finitely generated; (b) Γ has no nontrivial elements of finite order; (c) Γ is discontinuous.

If Γ is a Schottky group, then every element $\gamma \in \Gamma$, $\gamma \neq 1$, is hyperbolic (i.e., it is conjugate to a homothety $x \mapsto qx$ with $q \in k$ and $0 < |q| < 1$), and Γ is a free group. Furthermore, the set Σ_Γ lies in $\mathbf{P}^1(k)$, and if the rank of Γ is greater than 1, then Σ_Γ is a perfect set (i.e., Σ_Γ coincides with its set of limit points). If the rank of Γ is equal to 1, then Σ_Γ consists of two points.

Every Schottky group can be constructed in the following way (**[GP]**, §I.4). Let $B_1, C_1, \dots, B_g, C_g$ be $2g$ open discs in $\mathbf{P}^1(\widehat{k}^a)$ with centers in k such that $r(B_i)r(C_i) \in |k^*|$, $1 \le i \le g$, and the corresponding closed discs $B_1^+, C_1^+, \dots, B_g^+, C_g^+$ are disjoint. There exist elements $\gamma_1, \dots, \gamma_g \in \mathrm{PGL}_2(k)$ such that $\gamma_i(\mathbf{P}^1(\widehat{k}^a)\backslash B_i) = C_i^+$ and $\gamma_i(\mathbf{P}^1(\widehat{k}^a)\backslash B_i^+) = C_i$, $1 \le i \le g$. Then $\gamma_1, \dots, \gamma_g$ are free generators of a Schottky group $\Gamma \subset \mathrm{PGL}_2(k)$. Moreover, if $F = \mathbf{P}^1(\widehat{k}^a)\backslash \bigcup_{i=1}^n (B_i \cup C_i)$, then $\bigcup_{\gamma\in\Gamma} \gamma F = \mathbf{P}^1(\widehat{k}^a)\backslash\Sigma_\Gamma$.

Let Γ be a Schottky group of rank $g \ge 1$. Consider the action of Γ on the projective line $\mathbf{P}^1$. From the above facts it follows that Γ acts freely on the open set $\Omega_\Gamma = \mathbf{P}^1\backslash\Sigma_\Gamma$, and the quotient space Ω_Γ/Γ is compact. So Ω_Γ/Γ is endowed with the obvious k-analytic structure and is a proper k-analytic curve. By Corollary 3.4.14, the k-analytic curve Ω_Γ/Γ is associated with a smooth geometrically irreducible projective curve X_Γ over k, which is called the *Mumford curve* associated with Γ. Let Δ_Γ denote the skeleton of Ω_Γ. If $g \ge 2$, it is a locally finite tree having countably many vertices. The group Γ acts on Δ_Γ, and so Δ_Γ/Γ is a closed subset of X_Γ^{an}. From the proof of Theorem 4.4.1 it follows that $\Delta^{\mathrm{an}}(X_\Gamma) = \Delta_\Gamma/\Gamma$.

4.4.1. THEOREM. *The following properties of a smooth geometrically connected projective curve X of genus $g \ge 1$ are equivalent:*

(a) *X is a Mumford curve;*

(b) *the universal covering Ω of X^{an} is isomorphic to an open subset of $\mathbf{P}^1$ whose complement lies in $\mathbf{P}^1(k)$;*

(c) *X^{an} has $\widetilde{k}$-split totally degenerate reduction.*

PROOF. The implication (a) $\Rightarrow$ (b) is trivial.

(b) $\Rightarrow$ (a). The fundamental group $\Gamma = \pi_1(X^{\mathrm{an}})$ acts freely on Ω. We claim that $\mathrm{Aut}(\Omega) = \{\sigma \in \mathrm{PGL}_2(k) |\ \sigma(\Omega) = \Omega\}$. We first suppose that $\Sigma = \mathbf{P}^1\backslash\Omega \neq \mathbf{P}^1(k)$. Then we may assume that $\infty \in \Omega$. Let $\sigma \in \mathrm{Aut}(\Omega)$, and set $x = \sigma^{-1}(\infty) \in \Omega$. Then there exists a function $F(z) = \frac{az+b}{cz+d}$ with $\begin{pmatrix} a & b \\ c & d \end{pmatrix} \in$

$\mathrm{GL}_2(k)$ such that $F(x) = \infty$ and $F - \sigma$ is bounded in a neighborhood of the point x. Since F and σ are bounded outside of any neighborhood of $x \in \Omega$, it follows that $F - \sigma$ is bounded on all of Ω. Hence, $F - \sigma$ is a constant, and therefore, σ extends to an automorphism of $\mathbf{P}^1$. If $\Sigma = \mathbf{P}^1(k)$ (this can occur only if k is a local field), we can consider the situation over a finite extension K of k and, as previously, find that σ belongs to $\mathrm{PGL}_2(K)$. But since $\sigma(\mathbf{P}^1(k)) = \mathbf{P}^1(k)$, we have $\sigma \in \mathrm{PGL}_2(k)$. Thus, the action of Γ on Ω extends to an action on $\mathbf{P}^1$. It is clear that Γ is discontinuous. Since Γ is finitely generated and free, Γ is a Schottky group. It is also clear that Σ is the set of limit points of Γ, and hence $\Omega = \Omega_\Gamma$.

(a) $\Rightarrow$ (c). Let $X = X_\Gamma$ for a Schottky group $\Gamma \subset \mathrm{PGL}_2(k)$. If $g \geq 2$, then Δ_Γ is a graph, and all of its vertices are points of the form $p(E(a, r))$ with $a \in k$ and $r \in |k^*|$. If $g = 1$, then we choose a point $x \in \Delta_\Gamma$ of the form $p(E(a, r))$ with $a \in k$ and $r \in |k^*|$, and declare the points γx with $\gamma \in \Gamma$ to be vertices of Δ_Γ. In this way we endow Δ_Γ with a graph structure which is invariant under the action of Γ. Let $\tau\colon \Omega_\Gamma \to \Delta_\Gamma$ be the retraction map. For an edge ℓ of Δ_Γ, we set $V_\ell = \tau^{-1}(\ell)$. The covering $\mathscr{U} = \{V_\ell\}$ is a distinguished formal covering of Ω_Γ with $\Delta_{\mathscr{U}}(\Omega_\Gamma) = \Delta_\Gamma$. It is invariant under the action of Γ, and it has $\tilde{k}$-split totally degenerate reduction. First suppose that the endpoints of any edge ℓ of Δ_Γ are not equivalent under the action of Γ. Then, for any $V_\ell \in \mathscr{U}$ and $\gamma \in \Gamma$, $\gamma \neq 1$, one has $\gamma(V_\ell) \cap V_\ell = \varnothing$, and hence V_ℓ is isomorphic to its image in Ω_Γ/Γ. Thus, we obtain a distinguished formal covering of X_Γ^{an} with $\tilde{k}$-split totally degenerate reduction.

Note that in the construction of the covering $\mathscr{U}$ of Ω_Γ, the graph Δ_Γ could have additional vertices of the form $p(E(a, r))$, $a \in k$, $r \in |k^*|$. So if the valuation on k is not discrete, we can always add such vertices and assume that adjacent vertices of Δ_Γ are not equivalent under the action of Γ. If the valuation on k is discrete, we can take a ramified separable quadratic extension K of k and, as before, obtain a covering $\mathscr{U}$ of $(X_\Gamma \otimes K)^{\mathrm{an}}$ with $\tilde{K}$-split totally degenerate reduction. Let $\mathscr{V}$ be a distinguished formal covering of $(X_\Gamma \otimes K)^{\mathrm{an}}$ such that $\mathscr{U}$ refines $\mathscr{V}$ and the reduction $\widetilde{(X_\Gamma \otimes K)}_{\mathscr{V}}$ is stable. One can show that $\mathscr{V}$ is induced by a distinguished formal covering of X_Γ^{an} which has all of the required properties.

(c) $\Rightarrow$ (b). Let $\varphi\colon \Omega \to X^{\mathrm{an}}$ be the universal covering of X^{an}, let $\mathscr{U} = \{U_i\}_{i \in I}$ be a distinguished formal covering of X^{an} with $\tilde{k}$-split totally degenerate reduction $\widetilde{X}$, and let $\pi\colon X^{\mathrm{an}} \to \widetilde{X}$ be the reduction map. First, we claim that Ω admits a distinguished formal covering $\mathscr{V} = \{V_j\}_{j \in J}$ with $\tilde{k}$-split totally degenerate reduction $\widetilde{\Omega}$. (It is clear that all of the irreducible components of such a reduction are nonsingular.) Replacing the covering $\mathscr{U}$ by an equivalent one, we may suppose that every curve $\widetilde{U}_i$, $i \in I$, contains at most one singular point of $\widetilde{X}$. If $\widetilde{X}$ has no double points lying in only one component, then each U_i is simply connected, and hence $\varphi^{-1}(U_i) = \cup U_{i,\sigma}$, $\sigma \in \pi_1(X^{\mathrm{an}})$, is a disjoint union of affinoid subdomains of Ω which are isomorphic to U_i. In this case we can take $J = I \times \pi_1(X^{\mathrm{an}})$ and $V_j = U_{i,\sigma}$, where $j = (i, \sigma) \in J$. Now suppose that $\widetilde{X}$ contains double points lying in only one component. Let U_i be an element of $\mathscr{U}$ for which $\widetilde{U}_i$ contains such a point $\widetilde{y}$. Let $\widetilde{x}$ be its generic point, and set $x = \pi^{-1}(\widetilde{x})$. We note that, since U_i is distinguished, the valuation on the field $\mathscr{H}(x)$ takes values in $|k|$. The fibre $\pi^{-1}(\widetilde{y})$ is isomorphic to an open annulus $R = \{z \in \mathbf{A}^1 |\ \alpha < |T(z)| < 1\}$ with $\alpha \in |k^*|$. If the valuation on k is not discrete, we can choose $a \in k$ with $\alpha < |a| < 1$ and set $U_i' = U_i \backslash \{z \in R |\ |a| < |T(z)| < 1\}$, $U_i'' = U_i \backslash \{z \in R |\ \alpha < |T(z)| < |a|\}$. Then $U_i = U_i' \cup U_i''$ is a distinguished formal covering of U_i whose reduction consists of two nonsingular rational curves with a common $\widetilde{k}$-point. Replacing every such element $U_i \in \mathscr{U}$ by U_i' and U_i'', we obtain a distinguished formal covering of X^{an} whose reduction has no double points lying in only one component. If the valuation on k is discrete, we go through the same procedure with arbitrary $a \in k^a$; but in this case the formal covering $U_i = U_i' \cup U_i''$ is not distinguished. We have $\varphi^{-1}(U_i') = \cup U_{i,\sigma}'$ and $\varphi^{-1}(U_i'') = \cup U_{i,\sigma}''$, $\sigma \in \pi_1(X^{\mathrm{an}})$, where $U_{i,\sigma}'$ and $U_{i,\sigma}''$ are affinoid subdomains of Ω isomorphic to U_i' and U_i'', respectively. Supposing that $U_{i,\sigma}' \cap U_{i,\sigma}'' \neq \varnothing$, we set $U_{i,\sigma} = U_{i,\sigma}' \cup U_{i,\sigma}''$. The set $U_{i,\sigma}$ is an affinoid subdomain of Ω whose reduction consists of two nonsingular rational curves with a common $\widetilde{k}$-point. Moreover, $U_{i,\sigma}$ is distinguished. Indeed, since k is stable, it suffices to show that the values of the spectral norm on $U_{i,\sigma}$ lie in $|k|$. But these values are determined by the valuations on the fields $\mathscr{H}(x_\sigma)$, $x_\sigma \in \varphi^{-1}(x)$, which are isomorphic to $\mathscr{H}(x)$. In this way we obtain the required covering $\mathscr{V}$ of Ω.

We now note that, up to isomorphism, there is only a finite number of distinct fibres of the reduction $\Omega \to \widetilde{\Omega}$ over the double points of $\widetilde{\Omega}$, because they are isomorphic to the corresponding fibres of the reduction $X^{\mathrm{an}} \to \widetilde{X}$. Therefore, we can apply a result of Liu ([**Li**]). ∎

4.5. Picard's Theorem

4.5.1. THEOREM. *Every holomorphic map from the affine line to a smooth algebraic curve of genus* $g \geq 1$ *is constant.*

4.5.2. LEMMA. *Every nonconstant holomorphic map* $f: \mathbf{A}^1 \to \mathbf{A}^1$ *is surjective.*

PROOF. We may assume that the field k is algebraically closed and that $f(0) = 0$. From the elementary properties of Newton polygons it follows that $f(E_0(0, r)) = E_0(0, \|f\|_{E(0,r)})$, where $E_0(0, r) = E(0, r) \cap k = \{x \in k |\ |x| \leq r\}$. Since $E_0(0, r)$ is everywhere dense in $E(0, r)$, we have $f(E(0, r)) =$

$E(0, ||f||_{E(0,r)})$. It is also clear that $||f||_{E(0,r)} \to \infty$ as $r \to \infty$. Hence, $f(\mathbf{A}^1) = \mathbf{A}^1$. ∎

PROOF OF THEOREM 4.5.1. We may assume that the field k is algebraically closed, and the curve X is projective and irreducible. By the Stable Reduction Theorem, the k-analytic curve X^{an} admits a stable reduction $\widetilde{X}$. First suppose that all of the components of $\widetilde{X}$ are rational. By Theorem 4.4.1, X^{an} admits a uniformization $\varphi\colon \Omega \to X^{\mathrm{an}}$ by an open subset Ω of $\mathbf{P}^1$ whose complement contains at least two k-points. So we may suppose that Ω lies in $\mathbf{A}^1$ and does not coincide with it. Since $\mathbf{A}^1$ is contractible, there exists a commutative diagram of holomorphic maps:

$$\begin{array}{ccc} \mathbf{A}^1 & \xrightarrow{\quad f \quad} & X^{\mathrm{an}} \\ & {}_{h}\searrow \quad \nearrow_{\varphi} & \\ & \Omega & \end{array}$$

Since $\Omega \subset \mathbf{A}^1$ and $\Omega \neq \mathbf{A}^1$, it follows from Lemma 4.5.2 that the map h is constant, and therefore f is constant.

Now suppose that $\widetilde{X}$ contains a nonrational component $\mathscr{X}$. Let $\widetilde{x}$ be its generic point, and let x be the inverse image of $\widetilde{x}$ under the reduction map $\pi\colon X \to \widetilde{X}$. By Proposition 2.4.4(ii), $\widetilde{\mathscr{H}(x)} = \widetilde{k}(\widetilde{x})$. From this it follows that the point x does not lie in the image of f. Indeed, if $x = f(y)$, then $\widetilde{\mathscr{H}(x)}$ is embedded in $\widetilde{\mathscr{H}(y)}$. But $\widetilde{\mathscr{H}(y)}$ for $y \in \mathbf{A}^1$ coincides with $\widetilde{k}$ or with the field of rational functions over $\widetilde{k}$ (see §1.4.4).

Let $\widetilde{x}'$ and $\widetilde{x}''$ be two distinct closed points of $\mathscr{X}$ which are smooth in $\widetilde{X}$. One has $\overline{\pi^{-1}(\widetilde{x}')} = \pi^{-1}(\widetilde{x}') \cup \{x\}$ and $\overline{\pi^{-1}(\widetilde{x}'')} = \pi^{-1}(\widetilde{x}'') \cup \{x\}$. Since $\mathbf{A}^1$ is arcwise connected and $x \notin f(\mathbf{A}^1)$, the open subsets $\pi^{-1}(\widetilde{x}')$ and $\pi^{-1}(\widetilde{x}'')$ cannot both contain points of $f(\mathbf{A}^1)$. Hence, we can find a point $y \in X_0^{\mathrm{an}} = X_0$ and an open neighborhood $\mathscr{U}$ of y in X^{an} which does not meet $f(\mathbf{A}^1)$. By the Riemann-Roch Theorem, there exists a rational function h on X with a pole only at the point y. Taking the composition of $f\colon \mathbf{A}^1 \to X^{\mathrm{an}}$ with $h\colon X^{\mathrm{an}} \backslash \mathscr{U} \to \mathbf{A}^1$, we obtain a bounded holomorphic function on $\mathbf{A}^1$. Therefore, f is constant. ∎

Practically the same proof gives the following stronger fact.

4.5.3. THEOREM. *Let X be a smooth geometrically connected projective curve of genus ≥ 1, and let D be an open disc in $\mathbf{A}^1$. Then for any holomorphic map $D \to X^{\mathrm{an}}$, the image of D does not meet $\Delta^{\mathrm{an}}(X)$.* ∎

4.5.4. COROLLARY. *Let $\varphi\colon Y \to X$ be a morphism of smooth geometrically connected projective curves of genus ≥ 1. Then $\varphi^{-1}(\Delta^{\mathrm{an}}(X)) \subset \Delta^{\mathrm{an}}(Y)$.* ∎

4.6. A characterization of curves whose Betti number is equal to the genus

4.6.1. THEOREM. *Let X be a smooth, geometrically connected projective curve of genus $g \geq 1$. Then the Betti number $b(X^{\mathrm{an}})$ of X^{an} is equal to g*

if and only if X *is a Mumford curve or a principal homogeneous space over a Tate elliptic curve.*

Proof. The converse implication follows from Theorem 4.4.1 and the following lemma.

4.6.2. **Lemma.** *Let* X *be a principal homogeneous space over a Tate elliptic curve* E *with period* $q \in k^*$, $|q| < 1$.

(i) *There exists a unique minimal extension* K *of* k *such that* X *splits over* K. *Moreover,* K *is cyclic over* k, *and* $q \in N_{K/k}(K^*)$.

(ii) *The universal covering of* X^{an} *is isomorphic to* $\mathbf{G}^{\mathrm{an}}_{m,K}$, *and* $\pi_1(X^{\mathrm{an}}) \tilde{\to} \mathbf{Z}$.

(iii) *The action of* $\pi_1(X^{\mathrm{an}}) = \langle\gamma\rangle$ *on* $\mathbf{G}^{\mathrm{an}}_{m,K}$ *can be described as follows:* $\gamma(x) = \lambda \cdot {}^{\sigma}x$ *for some generator* σ *of* $G = \mathrm{Gal}(K/k)$ *and some* $\lambda \in K$ *with* $N_{K/k}(\lambda) = q$. *In particular, the canonical morphism* $(X \otimes K)^{\mathrm{an}} \to X^{\mathrm{an}}$ *is a local isomorphism.*

Proof. The group of principal homogeneous spaces over E is isomorphic to $H^1(\mathcal{G}, E(k^s))$, where k^s is the separable closure of k and $\mathcal{G} = \mathrm{Gal}(k^s/k)$. Using the exact sequence of $\mathcal{G}$-modules $0 \to \mathbf{Z} = \langle q\rangle \to k^{s*} \to E(k^s) \to 0$ and the equality $H^1(\mathcal{G}, k^{s*}) = 0$ (Hilbert's Theorem 90), one obtains an embedding $H^1(\mathcal{G}, E(k^s)) \hookrightarrow H^2(\mathcal{G}, \mathbf{Z})$. From the exact sequence $0 \to \mathbf{Z} \to \mathbf{Q} \to \mathbf{Q}/\mathbf{Z} \to 0$, one obtains an isomorphism $H^1(\mathcal{G}, \mathbf{Q}/\mathbf{Z}) \tilde{\to} H^2(\mathcal{G}, \mathbf{Z})$. Since $H^1(\mathcal{G}, \mathbf{Q}/\mathbf{Z}) = \mathrm{Hom}(\mathcal{G}, \mathbf{Q}/\mathbf{Z})$, we obtain an embedding $j\colon H^1(\mathcal{G}, E(k^s)) \hookrightarrow \mathrm{Hom}(\mathcal{G}, \mathbf{Q}/\mathbf{Z})$. This embedding commutes with extension of the ground field. In particular, if c is an element of $H^1(\mathcal{G}, E(k^s))$ corresponding to X, then the field K in (i) is the field corresponding to the kernel of the homomorphism $j(c)\colon \mathcal{G} \to \mathbf{Q}/\mathbf{Z}$. We see that K is cyclic, and c comes from the group $H^1(G, E(K))$.

By construction, we have an isomorphism

$$j\colon H^1(G, E(K)) \tilde{\to} \mathrm{Hom}(G, \mathbf{Q}/\mathbf{Z}).$$

Let τ be a generator of G, and set $n = [K: k]$. Then $H^1(G, E(K)) = \{\alpha \in K^* | N_{K/k}(\alpha) \in \langle q\rangle\}/\{({}^{\tau}\beta/\beta)q^t | \beta \in K^*, t \in \mathbf{Z}\}$. Since $H^1(G, K^*) = 0$, we obtain an isomorphism $H^1(G, E(K)) \tilde{\to} (N_{K/k}(K^*) \cap \langle q\rangle)/\langle q^n\rangle$. Since the order of $H^1(G, E(K))$ is equal to n, we have $q \in N_{K/k}(K^*)$.

Let α be an element of K with $N_{K/k}(\alpha) = q$. Since the element c is a generator of $H^1(G, E(K))$, it can be represented by a cocycle $\{c_{\tau^i} = \alpha^{\ell} \cdot {}^{\tau}\alpha^{\ell} \cdot \ldots \cdot {}^{\tau^{i-1}}\alpha^{\ell}\}_{0 \le i \le n-1}$ for some $1 \le \ell \le n-1$ prime to n. Let t be the integer with $1 \le t \le n-1$ and $\ell t = 1 + sn$, and set $\sigma = \tau^t$ and $\lambda = (\alpha^{\ell} \cdot {}^{\tau}\alpha^{\ell} \cdot \ldots \cdot {}^{\tau^{t-1}}\alpha^{\ell})/q^s$. The element σ is a generator of G, and $N_{K/k}(\lambda) = q$. It follows that c can also be represented by the cocycle $\{c'_{\sigma^i} = \lambda \cdot {}^{\sigma}\lambda \cdot \ldots \cdot {}^{\sigma^{i-1}}\lambda\}_{0 \le i \le n-1}$. This gives a representation of X as the quotient of $E \otimes K$ by the action of G in which σ sends x to $\lambda \cdot {}^{\sigma}x$. The action of σ lifts to an action on the universal covering $\mathbf{G}_{m,K}^{\mathrm{an}}$ of $(E \otimes K)^{\mathrm{an}}$ as follows: $x \mapsto \lambda \cdot {}^{\sigma}x$. If γ denotes this automorphism, then $\gamma^n = q$. Hence $\mathbf{G}_{m,K}^{\mathrm{an}}$ is the universal covering of X^{an}. ∎

4.6.3. Lemma. *Let X be a smooth geometrically connected projective curve of genus $g \geq 1$ with $b(X^{\mathrm{an}}) = g$. Suppose that the universal covering $\mathcal{U}$ of X^{an} is geometrically connected. (That is, $\mathcal{U} \otimes K$ is connected for any finite extension K of k.) Then X is a Mumford curve.*

Proof. By the Stable Reduction Theorem, there exists a finite Galois extension K over k such that the K-analytic space $(X \otimes K)^{\mathrm{an}}$ has $\widetilde{K}$-split stable reduction. Since the morphism $\varphi \otimes K \colon \mathcal{U} \otimes K \to (X \otimes K)^{\mathrm{an}}$ is a local isomorphism and $\mathcal{U} \otimes K$ is connected, it is the universal covering of $(X \otimes K)^{\mathrm{an}}$.

By Theorem 4.4.1, $\mathcal{U} \otimes K$ is isomorphic to $\Omega_\Gamma \subset \mathbf{P}^1 \otimes K$ for some Schottky group $\Gamma \subset \mathrm{PGL}_2(K)$ isomorphic to $\pi_1(X^{\mathrm{an}})$. We claim that the group $G = \mathrm{Gal}(K/k)$ acts trivially on Δ_Γ. Indeed, since the actions of Γ and G on Δ_Γ commute, it follows that the point gx_0 is also fixed under G for any point $x_0 \in \Delta_\Gamma$ fixed under G and any $g \in \Gamma$. Since Δ_Γ is the union of ℓ_{x_0, gx_0} over all $g \in \Gamma$, we conclude that G acts trivially on Δ_Γ. In particular, the set $\Sigma_\Gamma = (\mathbf{P}^1 \otimes K) \backslash \Omega_\Gamma$ comes from $\mathbf{P}^1(K)^G = \mathbf{P}^1(k)$. We may assume that $\{0, \infty\} \subset \Sigma_\Gamma$.

Furthermore, we claim that for any invertible function $f \in \mathcal{O}(\Omega_\Gamma)^*$ there exists a constant $a \in K^*$ such that $af \in (\mathcal{O}(\Omega_\Gamma)^*)^G = \mathcal{O}(\mathcal{U})^*$. Indeed, since G acts trivially on Δ_Γ, $|({}^{\sigma}f)(x)| = |f(x)|$ for all $\sigma \in G$ and $x \in \Delta_\Gamma$. By Remark 4.2.4(iii), the function ${}^{\sigma}f/f$ is constant on Ω_Γ, i.e., ${}^{\sigma}f = fb_\sigma$, where $\{b_\sigma\}_{\sigma \in G}$ is a 1-cocycle of G in K^*. By Hilbert's Theorem 90, there exists $a \in K^*$ such that $b_\sigma = a/{}^{\sigma}a$, $\sigma \in G$. Hence, $af \in \mathcal{O}(\mathcal{U})^*$.

We can now choose a function $z \in \mathcal{O}(\mathcal{U})^*$ which comes from the coordinate function of $\mathbf{P}^1 \otimes K$. The corresponding morphism $z \colon \mathcal{U} \to \mathbf{P}^1$ is an open immersion because it becomes an open immersion after tensoring with K. ∎

We now prove the direct implication of the theorem. Let $\pi \colon \mathcal{U} \to X^{\mathrm{an}}$ be the universal covering of X^{an}. First suppose that $g > 1$. By Lemma 4.6.3, it suffices to show that $\mathcal{U} \otimes K$ is connected for any finite extension K of k. Let $\mathcal{U}_1$ be a connected component of $\mathcal{U} \otimes K$. If Γ_1 is the stabilizer of $\mathcal{U}_1$ in $\pi_1(X^{\mathrm{an}})$, then $\mathcal{U}_1/\Gamma_1 \tilde{\to} (X \otimes K)^{\mathrm{an}}$. Since the group $\pi_1(X^{\mathrm{an}})$ is free of rank > 1, any proper subgroup of it is of greater rank. But $g = g(X \otimes K) = \mathrm{rk}(\Gamma_1)$. Hence, $\mathcal{U} \otimes K$ is connected.

Now suppose that $g = 1$. Let K be the algebraic closure of k in $\mathcal{O}(\mathcal{U})$. Since $\mathcal{U}$ is irreducible, K is a field which is finite over k. The group $\Gamma = \pi_1(X^{\mathrm{an}})$ acts on K, and $K^\Gamma = k$, because X is geometrically irreducible. Hence, K is a cyclic extension of k. If Γ_1 denotes the subgroup of Γ

consisting of elements which act trivially on K, then $\Gamma/\Gamma_1 \tilde{\to} G = \mathrm{Gal}(K/k)$. The K-analytic space $\mathscr{U}/\Gamma_1$ is proper, and therefore it is associated with a smooth geometrically connected projective curve of genus 1 over K. Since the K-analytic space $\mathscr{U}$ is geometrically connected, we can apply Lemma 4.6.3. It follows that $\mathscr{U}$ is isomorphic to $\mathbf{G}^{\mathrm{an}}_{m,K}$. The group Γ acts on it. All of the automorphisms of the k-analytic space $\mathbf{G}^{\mathrm{an}}_{m,K}$ are easily described. It follows that one can choose a generator γ of Γ which acts on $\mathbf{G}^{\mathrm{an}}_{m,K}$ as follows: $x \mapsto \lambda \cdot {}^{\sigma}x$ for some generator σ of G and some $\lambda \in K^*$ with $|\lambda| < 1$. If E is a Tate elliptic curve with period $q = N_{K/k}(\lambda)$, then X is the principal homogeneous space over E corresponding to the cocycle $\{c_{\sigma^i} = \lambda \cdot {}^{\sigma}\lambda \cdot \ldots \cdot {}^{\sigma^{i-1}}\lambda\}_{0 \le i \le n-1}$, $n = [K : k]$, in $H^1(G, E(K))$. ∎

CHAPTER 5

Analytic Groups and Buildings

All analytic spaces considered in this section are assumed to be separated.

5.1. Subgroups and homogeneous spaces of analytic groups. A *k-analytic group* is a k-analytic space G which is a group object in the category of k-analytic spaces. This means that there are three morphisms $m: G \times G \to G$ (multiplication), $e: \mathscr{M}(k) \to G$ (identity) and $i: G \to G$ (inversion) which satisfy the natural conditions (see **[M1]**, §11). Homomorphisms of k-analytic groups are defined in the obvious way. For example, the k-analytic spaces associated with group schemes over k are k-analytic groups. An *action* of a k-analytic group G on a k-analytic space X is a morphism $\mu: G \times X \to X$ satisfying the natural conditions (see **[M1]**, §12). For example, there is a left (resp. right) action of G on G defined by $(g, x) \mapsto gx$ (resp. $(g, x) \mapsto xg^{-1}$) and a two-sided action of $G \times G$ on G defined by $(g, h, x) \mapsto gxh^{-1}$. The restriction of the latter action to the diagonal subgroup $G \subset G \times G$ is called the action by inner automorphisms.

For a k-analytic group G, the image of $\mathscr{M}(k)$ under e is denoted by e, and the image of an element $g \in G$ (resp. a subset $H \subset G$) under i is denoted by g^{-1} (resp. H^{-1}). Let G act on a k-analytic space X. For $g \in G$ and $x \in X$, we denote by gx the subset of X consisting of all elements y such that there exists $w \in G \times X$ with $p_1(w) = g$, $p_2(w) = x$ and $\mu(w) = y$, i.e., $gx = \mu(p^{-1}(g, x))$. Here (and later) p_1 and p_2 denote the projections of $G \times X$ onto the factors, and $p = (p_1, p_2): |G \times X| \to |G| \times |X|$. If $H \subset G$ and $Y \subset X$, then HY denotes the union of the subsets gx over all $g \in H$ and $x \in Y$. In the case of the action of G on G by inner automorphisms, the sets gx and HY are denoted by ${}^g x$ and ${}^H Y$, respectively.

5.1.1. PROPOSITION. (i) *$(gh)x = g(hx)$ and $ex = x$. If $y \in gx$, then $x \in g^{-1}y$.*

(ii) *Given an open neighborhood $\mathscr{W}$ of the set gx, there exist open neighborhoods $\mathscr{U}$ of g and $\mathscr{V}$ of x such that $g'x' \subset \mathscr{W}$ for all $g' \in \mathscr{U}$ and $x' \in \mathscr{V}$.*

(iii) *If Y is compact (resp. closed) and H is compact, then HY is compact (resp. closed).*

If K is a valuation field over k, we denote by π_K (or simply by π) the canonical map $X' = X\widehat{\otimes}K \to X$.

5.1.2. LEMMA. *Let K be a valuation field over k, and let $H \subset G$ and $Y \subset X$. Then:*
(i) $\pi^{-1}(HY) = \pi^{-1}(H)\pi^{-1}(Y)$;
(ii) *if $g \in G$ and $g' \in \pi^{-1}(g)$ (resp. $x \in X$ and $x' \in \pi^{-1}(x)$), then $gY = \pi(g'\pi^{-1}(Y))$ (resp. $Hx = \pi(\pi^{-1}(H)x')$).*

PROOF. (i) The inclusion $\pi^{-1}(H)\pi^{-1}(Y) \subset \pi^{-1}(HY)$ is trivial. Let $y \in HY$, $y' \in \pi^{-1}(y)$ and let w be an element of $G \times X$ with $p_1(w) \in H$, $p_2(w) \in Y$ and $\mu(w) = y$. To establish the reverse inclusion, it suffices to show that there exists $w' \in G' \times X'$ with $\pi(w') = w$ and $\mu(w') = y'$. This is clear because the set $\pi^{-1}(w) \cap \mu^{-1}(y')$ coincides with the spectrum of the algebra $\mathscr{H}(w)\widehat{\otimes}_{\mathscr{H}(y)}\mathscr{H}(y')$, which is not zero.

(ii) The inclusion $\pi(g'\pi^{-1}(Y)) \subset gY$ is trivial. To establish the reverse inclusion, it suffices to show that, for a point $w \in G \times X$ with $p_1(w) = g$ and $p_2(w) \in Y$, there exists a point $w' \in G' \times X'$ with $\pi(w') = w$ and $p_1(w') = g'$. As above, $\pi^{-1}(w) \cap p_1^{-1}(g')$ coincides with the spectrum of the algebra $\mathscr{H}(w)\widehat{\otimes}_{\mathscr{H}(g)}\mathscr{H}(g')$, which is not zero. ∎

PROOF OF PROPOSITION. (i) The equality $ex = x$ is obvious. The equality $(gh)x = g(hx)$ is clearly true if $g, h \in G(k)$. Let K be a valuation field over k such that there exist $g' \in \pi^{-1}(g) \cap G'(K)$ and $h' \in \pi^{-1}(h) \cap G'(K)$ (in the notation of Lemma 5.1.2). If $g \in G(k)$, then $\pi^{-1}(g) = \{g'\}$ and:

$$\begin{aligned} g(hx) &= g\pi(h'\pi^{-1}(x)) = \pi(g'(h'\pi^{-1}(x))) \\ &= \pi((g'h')\pi^{-1}(x)) = g(hx). \end{aligned}$$

In the general case we have:

$$\begin{aligned} (gh)x &= \pi((g'h')\pi^{-1}(x)) = \pi(g'(h'\pi^{-1}(x))) \\ &\subset \pi(g'\pi^{-1}(hx)) = g(hx); \end{aligned}$$

$$\begin{aligned} g(hx) &= \pi(g'\pi^{-1}(hx)) = \pi(g'(\pi^{-1}(h)\pi^{-1}(x))) \\ &= \pi((g'\pi^{-1}(h))\pi^{-1}(x)) \subset \pi(\pi^{-1}(gh)\pi^{-1}(x)) = (gh)x. \end{aligned}$$

Let $y \in gx$. If $g \in G(k)$, then $y = gx$ and $x = g^{-1}y$. In the general case, there exists $y' \in \pi^{-1}(y)$ such that $y' = g'x'$ with $g' \in \pi^{-1}(g) \cap G'(K)$. Therefore, $x' = g'^{-1}y'$ and $x \in g^{-1}y$.

(ii) Since $gx = \mu(p^{-1}(g, x))$, the assertion follows from the fact that the set $p^{-1}(g, x)$ is compact and coincides with the intersection of open subsets of the form $\mathscr{U} \times \mathscr{V}$ over all open neighborhoods $\mathscr{U}$ of g and $\mathscr{V}$ of x.

(iii) If H and Y are compact, then $HY = \mu(p^{-1}(H \times Y))$ is compact. Suppose that H is compact and Y is closed, and let $x \notin HY$. Then $H^{-1}x \cap Y = \varnothing$. By (ii), for each point $g \in H$ there exist open neighborhoods $\mathscr{U}_g$

of g and $\mathscr{V}_g$ of x such that $g'^{-1}x' \cap Y = \varnothing$ for all $g' \in \mathscr{U}_g$ and $x' \in \mathscr{V}_g$. Since H is compact, it lies in a finite union $\bigcup_{i=1}^n \mathscr{U}_{g_i}$. We set $\mathscr{V} = \bigcap_{i=1}^n \mathscr{V}_{g_i}$. Then $g^{-1}x' \cap Y = \varnothing$ for all $g \in H$ and $x' \in \mathscr{V}$, i.e., $\mathscr{V} \cap HY = \varnothing$. ∎

5.1.3. COROLLARY. *In the case of the action of G on G by inner automorphisms, one has*

(i) ${}^{(gh)}x = {}^g({}^hx)$ *and* ${}^ex = x$. *If* $y \in {}^gx$, *then* $x \in {}^{g^{-1}}y$.
(ii) ${}^ge = e$, ${}^g(x^{-1}) = ({}^gx)^{-1}$ *and* ${}^g(xy) \subset {}^gx{}^gy$.
(iii) $gx \subset {}^gxg$ *and* $xg \subset g\,{}^{g^{-1}}x$.

PROOF. (i) follows immediately from Proposition 5.1.1. The other statements are verified in the same way as Proposition 5.1.1, using the following fact which follows from Lemma 5.1.2. Let K be a valuation field over k such that there exists $g' \in \pi^{-1}(x) \cap G'(K)$. Then for any subset $H \subset G$, ${}^gH = \pi(g'\pi^{-1}(H)g'^{-1})$. ∎

5.1.4. EXAMPLE. Let $\mathrm{G_a}$ (resp. $\mathrm{G_m}$) be the additive (resp. multiplicative) group over k. Then $\mathrm{G_a^{an}} = \mathbf{A}^1$, $\mathrm{G_m^{an}} = \mathbf{A}^1\backslash\{0\}$, and they act on $\mathbf{A}^1$. The set $p(E(0, r)) + p(E(0, r'))$ coincides with $p(E(0, \max(r, r')))$ for $r \neq r'$ and with $E(0, r)$ for $r = r'$. Furthermore, $p(E(0, r))p(E(0, r')) = E(0, rr')\backslash D(0, rr')$.

A subset $H \subset G$ is called a *subgroup* if $H^{-1}H \subset H$, i.e., (a) $e \in H$; (b) $g \in H \Rightarrow g^{-1} \in H$; (c) $g, h \in H \Rightarrow gh \subset H$. For example, let $\varphi: H \to G$ be a homomorphism of k-analytic groups. If φ is a closed immersion, its image is a subgroup called a *closed k-analytic subgroup* of G. If φ identifies H with an affinoid domain in G, H is called an *affinoid subgroup* of G. For example, the closed discs $E(0, t)$ for $t > 0$ are affinoid subgroups of $\mathrm{G_a^{an}}$. The sets $\mathrm{G_{m,1}^{an}} = E(0, 1)\backslash D(0, 1)$ and $\mathrm{G_{m,t}^{an}} = E(1, t)$ for $0 < t < 1$ are affinoid subgroups of $\mathrm{G_m^{an}}$. From Lemma 5.1.2(i) it follows that if H is a subgroup of G, then $H' = \pi^{-1}(H)$ is a subgroup of G' (in the notation of Lemma 5.1.2). In this case the set of K-points $H'(K)$ is a subgroup of the abstract group $G'(K)$.

Let G act on a k-analytic space X, and let H be a subgroup of G. A set of the form Hx, $x \in X$, is called the *orbit of x under H*. From Proposition 5.1.1(i) it follows that two orbits having nonempty intersection coincide. So we can consider the set of orbits $H\backslash X$. We endow it with the quotient topology. For example, G acts on G on the left (resp. right), and $G \times G$ acts on G on both sides. In this way we obtain the set of *left* (resp. *right*, resp. *double*) *cosets* $H\backslash G$ (resp. G/H, resp. $H\backslash G/H$). Furthermore, for a closed subset $F \subset G$, one can consider the sets $H\backslash HF$ and FH/H, which are closed subsets of $H\backslash G$ and G/H, respectively.

5.1.5. PROPOSITION. *Let a k-analytic group G act on a k-analytic space X, and let H be a compact subgroup of G.*

(i) *The space* $H\backslash X$ *is locally compact, and the canonical map* $X \to H\backslash X$ *is proper.*

(ii) *If* K *is a finite Galois extension of* k, *then* $(H'\backslash X')/\mathrm{Gal}(K/k) \tilde{\to} H\backslash X$, *where* $H' = \pi^{-1}(H) \subset X' = X\widehat{\otimes}K$.

(iii) *For any closed subgroup* $F \subset G$, *there are homeomorphisms* $F/(F \cap H) \tilde{\to} FH/H$ *and* $(F \cap H)\backslash F \tilde{\to} H\backslash HF$.

PROOF. (i) follows from the following general fact. Let R be an equivalence relation on a locally compact space X such that the R-saturation of any compact subset is compact. Then the set of equivalence classes X/R is locally compact, and the canonical map $X \to X/R$ is proper.

(ii) By Lemma 5.1.2(i), $\pi^{-1}(Hx) = H'\pi^{-1}(x)$. Since the Galois group acts transitively on the set $\pi^{-1}(x)$, our map is injective. Its surjectivity is obvious.

(iii) The fact that the map $F/(F \cap H) \to FH/H$: $f(F \cap H) \mapsto fH$ is bijective and continuous is easy. Let $\mathcal{U}$ be an open neighborhood of the set $f(F \cap H)$ in F with $\mathcal{U}(F \cap H) = \mathcal{U}$. Since $fH \cap F = f(F \cap H)$ and the compact set fH coincides with the intersection of its open neighborhoods $\mathcal{W}$ in FH with $\mathcal{W}H = \mathcal{W}$, it follows that there exists such a $\mathcal{W}$ with $\mathcal{W} \cap F \subset \mathcal{U}$. It follows that $\mathcal{W} = (\mathcal{W} \cap F)H \subset \mathcal{U}H$. That is, the set $\mathcal{U}H$ is open in FH. ∎

5.1.6. COROLLARY. *Let* H *be a compact subgroup of a* k*-analytic group* G. *Then the natural* (*left*) *action of the group* $G(k)$ *on the locally compact space* G/H *is proper in the sense that the images and inverse images of relatively compact subsets under the mapping* $G(k) \times G/H \to G/H \times G/H$: $(g, x) \mapsto (gx, x)$ *are relatively compact.* ∎

5.1.7. REMARK. (i) The *normalizer* (resp. *centralizer*) of a subset $H \subset G$ is the set $N(H) = \{g \in G | {}^gH \subset H$ and ${}^{g^{-1}}H \subset H\}$ (resp. $C(H) = \{g \in G | {}^gx = {}^{g^{-1}}x = x$ for all $x \in H\}$). If $g \in N(H)$, then ${}^gH = {}^{g^{-1}}H = H$ and $gH = Hg$. It is easy to see that $N(H)$ is a subgroup of G, and $C(H)$ is a closed subgroup of G. If $H = \{x\}$ is a point, then $C(x) = N(x)$. The centralizer of G is called the *center* of G and is denoted by $Z(G)$. A subgroup $H \subset G$ is called a *normal divisor* if $N(H) = G$.

(ii) A normal divisor $H \subset G$ is said to be *thick* if $gg^{-1} \subset H$ for all $g \in G$. The product of two cosets of a thick normal divisor H is a coset. Indeed, if $z \in xy$, then $xy \subset zy^{-1}y \subset zH$, i.e., $xHyH = xyH = zH$. In particular, G/H is an abstract group. The intersection of all thick normal divisors is a thick normal divisor. It is called the *kernel* of G. For example, the kernel of $\mathrm{G}_{\mathrm{m}}^{\mathrm{an}}$ is $\mathrm{G}_{\mathrm{m},1}^{\mathrm{an}}$, and there is an isomorphism of locally compact groups $\mathrm{G}_{\mathrm{m}}^{\mathrm{an}}/\mathrm{G}_{\mathrm{m},1}^{\mathrm{an}} \tilde{\to} \mathbf{R}_+^*$.

A k-analytic space X is said to be *Stein* if there exists an increasing sequence of affinoid domains $X_1 \subset X_2 \subset \dots$ such that each X_n is a Weierstrass domain in X_{n+1} and $X = \bigcup_{n=1}^{\infty} X_n$. For example, k-affinoid spaces and k-analytic spaces associated with affine schemes of finite type over k are Stein. The property of being Stein is preserved under extension of the ground field, and the product of two Stein spaces is a Stein space. The Weierstrass subdomains of the affinoid domains X_n are called *Weierstrass domains in* X. It is clear that this notion does not depend on the choice of the sequence $\{X_n\}_{n\geq 1}$.

Let X be a Stein space, and let Σ be a relatively compact subset of X. Then Σ is contained in some Weierstrass domain V. As in §2.6, we define the compact set $\Sigma^h = \{x \in X \,|\, |f(x)| \leq \|f\|_\Sigma \text{ for all } f \in \mathscr{A}_V\}$ (the holomorphically convex envelope of Σ) and the Banach k-algebra $\mathscr{P}(\Sigma)$ which is the completion of $\mathscr{A}_V$ with respect to the seminorm $\| \ \|_\Sigma$. These objects do not depend on V, and one has $\mathscr{M}(\mathscr{P}(\Sigma)) \tilde{\to} \Sigma^h$.

5.1.8. PROPOSITION. *Let a Stein group G act on a Stein space X, and let $A \subset G$ and $\Sigma \subset X$ be relatively compact subsets. Then $(A^{-1})^h = (A^h)^{-1}$ and $A^h\Sigma^h \subset (A\Sigma)^h$.*

PROOF. The first equality is trivial. Let U (resp. V) be a Weierstrass domain in G (resp. X) containing A (resp. Σ), and let W be a Weierstrass domain in X containing the set UV. For every function $f \in \mathscr{A}_W$, one has the function $\mu^*(f)$ defined on the Weierstrass domain $U \times V$ in $G \times X$. Let $g \in A^h$, $x \in \Sigma^h$ and $y \in gx$. Take a point $z \in G \times X$ with $p_1(z) = g$, $p_2(z) = x$, and $\mu(z) = y$. This point is contained in $p^{-1}(A^h \times \Sigma^h) = p^{-1}(A \times \Sigma)^h \subset U \times V$, and for any $f \in \mathscr{A}_W$ one has:

$$|f(y)| = |\mu^*(f)(z)| \leq \|\mu^*(f)\|_{p^{-1}(A\times\Sigma)^h} \leq \|f\|_{(A\Sigma)^h} = \|f\|_{A\Sigma}. \qquad \blacksquare$$

5.1.9. COROLLARY. *Let G be a Stein group. Then*:

(i) *The holomorphically convex envelope of a relatively compact subgroup is a compact subgroup.*

(ii) *Every compact subgroup of G is holomorphically convex.*

PROOF. (i) follows immediately from Proposition 5.1.8. (ii) Let H be a compact subgroup of G, and let $x \in H^h \backslash H$. Then $H^h = H^h x \subset H^h\{x\}^h \subset (Hx)^h \subset H^h$. But this contradicts the equality $H \cap Hx = \varnothing$. ∎

5.2. Peaked points and the $*$-multiplication

5.2.1. DEFINITION. A commutative Banach k-algebra with identity $\mathscr{A}$ is said to be *peaked* (*over* k) if, for any valuation field K over k, the norm on the algebra $\mathscr{A} \widehat{\otimes} K$ is multiplicative.

5.2.2. LEMMA. *Let $\mathscr{A}$ be a commutative Banach k-algebra with identity whose norm is multiplicative, and let x be the corresponding point of $\mathscr{M}(\mathscr{A})$. Then the algebra $\mathscr{A}$ is peaked if and only if the field $\mathscr{H}(x)$ is peaked.*

PROOF. Consider the commutative diagram of isometric embeddings:

$$\begin{array}{ccc} \mathscr{A} & \longrightarrow & \mathscr{A}\widehat{\otimes}K \\ \downarrow & & \downarrow \\ \mathscr{H}(x) & \longrightarrow & \mathscr{H}(x)\widehat{\otimes}K \end{array}$$

It follows that if the norm on $\mathscr{H}(x)\widehat{\otimes}K$ is multiplicative, then the same is true for the algebra $\mathscr{A}\widehat{\otimes}K$. Conversely, suppose that the norm on $\mathscr{A}\widehat{\otimes}K$ is multiplicative. Because the fraction field $Q(\mathscr{A})$ of $\mathscr{A}$ is everywhere dense in $\mathscr{H}(x)$, it follows that the subalgebra M of $\mathscr{H}(x)\widehat{\otimes}K$ generated by $Q(\mathscr{A})$ and $\mathscr{A}\widehat{\otimes}K$ is everywhere dense. Since M consists of the products fg with $f \in Q(\mathscr{A})$ and $g \in \mathscr{A}\widehat{\otimes}K$, and $||fg|| = ||f||\ ||g||$ for $f \in \mathscr{H}(x)$ and $g \in \mathscr{H}(x)\widehat{\otimes}K$, the norm on $\mathscr{H}(x)\widehat{\otimes}K$ is multiplicative. ∎

5.2.3. COROLLARY. *Let K be a peaked valuation field, and let k' be a valuation field over k. Let K' denote the completion of the fraction field of the algebra $K\widehat{\otimes}k'$ with respect to the norm that is obtained by extending the norm on $K\widehat{\otimes}k'$. Then K' is peaked over k'. If k' is peaked over k, then K' is peaked over k.* ∎

A point x of a k-analytic space is said to be *peaked* if the field $\mathscr{H}(x)$ is peaked. The set of peaked points is denoted by X_p. If K is a valuation field (resp. a peaked valuation field) over k, one can construct a map $\sigma = \sigma_K\colon X_p \to X\widehat{\otimes}K$ (resp. $X \to X\widehat{\otimes}K$) as follows: a point $x \in X_p$ (resp. $x \in X$) goes to the point in $\pi_K^{-1}(x)\widetilde{\to}\mathscr{M}(\mathscr{H}(x)\widehat{\otimes}K)$ corresponding to the multiplicative norm on the algebra $\mathscr{H}(x)\widehat{\otimes}K$. From Corollary 5.2.3 it follows that $\sigma(X_p) \subset (X\widehat{\otimes}K)_p$.

5.2.4. COROLLARY. *Let X be a Stein space, and let K be a valuation field over k. Then for any point $x \in X_p$ one has $\mathscr{P}(\{x\})\widehat{\otimes}K\widetilde{\to}\mathscr{P}(\{\sigma_K(x)\})$. In particular, $\pi^{-1}(\{x\}^h) = \{\sigma_K(x)\}^h$.*

PROOF. By Lemma 5.2.2, the algebra $\mathscr{P}(\{x\})$ is peaked, and therefore the norm on $\mathscr{P}(\{x\})\widehat{\otimes}K$ is multiplicative. It follows that the homomorphism under consideration is isometric. Since its image is everywhere dense, it is an isomorphism. ∎

From Lemma 5.2.2 it follows that points of $\mathbf{A}^n$ which are associated with closed discs with center at zero are peaked. Actually, all of the peaked points which occur in this section are of that type. In the next section we need a more general class of peaked points which are obtained as follows.

5.2.5. PROPOSITION. *Let $X = \mathscr{M}(\mathscr{A})$ be a distinguished strictly k-affinoid space (see §4.3) such that its reduction $\widetilde{X} = \mathrm{Spec}(\widetilde{\mathscr{A}})$ is geometrically reduced and irreducible. Then the point $x \in X$ corresponding to the spectral norm on $\mathscr{A}$ is peaked.*

PROOF. We may assume that $\mathscr{A}$ is equipped with the spectral norm (which is multiplicative, by hypothesis). We have to verify that, for any valuation field K over k, the norm on the algebra $\mathscr{A}\widehat{\otimes}K$ is multiplicative. Note that the case of trivial valuation is easily reduced to the case of nontrivial valuation after tensoring by the field K_r for some $0 < r < 1$. By a result of Bosch ([**Bo1**], 6.4), the K-affinoid space $X\widehat{\otimes}K$ is distinguished, and its reduction is irreducible. Hence, the spectral seminorm on $\mathscr{A}\widehat{\otimes}K$ is equivalent to the norm and is multiplicative. We must verify that the norm on $\mathscr{A}\widehat{\otimes}K$ coincides with the spectral norm. It suffices to show that if $\varphi \in (\mathscr{A}\widehat{\otimes}K)^\circ$, then $||\varphi|| \le 1$. By the same result of Bosch, there exists an element $\psi = \sum_{i=1}^n f_i \otimes \lambda_i \in \mathscr{A} \otimes K$ with $f_i \in \mathscr{A}^\circ$, $\lambda_i \in K^\circ$ and $||\varphi - \psi|| < \rho(\varphi)$. (We assume that $\varphi \ne 0$.) Hence, $||\varphi|| = ||\psi|| \le 1$. ∎

Recall (see [**BGR**], 2.8.2/1) that a Banach space B over k is said to be *of countable type* if B contains an everywhere dense subspace of at most countable dimension. For example, any k-affinoid algebra is of countable type. Furthermore, if a commutative Banach k-algebra $\mathscr{A}$ is of countable type, then the fields $\mathscr{H}(x)$ for $x \in \mathscr{M}(\mathscr{A})$ are of countable type. In particular, the fields $\mathscr{H}(x)$ for points of k-analytic spaces are of countable type.

5.2.6. LEMMA. *Let $\mathscr{A}$ be a commutative Banach k-algebra, and let B be a Banach space of countable type. Let φ be an element of $\mathscr{A}\widehat{\otimes}B$, and for $x \in X = \mathscr{M}(\mathscr{A})$ let $\nu_x(\varphi)$ be the image of φ in $\mathscr{H}(x)\widehat{\otimes}B$. Then the real valued function $x \mapsto ||\nu_x(\varphi)||$ on X is continuous.*

PROOF. We have to verify that, for any $c > 0$, the sets $\mathscr{W}_{<c} = \{x \in X \,|\, ||\nu_x(\varphi)|| < c\}$ and $\mathscr{W}_{>c} = \{x \in X \,|\, ||\nu_x(\varphi)|| > c\}$ are open in X.

Let $x \in \mathscr{W}_{<c}$. Take $\alpha > 1$ with $||\nu_x(\varphi)|| < \alpha^{-1}c$. By [**BGR**], 2.8.2/4, there exists an α-cartesian basis of B, i.e., a linearly independent family $\{b_i\}_{i\in I}$ of elements of B which generate an everywhere dense subspace in B and for which $||\sum_{i=1}^n a_i b_i|| \ge \alpha^{-1} \max_{1\le i\le n}\{|a_i|\,||b_i||\}$ for all $a_i \in k$. It follows that for every Banach space V the family of elements of the form $\sum_{i=1}^n v_i \otimes b_i$, $v_i \in V$, is everywhere dense in $V\widehat{\otimes}B$, and $||\sum_{i=1}^n v_i \otimes b_i|| \ge \alpha^{-1} \max_{1\le i\le n}\{||v_i||\,||b_i||\}$. In particular, there exists an element $\sum_{i=1}^n f_i \otimes b_i \in \mathscr{A}\widehat{\otimes}B$ with $||\varphi - \sum_{i=1}^n f_i \otimes b_i|| < \alpha^{-1}c$. Let $\mathscr{U}$ denote the open set $\{y \in X \,|\, \max_{1\le i\le n}\{|f_i(y)|\,||b_i||\} < c\}$. We claim that $x \in \mathscr{U} \subset \mathscr{W}_{<c}$. Indeed, one has:

$$||\sum_{i=1}^n f_i(x) \otimes b_i|| \le \max\{||\nu_x(\varphi) - \sum_{i=1}^n f_i(x) \otimes b_i||, ||\nu_x(\varphi)||\} < \alpha^{-1}c,$$

and hence $\max_{1\le i\le n}\{|f_i(x)|\,||b_i||\} \le \alpha||\sum_{i=1}^n f_i(x) \otimes b_i|| < c$, i.e. $x \in \mathscr{U}$. Let $y \in \mathscr{U}$. Then $||\sum_{i=1}^n f_i(y) \otimes b_i|| \le \max_{1\le i\le n}\{|f_i(y)|\,||b_i||\} < c$, and

$$||\nu_y(\varphi)|| \le \max\{||\nu_y(\varphi) - \sum_{i=1}^n f_i(y) \otimes b_i||, ||\sum_{i=1}^n f_i(y) \otimes b_i||\} < c,$$

i.e., $y \in \mathscr{W}_{<c}$.

Now let $x \in \mathscr{W}_{>c}$. Take $\alpha > 1$ with $\|\nu_x(\varphi)\| > \alpha c$. As above, we take an α-cartesian basis of B and an element $\sum_{i=1}^n f_i \otimes b_i \in \mathscr{A}\widehat{\otimes}B$ with $\|\varphi - \sum_{i=1}^n f_i \otimes b_i\| < c$. We verify that for the open set

$$\mathscr{U} = \left\{y \in X \middle| \max_{1\le i\le n}\{|f_i(y)|\ \|b_i\|\} > \alpha c\right\},$$

one has $x \in \mathscr{U} \subset \mathscr{W}_{>c}$. Indeed, since $\|\nu_x(\varphi) - \sum_{i=1}^n f_i(x) \otimes b_i\| < c$ and $\|\nu_x(\varphi)\| > \alpha c$, it follows that

$$\max_{1\le i\le n}\{|f_i(x)|\ \|b_i\|\} \ge \|\sum_{i=1}^n f_i(x) \otimes b_i\| = \|\nu_x(\varphi)\| > \alpha c.$$

Let $y \in \mathscr{U}$. Then $\|\sum_{i=1}^n f_i(y)\otimes b_i\| \ge \alpha^{-1}\max_{1\le i\le n}\{|f_i(y)|\ \|b_i\|\} > c$. Since $\|\nu_y(\varphi) - \sum_{i=1}^n f_i(y) \otimes b_i\| < c$, we have $\|\nu_y(\varphi)\| = \|\sum_{i=1}^n f_i(y) \otimes b_i\| > c$. ∎

5.2.7. COROLLARY. *For a k-analytic space X and a valuation field (resp. peaked valuation field) K of countable type over k, the map $\sigma: X_p \to X\widehat{\otimes}K$ (resp. $\sigma: X \to X\widehat{\otimes}K$) is continuous (the set X_p is endowed with the induced topology).*

PROOF. We may assume that $X = \mathscr{M}(\mathscr{A})$ is a k-affinoid space, and we must show that for any $\varphi \in \mathscr{A}\widehat{\otimes}K$ the real valued function $x \mapsto |\varphi(\sigma(x))|$ on X_p (resp. X) is continuous. Since $|\varphi(\sigma(x))| = \|\nu_x(\varphi)\|$ for $x \in X_p$ (resp. $x \in X$), the required fact follows from Lemma 5.2.6. ∎

Let a k-analytic group G act on a k-analytic space X, and let $g \in G$ and $x \in X$. Suppose that at least one of the points g, x is peaked. Then the norm on the algebra $\mathscr{H}(g)\widehat{\otimes}\mathscr{H}(x)$ is multiplicative. Let $w_{g,x}$ be the point of $p^{-1}(g, x) = \mathscr{M}(\mathscr{H}(g)\widehat{\otimes}\mathscr{H}(x))$ corresponding to this norm, and set $g * x = \mu(w_{g,x})$. If $g \in G(k)$ or $x \in X(k)$, then $g * x = gx$.

5.2.8. PROPOSITION. (i) *If $g \in G_p$ and $x \in X_p$, then $g * x \in X_p$, and, moreover, $\sigma_K(g * x) = \sigma_K(g) * \sigma_K(x)$ for any valuation field K over k.*

(ii) *Given a commutative diagram*

$$\begin{array}{ccccc} G & \times & X & \to & X \\ \downarrow\varphi & & \downarrow\psi & & \downarrow\psi \\ G' & \times & X' & \to & X' \end{array},$$

one has $\psi(g * x) = \varphi(g) * \psi(x)$.

(iii) *If at least two of the points g, h, x are peaked, then $(g * h) * x = g * (h * x)$.*

(iv) *If X is Stein, then $gx \subset \{g * x\}^h$.*

(v) *If g (resp. x) is peaked, then the map $X \to X: x \mapsto g * x$ (resp. $G \to X: g \mapsto g * x$) is continuous.*

PROOF. (i) By Corollary 5.2.3, the point $w_{g,x}$ is peaked. Since $\mathscr{H}(g * x)$ is a subfield of $\mathscr{H}(w_{g,x})$, we have $g * x \in X_p$. The second assertion follows from the equality $\sigma(w_{g,x}) = w_{\sigma(g),\sigma(x)}$, which is easily verified.

(ii) First we remark that if $g \in G_p$ (resp. $x \in X_p$), then $\varphi(g) \in G'_p$ (resp. $\psi(x) \in X'_p$). Next we note that the norm on the algebra $\mathscr{H}(g)\widehat{\otimes}\mathscr{H}(x)$ induces the norm on the algebra $\mathscr{H}(\varphi(g))\widehat{\otimes}\mathscr{H}(\psi(x))$. Thus, $(\varphi, \psi)(w_{g,x}) = w_{\varphi(g),\psi(x)}$.

(iii) By (i), at least one of the points $w_{g,h}, x$ (resp. $g, w_{h,x}$) is peaked. If w' (resp. w'') is the point of $G \times G \times X$ corresponding to the first (resp. second) pair of points, then $w' = w''$, because both points correspond to the (multiplicative) norm on the algebra $(\mathscr{H}(g)\widehat{\otimes}\mathscr{H}(h))\widehat{\otimes}\mathscr{H}(x) = \mathscr{H}(g)\widehat{\otimes}(\mathscr{H}(h)\widehat{\otimes}\mathscr{H}(x))$. Hence, $(g * h) * x = \mu(w') = \mu(w'') = g * (h * x)$.

(iv) Let $y \in gx$, and let u be a point of $p^{-1}(g, x) = \mathscr{M}(\mathscr{H}(g)\widehat{\otimes}\mathscr{H}(x))$ with $\mu(u) = y$. Then $|\varphi(u)| \le |\varphi(w_{g,x})|$ for all $\varphi \in \mathscr{H}(g)\widehat{\otimes}\mathscr{H}(x)$. In particular, if $f \in \mathscr{O}(X)$, then $|f(y)| = |\mu^*(f)(u)| \le |\mu^*(f)(w_{g,x})| = |f(g * x)|$.

(v) follows from Corollary 5.2.7. ∎

5.2.9. Corollary. *Let g be a peaked point in a Stein group G. Then the set $\{g\}^h$ is a subgroup of G if and only if $g^{-1} = g$ and $g * g = g$.*

Proof. The statement follows from Propositions 5.1.8 and 5.2.8(iv). ∎

A k-analytic space X is said to be *holomorphically separated* if, for every pair of distinct points x and y, there exists a function $f \in \mathscr{O}(X)$ with $|f(x)| \ne |f(y)|$. Let X be such a space, and let Σ be a compact subset in it. We say that a point $x \in \Sigma$ is a *maximal point of* Σ if $|f(y)| \le |f(x)|$ for all $f \in \mathscr{O}(X)$ and $y \in \Sigma$. If such a point exists, it is unique, and we say that Σ has a maximal point. The following statement is verified in the same way as Proposition 5.2.8(iv).

5.2.10. Proposition. *Let a Stein group G act on a k-analytic space X, and let H be a compact subgroup of G such that $H = \{g\}^h$ for some peaked point $g \in G$. Suppose that every orbit Hx is contained in a holomorphically separated open subset of X. Then the point $g * x$ is a maximal point of the orbit Hx in any holomorphically separated open neighborhood of Hx. In particular, the point $g * x$ depends only on the orbit Hx.* ∎

5.2.11. Corollary. *In the situation of Proposition 5.2.10, the map $x \mapsto g * x$ induces a homeomorphism of $H\backslash X$ with a closed subset of X.* ∎

5.2.12. Example. Let T be a split torus over k, let $\widehat{T}$ be its character group, and let $\Lambda = \mathrm{Hom}(\widehat{T}, \mathbf{R}^*_+)$. Every point $t \in T^{\mathrm{an}}$ determines a homomorphism $\psi(t)\colon \widehat{T} \to \mathbf{R}^*_+$ which sends a character $\chi \in \widehat{T}$ to $|\chi(t)|$. The set $T_1^{\mathrm{an}} = \{t \in T^{\mathrm{an}} \mid |\chi(t)| = 1 \text{ for all } \chi \in \widehat{T}\}$ is a thick affinoid subgroup of T^{an} (and actually is the kernel of T^{an}), and there is an isomorphism of locally compact groups $\psi\colon T^{\mathrm{an}}/T_1^{\mathrm{an}} \xrightarrow{\sim} \Lambda$ (see Remark 5.1.7(ii)). For $\lambda \in \Lambda$, let $\mathbf{t}_\lambda$ denote the point of T^{an} corresponding to the following multiplicative norm on $A(T)$ (the ring of regular functions on T):

$$|\sum_{\chi\in\widehat{T}} a_\chi \chi| = \max_{\chi\in\widehat{T}} |a_\chi|\lambda(\chi).$$

The point $\mathbf{t}_\lambda$ is peaked. Indeed, after fixing a basis $\chi_1, \dots, \chi_l$ of $\widehat{T}$, we can naturally embed T^{an} in $\mathbf{A}^l$. Under this embedding the point $\mathbf{t}_\lambda$ corresponds to the maximal point of the closed disc of radius $(\lambda(\chi_1), \cdots, \lambda(\chi_l))$ with center at zero. One has $\psi(\mathbf{t}_\lambda) = \lambda$, i.e., the correspondence $\lambda \mapsto \mathbf{t}_\lambda$ is inverse to ψ. It is easy to verify that:

(i) $T_1^{\mathrm{an}} = \{\mathbf{t}_0\}^h$;

(ii) if $t \in T^{\mathrm{an}}$, then $tT_1^{\mathrm{an}} = \{\mathbf{t}_{\psi(t)}\}^h$ and $t * \mathbf{t}_\lambda = \mathbf{t}_{\lambda+\psi(t)}$; in particular, $\mathbf{t}_\lambda * \mathbf{t}_\mu = \mathbf{t}_{\lambda+\mu}$, i.e., the set $\{\mathbf{t}_\lambda\}_{\lambda\in\Lambda}$ is a group with respect to $*$-multiplication.

5.2.13. PROPOSITION. *Let T^{an} act on a Stein space X, and let x be a peaked point of X such that $\mathbf{t}_0 * x = x$.*

(i) *If $t \in T^{\mathrm{an}}$, then $t\{x\}^h = \{\mathbf{t}_{\psi(t)} * x\}^h$;*

(ii) *if X is geometrically reduced and $\{x\}^h$ is an affinoid domain, then all $\{\mathbf{t}_\lambda * x\}^h$, $\lambda \in \Lambda$, are affinoid domains.*

PROOF. (i) If $t \in T(k)$, then $tx = t * (\mathbf{t}_0 * x) = \mathbf{t}_{\psi(y)} * x$ and $t\{x\}^h = \{tx\}^h = \{\mathbf{t}_{\psi(t)} * x\}^h$. For arbitrary t, let $K = \mathscr{H}(t)$ and take a point $t' \in \pi^{-1}(t) \cap T(K)$. If $x' = \sigma(x)$, then by Proposition 5.2.8(i) we have $\mathbf{t}'_0 * x' = x'$ and, by Corollary 5.2.4, $\pi^{-1}(\{x\}^h) = \{x'\}^h$ and $\pi^{-1}(\{\mathbf{t}_\lambda * x\}^h) = \{\mathbf{t}'_\lambda * x'\}^h$. We then have

$$t\{x\}^h = \pi(t'\{x'\}^h) = \pi(\{\mathbf{t}_{\psi(t)} * x'\}^h) = \{\mathbf{t}_{\psi(t)} * x\}^h .$$

(ii) If $\lambda \in \mathrm{Hom}(\widehat{T}, |k^*|)$, then $\lambda = \psi(t)$ for some $t \in T(k)$, and therefore $\{\mathbf{t}_\lambda * x\}^h = t\{x\}^h$ is an affinoid domain. We claim that the algebra $\mathscr{P}(\{\mathbf{t}_\lambda * x\})$ is k-affinoid for all $\lambda \in \Lambda$. If $\lambda \in \mathrm{Hom}(\widehat{T}, \sqrt{|k^*|})$, we take a finite Galois extension K over k such that $\lambda \in \mathrm{Hom}(\widehat{T}, |K^*|)$. From Corollary 5.2.4 it follows that $\mathscr{P}(\{\mathbf{t}_\lambda * x\}) = \mathscr{P}(\{\mathbf{t}'_\lambda * x'\})^{\mathrm{Gal}(K/k)}$, and hence the algebra $\mathscr{P}(\{\mathbf{t}_\lambda * x\})$ is k-affinoid, by Corollary 2.1.14(ii). In the general case it suffices to show that if $r \notin \sqrt{|k^*|}$ and the algebra $\mathscr{P}(\{\mathbf{t}'_\lambda * x'\})$ is K_r-affinoid, then the algebra $\mathscr{P}(\{\mathbf{t}_\lambda * x\})$ is k-affinoid. This follows from Corollary 2.1.8. It remains to verify the following simple fact. Let V be a holomorphically convex subset of a k-affinoid space $X = \mathscr{M}(\mathscr{A})$ such that, for some valuation field $K \supset k$, the set $V' = \pi_K^{-1}(V)$ is an affinoid domain in the K-affinoid space $X' = \mathscr{M}(\mathscr{A}')$, $\mathscr{A}' = \mathscr{A} \widehat{\otimes} K$, and $\mathscr{A}'_{V'} = \mathscr{P}(V') = \mathscr{P}(V)\widehat{\otimes} K$. Then V is an affinoid domain in X, and $\mathscr{A}_V = \mathscr{P}(V)$. Indeed, let $\varphi: \mathscr{A} \to \mathscr{B}$ be a bounded homomorphism to an affinoid k-algebra $\mathscr{B}$ such that the image of $\mathscr{M}(\mathscr{B})$ in X lies in V. Then φ induces a homomorphism $\varphi': \mathscr{A}' \to$

$\mathcal{B}\widehat{\otimes}K$, and the image of $\mathcal{M}(\mathcal{B}\widehat{\otimes}K)$ in X' lies in V'. Hence φ' extends to a homomorphism $\mathcal{A}'_{V'} \to \mathcal{A}\widehat{\otimes}K$. That is, φ' is continuous with respect to the seminorm $\| \ \|_{V'}$. Since the canonical homomorphism $\mathcal{A} \to \mathcal{A}'$ is isometric, φ is continuous with respect to the seminorm $\| \ \|_V$. That is, φ extends to a homomorphism $\mathcal{P}(V) \to \mathcal{B}$. ∎

5.3. A class of affinoid subgroups of a Chevalley group. The notation introduced here will be used until the end of this section.

Suppose that $\mathcal{G}$ is a reductive Chevalley group scheme over $\mathbf{Z}$ (**[Ch]**, **[DeG]**), $\mathcal{Z}$ is the center of $\mathcal{G}$, $\mathcal{T}$ is a maximal torus in $\mathcal{G}$, $\Phi \subset \widehat{\mathcal{T}}$ is the set of roots of $\mathcal{G}$ with respect to $\mathcal{T}$, $\mathcal{N}$ is the normalizer of $\mathcal{T}$ in $\mathcal{G}$, W is the Weyl group of the root system Φ, which we identify with $\mathcal{N}(\mathbf{Z})/\mathcal{T}(\mathbf{Z})$, and $\mathcal{U}_a$ is the unipotent subgroup of $\mathcal{G}$ normalized by $\mathcal{T}$, which corresponds to the root $a \in \Phi$. We fix a system of representatives $\{n_w\}$ of elements $w \in W$ in $\mathcal{N}(\mathbf{Z})$ and a system of isomorphisms $X_b : \mathcal{U}_b \xrightarrow{\sim} \mathbf{G}_a$, $b \in \Phi$.

Next, we fix a system of simple roots in Φ and a linear ordering on the set Φ_+ (resp. Φ_-) of positive (resp. negative) roots, and we set $\mathcal{U}_+ = \prod_{a\in\Phi_+} \mathcal{U}_a$, $\mathcal{U}_- = \prod_{a\in\Phi_-} \mathcal{U}_a$. One has $A(\mathcal{U}_+) = \mathbf{Z}[X_a]_{a\in\Phi_+}$ and $A(\mathcal{U}_-) = \mathbf{Z}[X_a]_{a\in\Phi_-}$, where $A(\mathcal{H})$ denotes the algebra of regular functions on an affine scheme $\mathcal{H}$. The morphism

$$\mathcal{U}_- \times \mathcal{T} \times \mathcal{U}_+ \to \mathcal{G}$$

induced by the multiplication on $\mathcal{G}$ is an open immersion. Its image $\mathbf{\Omega}$ (the "grosse cellule" of $\mathcal{G}$) does not depend on the ordering chosen on Φ_+ and Φ_-. One has $A(\mathbf{\Omega}) = A(\mathcal{G})[d^{-1}]$, where d is an element of $A(\mathcal{G})$ which takes the value 1 on the identity of $\mathcal{G}$, and $\mathcal{G} = \bigcup_{w\in W} \mathbf{\Omega} n_w$ (see **[Ch]**, **[DeG]**, Exp. XXII).

Now let k be a non-Archimedean field, and set $G = \mathcal{G}\otimes_{\mathbf{Z}} k$, $Z = \mathcal{Z}\otimes_{\mathbf{Z}} k$, $T = \mathcal{T} \otimes_{\mathbf{Z}} k$, $N = \mathcal{N} \otimes_{\mathbf{Z}} k$, $U_a = \mathcal{U}_a \otimes_{\mathbf{Z}} k$, $U_\pm = \mathcal{U}_\pm \otimes_{\mathbf{Z}} k$, and $\Omega = \mathbf{\Omega} \otimes_{\mathbf{Z}} k$. We identify T^{an} with its image in G^{an}, and we use the notation in Example 5.2.12. We have the following natural action of the group $N(k)$ on $\Lambda = \mathrm{Hom}(\widehat{T}, \mathbf{R}_+^*)$ by affine transformations. (Note that Λ is a vector space over $\mathbf{R}$.) If $n = tn_w$, where $t \in T(k)$ and $w \in W$, then $n\lambda = w(\lambda) + \psi(t)$, where $w(\lambda)$ is the image of λ under the action of w on Λ which is induced by its natural action on $\widehat{T}$. One has ${}^{n_w}\mathbf{t}_\lambda = \mathbf{t}_{w(\lambda)}$.

Let $\chi_1, \ldots, \chi_l$ be a fixed basis of $\widehat{T} = \widehat{\mathcal{T}}$. Then

$$A(\Omega) = k[\chi_i, \chi_i^{-1}, X_a]_{1\le i\le l, a\in\Phi}.$$

Therefore every element of $A(\Omega)$ can be written in the form $\sum \alpha_{\mu,\nu}\chi^\mu X^\nu$, where the sum is taken over all $\mu = (\mu_i)_{1\le i\le l}$, $\nu = (\nu_a)_{a\in\Phi}$ with $\nu_a \ge 0$ and $\chi^\mu = \prod_{1\le i\le l} \chi_i^{\mu_i}$, $X^\nu = \prod_{a\in\Phi} X_a^{\nu_a}$. Let $\mathbf{p}$ be the point of $\Omega^{\mathrm{an}} \subset G^{\mathrm{an}}$

corresponding to the following multiplicative norm on $A(\Omega)$:

$$|\sum \alpha_{\mu,\nu}\chi^{\mu}X^{\nu}| = \max|\alpha_{\mu,\nu}|,$$

and denote by $\mathbf{P}$ (resp $\mathbf{V}$) the holomorphically convex envelope of the point $\mathbf{p}$ in G^{an} (resp. Ω^{an}). One has $\mathbf{V} = \{x \in \Omega^{\mathrm{an}} | |\chi_i(x)| = 1, |X_a(x)| \le 1, 1 \le i \le l, a \in \Phi\}$ (in particular, $\mathbf{V}$ is an affinoid domain in Ω^{an}) and $T_1^{\mathrm{an}} \subset \mathbf{V} \subset \mathbf{P}$. It is clear that the point $\mathbf{p}$ (and hence $\mathbf{P}$ and $\mathbf{V}$) does not depend on the chosen basis of $\widehat{T}$.

5.3.1. PROPOSITION. (i) $\mathbf{P}$ *is an affinoid subgroup of* G^{an}; *the centralizer* (*resp. normalizer*) *of* $\mathbf{P}$ *coincides with* Z^{an} (*resp.* $Z^{\mathrm{an}}\mathbf{P}$);

(ii) *the point* $\mathbf{p}$ *is peaked, and* $\mathbf{t}_0 * \mathbf{p} = \mathbf{p} * \mathbf{t}_0 = \mathbf{p}$;

(iii) $\mathbf{P} = \bigcup_{w\in W} \mathbf{V}n_w$.

PROOF. The point $\mathbf{p}$ is peaked, by Lemma 5.2.2. The equality in (ii) follows from Proposition 5.2.10 and the fact that $\mathbf{P}$ is a subgroup, because $T_1^{\mathrm{an}} \subset \mathbf{P}$. We now remark that all of the objects that have been introduced are defined over a smaller field, namely, the closure of the field generated by the image of $\mathbf{Z}$ in k, and it suffices to verify (i) and (iii) for the smaller field. So we may assume that the valuation on k is discrete.

5.3.2. LEMMA. *Suppose that the valuation on* k *is discrete, let* $\mathscr{X} = \mathrm{Spec}(A)$ *be an affine scheme which is flat and of finite type over* k°, *and let* $X = \mathrm{Spec}(A \otimes_{k^\circ} k)$ *be its generic fibre. Suppose that the ring* $\widetilde{A} = A \otimes_{k^\circ} \widetilde{k}$ *has no zero divisors. If the valuation on* k *is nontrivial, we set* $\mathscr{A} = \varprojlim A/(\alpha^n) \otimes_{k^\circ} k$ *for some* $\alpha \in k$ *with* $0 < |\alpha| < 1$. *Otherwise, we set* $\mathscr{A} = A$ *and endow it with the trivial norm.*

(i) $\mathscr{A}$ *is a strictly* k*-affinoid algebra, and the* k*-affinoid space* $\mathscr{X}^{\mathrm{an}} = \mathscr{M}(\mathscr{A})$ *is a Weierstrass domain in* X^{an}.

(ii) $\widetilde{\mathscr{A}} = \widetilde{A}$.

(iii) *If* $\mathscr{Y} = \mathrm{Spec}(A[d^{-1}])$, *where the element* $d \in A$ *has nonzero image in* $\widetilde{A}$, *then* $\mathscr{Y}^{\mathrm{an}} = \mathscr{X}^{\mathrm{an}}(d^{-1})$.

PROOF. The case of trivial valuation is simple, and so we assume that the valuation on k is nontrivial. It is clear that $\mathscr{A}$ does not depend on α.

(i) Let $A = k^\circ[T_1, \cdots, T_n]/\mathbf{a}$. Since A is flat over k°, there is an exact sequence:

$$0 \to \mathbf{a}/(\alpha^n)\mathbf{a} \to k^\circ/(\alpha^n)[T_1, \cdots, T_n] \to A/(\alpha^n) \to 0.$$

Taking the projective limit and tensoring by k, we obtain an exact sequence:

$$0 \to \mathbf{a}k\{T_1, \cdots, T_n\} \to k\{T_1, \cdots, T_n\} \to \mathscr{A} \to 0.$$

Therefore, $\mathscr{A}$ is endowed with a strictly k-affinoid algebra structure, which is unique, by [**BGR**], 6.1.3/2. Since the image of $A \otimes_{k^\circ} k$ in $\mathscr{A}$ is everywhere dense, $\mathscr{X}^{\mathrm{an}} = \mathscr{M}(\mathscr{A})$ is a Weierstrass domain in X^{an}.

(ii) Let A' be the image of $k^\circ\{T_1, \cdots, T_n\}$ in $\mathscr{A}$. It is easy to see that $A'/\mathbf{m}A' = \tilde{A}$, where $\mathbf{m}$ is the maximal ideal in k°. Since $\mathbf{m}$ is principal, it follows that for each element $f \in \mathscr{A}$ there exists $\alpha \in k$ with $\alpha f \in A'\backslash\mathbf{m}A'$. Since the ring $\tilde{A}$ has no zero divisors, the function $f \mapsto |\alpha|^{-1}$ is a multiplicative norm on $\mathscr{A}$. Let x be the corresponding point in $\mathscr{X}^{\mathrm{an}}$. Since $\mathscr{A}^\circ \subset \{f \in \mathscr{A} \,|\, |f(x)| \le 1\} = A'$ and $\mathscr{A}^{\circ\circ} \subset \{f \in \mathscr{A} \,|\, |f(x)| < 1\} = \mathbf{m}\mathscr{A}'$, we have $\mathscr{A}^\circ = A'$ and $\mathscr{A}^{\circ\circ} = \mathbf{m}A'$, i.e., $\tilde{\mathscr{A}} = \tilde{A}$.

(iii) The ring $B = A[d^{-1}]$ is flat over k°, and $\tilde{B} = \tilde{A}[\tilde{d}^{-1}]$ has no zero divisors. Let $\mathscr{B} = \varprojlim B/(\alpha^n) \otimes_{k^\circ} k$. As in (i), from the exact sequence $0 \to (Td-1) \to A[T] \to B \to 0$ one obtains an exact sequence $0 \to (Td-1) \to \mathscr{A}\{T\} \to \mathscr{B} \to 0$. The required statement follows. ∎

We apply the lemma to the affine schemes $\mathscr{G}_{k^\circ} = \mathscr{G} \otimes_{\mathbf{Z}} k^\circ$ and $\Omega_{k^\circ} = \mathbf{\Omega} \otimes_{\mathbf{Z}} k^\circ$. It follows that $\mathscr{G}_{k^\circ}^{\mathrm{an}}$ is an affinoid subgroup of G^{an}, and $\mathbf{\Omega}_{k^\circ}^{\mathrm{an}} = \mathscr{G}_{k^\circ}^{\mathrm{an}}(d^{-1})$. It also follows that the Shilov boundary of $\mathscr{G}_{k^\circ}^{\mathrm{an}}$ coincides with that of $\Omega_{k^\circ}^{\mathrm{an}}$. Since $\Omega_{k^\circ}^{\mathrm{an}} = \mathbf{V}$ and $\mathscr{G}_{k^\circ}^{\mathrm{an}}$ is a Weierstrass domain in G^{an}, we see that $\mathscr{G}_{k^\circ}^{\mathrm{an}}$ coincides with the holomorphically convex envelope of the point $\mathbf{p}$ in G^{an}, i.e., with $\mathbf{P}$. To prove (i), we use the following commutative diagram:

$$\begin{array}{ccccccccc} 0 & \to & \mathscr{G}(k^\circ)/\mathscr{Z}(k^\circ) & \to & \mathrm{Aut}(\mathscr{G}_{k^\circ}) & \to & \mathrm{Aut\,ext}(\mathscr{G}_{k^\circ}) & \to & 0 \\ & & \downarrow & & \downarrow & & \downarrow & & \\ 0 & \to & G(k)/Z(k) & \to & \mathrm{Aut}(G) & \to & \mathrm{Aut\,ext}(G) & \to & 0 \end{array}.$$

From [**DeG**], Exp. XXIV, §1, it follows that the first two vertical homomorphisms are injective and the third one is bijective. If an element $g \in G(k)$ normalizes $\mathbf{P}$, it induces an automorphism of the scheme $\mathscr{G}_{k^\circ}$ which must be inner. That is, $g \in Z(k)\mathscr{G}(k^\circ) = Z(k)\mathbf{P}(k)$. For the same reason, if $g \in G(k)$ centralizes $\mathbf{P}$, then $g \in Z(k)$. Since this is true over all extensions of k, the statement (i) follows. The equality (iii) follows from the equality $\mathscr{G}_{k^\circ} = \bigcup_{w \in W} \Omega_{k^\circ} n_w$. ∎

By Proposition 5.3.1, we can apply the results of the previous subsection to G^{an} and its subgroup $\mathbf{P}$. First of all, Corollary 5.2.11 gives the following fact.

5.3.3. COROLLARY. *The mapping $g \mapsto g * \mathbf{p}$ induces a homeomorphism of $G^{\mathrm{an}}/\mathbf{P}$ with a closed subset of G^{an}.* ∎

Next, we apply Proposition 5.2.13 to the left action and the action by inner automorphisms of T^{an} on G^{an} and Ω^{an}. For $\lambda \in \Lambda$, we set $\mathbf{p}_\lambda = \mathbf{t}_\lambda * \mathbf{p} * \mathbf{t}_\lambda^{-1}$ and denote by $\mathbf{P}_\lambda$ (resp. $\mathbf{V}_\lambda$) the holomorphically convex envelope of the point $\mathbf{p}_\lambda$ in G^{an} (resp. Ω^{an}). For example, $\mathbf{p}_0 = \mathbf{p}$, $\mathbf{P}_0 = \mathbf{P}$, and $\mathbf{V}_0 = \mathbf{V}$.

5.3.4. COROLLARY. (i) $\mathbf{t}_\lambda\mathbf{P}$ *is a Weierstrass domain in G^{an}, and $\mathbf{t}_\lambda\mathbf{P} = \{\mathbf{t}_\lambda * \mathbf{p}\}^h = t\mathbf{P}$ for any $t \in T^{\mathrm{an}}$ with $\psi(t) = \lambda$. If $n \in N(k)$, then $n\mathbf{t}_\lambda * \mathbf{p} = \mathbf{t}_{n\lambda} * \mathbf{p}$.*

(ii) $\mathbf{P}_\lambda$ *is an affinoid subgroup of* G^{an}, *and* $\mathbf{P}_\lambda = t\mathbf{P}t^{-1} = {}^t\mathbf{P}$ *for any* $t \in T^{\text{an}}$ *with* $\psi(t) = \lambda$. *The centralizer (resp. normalizer) of* $\mathbf{P}_\lambda$ *coincides with* Z^{an} *(resp.* $Z^{\text{an}}\mathbf{P}_\lambda$*). If* $n \in N(k)$, *then* $n\mathbf{P}_\lambda n^{-1} = \mathbf{P}_{n\lambda}$.

(iii) $\mathbf{V}_\lambda$ *is an affinoid domain in* Ω^{an}, *and* $\mathbf{V}_\lambda = t\mathbf{V}_\lambda t^{-1} = {}^t\mathbf{V}_\lambda$ *for any* $t \in T^{\text{an}}$ *with* $\psi(t) = \lambda$.

PROOF. It suffices to verify that $n\mathbf{t}_\lambda\mathbf{P} = \mathbf{t}_{n\lambda}\mathbf{P}$ and that $\mathbf{P}_\lambda$ is a subgroup of G^{an}. If $n = tn_w$ for $t \in T(k)$ and $w \in W$, then:

$$n\mathbf{t}_\lambda\mathbf{P} = t^{n_w}\mathbf{t}_\lambda n_w\mathbf{P} = \mathbf{t}_{w(\lambda)+\psi(t)}\mathbf{P} = \mathbf{t}_{n\lambda}\mathbf{P}.$$

Furthermore, $\mathbf{P}_\lambda$ is obviously a subgroup if $\lambda \in \text{Hom}(\widehat{T}, |k^*|)$. In the general case we take a larger valuation field $K \supset k$ such that there exists $t' \in T(K)$ with $\psi(t') = \lambda$. Then $\pi^{-1}(\mathbf{P}_\lambda) = t'\mathbf{P}'t'^{-1}$ is a subgroup of $(G\widehat{\otimes}K)^{\text{an}}$, and hence $\mathbf{P}_\lambda$ is a subgroup of G^{an}. ∎

Finally, we apply Proposition 5.2.13 to the two-sided action of $T^{\text{an}} \times T^{\text{an}} = (T \times T)^{\text{an}}$ on G^{an} and Ω^{an}.

5.3.5. COROLLARY. *Let* $\lambda, \lambda' \in \Lambda$. *Then* $\mathbf{t}_{\lambda'}\mathbf{P}\mathbf{t}_\lambda^{-1}$ *(resp.* $\mathbf{t}_{\lambda'}\mathbf{V}\mathbf{t}_\lambda^{-1}$*) is an affinoid domain which coincides with the holomorphically convex envelope of the point* $\mathbf{t}_{\lambda'} * \mathbf{p} * \mathbf{t}_\lambda^{-1}$ *in* G^{an} *(resp.* Ω^{an}*). Moreover,* $\mathbf{t}_{\lambda'}\mathbf{P}\mathbf{t}_\lambda^{-1} = t'\mathbf{P}t^{-1}$ *(resp.* $\mathbf{t}_{\lambda'}\mathbf{V}\mathbf{t}_\lambda^{-1} = t'\mathbf{V}t^{-1}$*) for all* $t, t' \in T^{\text{an}}$ *with* $\psi(t) = \lambda$ *and* $\psi(t') = \lambda'$. ∎

5.3.6. COROLLARY. *If* $f = \sum \alpha_{\mu.\nu}\chi^\mu X^\nu \in A(\Omega)$, *then*:

$$|f(\mathbf{t}_{\lambda'} * \mathbf{p} * \mathbf{t}_\lambda^{-1})| = \max|\alpha_{\mu,\nu}|\prod_{i=1}^{l}(\lambda' - \lambda)(\chi_i)^{\mu_i}\prod_{a\in\Phi_-}\lambda'(a)^{\nu_a}\prod_{a\in\Phi_+}\lambda(a)^{\nu_a}.$$

In particular, the set $\mathbf{t}_{\lambda'}\mathbf{V}\mathbf{t}_\lambda^{-1}$ *consists of all points* $x \in \Omega^{\text{an}}$ *with* $|\chi_i(x)| = (\lambda' - \lambda)(\chi_i)$ *for* $1 \le i \le l$, $|X_a(x)| \le \lambda'(a)$ *for* $a \in \Phi_-$, *and* $|X_a(x)| \le \lambda(a)$ *for* $a \in \Phi_+$.

PROOF. First suppose that there exist $t, t' \in T(k)$ with $\psi(t) = \lambda$, $\psi(t') = \lambda'$. By Corollary 5.3.5, $|f(\mathbf{t}_{\lambda'} * \mathbf{p} * \mathbf{t}_\lambda^{-1})| = \|f\|_{t'\mathbf{V}_{t^{-1}}}$. It is easy to see that $\|f\|_{t'\mathbf{V}_{t^{-1}}}$ is equal to the right side of the formula. In the general case we take a larger valuation field $K \supset k$ such that $\lambda, \lambda' \in \text{Hom}(\widehat{T}, |K^*|)$, and we note that $\sigma_K(\mathbf{t}_{\lambda'} * \mathbf{p} * \mathbf{t}_\lambda^{-1}) = \mathbf{t}'_{\lambda'} * \mathbf{p}' * \mathbf{t}'^{-1}_\lambda$, where $\mathbf{t}'_{\lambda'}$, $\mathbf{p}'$ and $\mathbf{t}'_\lambda$ are the corresponding points in $(G\widehat{\otimes}K)^{\text{an}}$. ∎

From Corollary 5.3.6 it follows that for $f = \sum \alpha_{\mu,\nu}\chi^\mu X^\nu \in A(\Omega)$:

$$|f(\mathbf{p}_\lambda)| = \max|\alpha_{\mu.\nu}|\prod_{\alpha\in\Phi}\lambda(a)^{\nu_a}$$

and that $\mathbf{V}_\lambda = \{x \in \Omega^{\text{an}} \mid |\chi_i(x)| = 1, |X_a(x)| \le \lambda(a), 1 \le i \le l, a \in \Phi\}$. In particular, if P_λ denotes the subgroup of $G(k)$ generated by $T_1^{\text{an}}(k)$ and by all subgroups $\{x \in U_a(k) \mid |X_a(x)| \le \lambda(a)\}$, $a \in \Phi$, then $\mathbf{V}_\lambda(k) \subset P_\lambda \subset \mathbf{P}_\lambda(k)$. We remark that $nP_\lambda n^{-1} = P_{n\lambda}$ for any $n \in N(k)$.

The following theorem gives a description of the set of k-rational points of the affinoid domains $\mathbf{t}_{\lambda'}\mathbf{P}\mathbf{t}_{\lambda}^{-1}$. Before stating it, we need some notation. For $\lambda, \lambda' \in \Lambda$ we set $N(k)_{\lambda,\lambda'} = \{n \in N(k) | n\lambda = \lambda'\}$ and $N(k)_{\lambda} = N(k)_{\lambda,\lambda}$. The latter set is the stabilizer of λ in $N(k)$. It is clear that $T_1^{\mathrm{an}}(k) \subset N(k)_{\lambda}$ and that $N(k)_{\lambda}/T_1^{\mathrm{an}}(k)$ maps injectively into the Weyl group $W = N(k)/T(k)$. We denote by W_{λ} the image of $N(k)_{\lambda}$ in W. Furthermore, $N(k)_{\lambda,\lambda'}N(k)_{\lambda} = N(k)_{\lambda,\lambda'}$, and so $N(k)_{\lambda,\lambda'}$ is a union of a certain (possibly empty) family of right cosets of the subgroup $N(k)_{\lambda}$ in $N(k)$, and $N(k)_{\lambda,\lambda'}/T_1^{\mathrm{an}}(k)$ maps injectively into W. Let $W_{\lambda,\lambda'}$ denote the image of $N(k)_{\lambda,\lambda'}$ in W. We remark that the group W_{λ} does not depend on the field k, but the set $W_{\lambda,\lambda'}$ does depend on it.

5.3.7. THEOREM. *Let $\{n_{w,\lambda,\lambda'}\}$ be a system of representatives of elements $w \in W_{\lambda,\lambda'}$ in $N(k)_{\lambda,\lambda'}$. Then:*

$$(\mathbf{t}_{\lambda'}\mathbf{P}\mathbf{t}_{\lambda}^{-1})(k) = \bigcup_{w\in W_{\lambda,\lambda'}} \mathbf{V}_{\lambda'}(k)n_{w,\lambda,\lambda'} .$$

In particular, $(\mathbf{t}_{\lambda'}\mathbf{P}\mathbf{t}_{\lambda}^{-1})(k)$ is nonempty if and only if there exists $n \in N(k)$ with $\lambda' = n\lambda$.

PROOF. It suffices to verify the equality for some system of representatives. By Proposition 5.3.1(iii), we have:

$$(\mathbf{t}_{\lambda'}\mathbf{P}\mathbf{t}_{\lambda}^{-1})(k) = \bigcup_{w\in W} (\mathbf{t}_{\lambda'}\mathbf{V}n_w\mathbf{t}_{\lambda}^{-1})(k) = \bigcup_{w\in W} (\mathbf{t}_{\lambda'}\mathbf{V}\mathbf{t}_{w(\lambda)}^{-1})(k)n_w .$$

From Corollary 5.3.6 it follows that:

$$(\mathbf{t}_{\lambda'}\mathbf{V}\mathbf{t}_{w(\lambda)}^{-1})(k) \subset \{x \in \Omega(k) | |\chi(x)| = (\lambda' - w(\lambda))(\chi) \text{ for all } \chi \in \widehat{T}\} .$$

The latter set is nonempty if and only if there exists $t \in T(k)$ with $\lambda' - w(\lambda) = \psi(t)$. This means that the element $n_{w,\lambda,\lambda'} = tn_w$ lies in $N(k)_{\lambda,\lambda'}$. Since $\mathbf{t}_{w(\lambda)}^{-1}n_w = \mathbf{t}_{\lambda'}^{-1}n_{w,\lambda,\lambda'}$, we have:

$$(\mathbf{t}_{\lambda'}\mathbf{P}\mathbf{t}_{\lambda}^{-1})(k) = \bigcup_{w\in W_{\lambda,\lambda'}} (\mathbf{t}_{\lambda'}\mathbf{V}\mathbf{t}_{w(\lambda)}^{-1}n_w)(k) = \bigcup_{w\in W_{\lambda,\lambda'}} \mathbf{V}_{\lambda'}(k)n_{w,\lambda,\lambda'} . \quad \blacksquare$$

5.3.8. COROLLARY. *Let $\{n_{w,\lambda}\}$ be a system of representatives of elements $w \in W_{\lambda}$ in $N(k)_{\lambda}$. Then:*

$$\mathbf{P}_{\lambda}(k) = \bigcup_{w\in W_{\lambda}} \mathbf{V}_{\lambda}(k)n_{w,\lambda} .$$

In particular, $\mathbf{P}_{\lambda}(k) = P_{\lambda}N(k)_{\lambda}$. ∎

Corollary 5.3.8 implies that the subgroups $\mathbf{P}_{\lambda}(k)$ are the ones which are connected with the Bruhat-Tits building of G (see the following subsection). By **[BT]**, 7.3.4, for any $\lambda, \lambda' \in \Lambda$ there is a Cartan decomposition $G(k) = \mathbf{P}_{\lambda'}(k)T(k)\mathbf{P}_{\lambda}(k)$. Since the equality is true over all extensions of k, we have the equality $G^{\mathrm{an}} = \mathbf{P}_{\lambda'}T^{\mathrm{an}}\mathbf{P}_{\lambda}$. If $\lambda' = \lambda$, this implies the following fact.

5.3.9. COROLLARY. $\mathbf{P}_\lambda$ *is a maximal compact subgroup of* G^{an}. ∎

5.4. Two embeddings of the building $\mathscr{B}(G)$ in G^{an}

5.4.1. DEFINITION. The Bruhat-Tits building $\mathscr{B}(G) = \mathscr{B}(G, k)$ of G over k is the quotient set of $G(k) \times \Lambda$ by the following equivalence relation: $(g, \lambda) \sim (g', \lambda')$ if and only if there exists $n \in N(k)$ such that $\lambda' = n\lambda$ and $g^{-1}g'n \in P_\lambda$.

The group $G(k)$ acts on $\mathscr{B}(G)$ through left multiplication on the first factor. From the definition it follows that the stabilizer of the point $(1, \lambda)$ coincides with $\widehat{P}_\lambda = P_\lambda N(k)_\lambda$, i.e., with the subgroup $\mathbf{P}_\lambda(k)$ (see Corollary 5.3.8). We identify Λ with its image in $\mathscr{B}(G)$ under the injective map $\lambda \mapsto (1, \lambda)$. If the valuation on k is nontrivial, $\mathscr{B}(G)$ coincides with the building defined by Bruhat and Tits in **[BT]**, 7.4.2. Suppose that the valuation on k is trivial. Then the group $G(k)$ fixes the point $0 \in \Lambda$ corresponding to the trivial homomorphism $\widehat{T} \to \mathbf{R}_+^*$. In particular, 0 lies in all subsets $g\Lambda$, $g \in G(k)$. It is easy to see that the set of all half-lines in Λ emanating from 0 is the Coxeter complex of W. From this it follows that the set of all half-lines in $\mathscr{B}(G)$ emanating from 0 is the spherical building of G. In other words, $\mathscr{B}(G)$ is the cone over the spherical building.

The subsets of the form $g\Lambda$, $g \in G(k)$, are called *apartments* of $\mathscr{B}(G)$. By **[BT]**, 7.4.10, the group $\widehat{P}_\lambda$ (i.e., $\mathbf{P}_\lambda(k)$) acts transitively on the set of apartments containing the point $\lambda \in \Lambda$. From the Cartan decomposition it follows that, given two points $\lambda, \lambda' \in \mathscr{B}(G)$, there is an apartment containing them both.

By **[BT]**, 7.4.9, the stabilizer of Λ in $G(k)$ coincides with $N(k)$. Therefore, if one fixes a scalar product on Λ invariant under the action of the Weyl group, one can endow all apartments with a Euclidean affine space structure such that, for any $g \in G(k)$, the mapping $\lambda \mapsto g\lambda$ is an isomorphism of Euclidean affine spaces $\Lambda \xrightarrow{\sim} g\Lambda$. In particular, given two points $\lambda, \lambda' \in \mathscr{B}(G)$, one can define the distance $d(\lambda, \lambda')$ between them and the geodesic segment $[\lambda, \lambda']$ joining them. Thus, the building $\mathscr{B}(G)$ is endowed with a metric that is invariant under the action of $G(k)$ and defines its topology. If the field k is maximally complete, the metric space $\mathscr{B}(G)$ is complete (**[BT]**, 7.5.4). If the field k is local or finite, the space $\mathscr{B}(G)$ is locally compact.

5.4.2. THEOREM. *The correspondence* $(g, \lambda) \mapsto g\mathbf{t}_\lambda * \mathbf{p}$ *induces a continuous embedding*

$$\Theta: \mathscr{B}(G) \longrightarrow G^{an}$$

that commutes with the action of $G(k)$. *If the field* k *is local or finite, then* Θ *is a homeomorphism of* $\mathscr{B}(G)$ *with a closed subset of* G^{an}.

PROOF. First of all we verify that the mapping Θ is well defined. Let $\lambda' = n\lambda$, where $n \in N(k)$ and $g^{-1}g'n \in P_\lambda$. Then:

$$g'\mathbf{t}_{\lambda'}\mathbf{P} = g'n\mathbf{t}_\lambda\mathbf{P} \subset g\mathbf{P}_\lambda\mathbf{t}_\lambda\mathbf{P} = g\mathbf{t}_\lambda\mathbf{P}.$$

It follows that $g'\mathbf{t}_{\lambda'}\mathbf{P} = g\mathbf{t}_\lambda\mathbf{P}$, and hence $g'\mathbf{t}_{\lambda'} * \mathbf{p} = g\mathbf{t}_\lambda * \mathbf{p}$.

Next we verify injectivity. Suppose that $g\mathbf{t}_\lambda * \mathbf{p} = g'\mathbf{t}_{\lambda'} * \mathbf{p}$. Then $g\mathbf{t}_\lambda\mathbf{P} = g'\mathbf{t}_{\lambda'}\mathbf{P}$, and therefore $g'^{-1}g\mathbf{P}_\lambda = \mathbf{t}_{\lambda'}\mathbf{P}\mathbf{t}_\lambda^{-1}$. Since $g'^{-1}g\mathbf{P}_\lambda(k) \neq \varnothing$, it follows by Theorem 5.3.7 that there exists $n' \in N(k)$ with $\lambda' = n'\lambda$. We have $g'^{-1}g\mathbf{P}_\lambda = n'\mathbf{t}_\lambda\mathbf{P}\mathbf{t}_\lambda^{-1} = n'\mathbf{P}_\lambda$. It follows that $n'^{-1}g'^{-1}g \in \mathbf{P}_\lambda(k)$. By Corollary 5.3.8, $n'^{-1}g'^{-1}g \in n''P_\lambda$ for some $n'' \in N(k)_\lambda$. We get $n = n'n'' \in N(k)$, $\lambda' = n\lambda$ and $g^{-1}g'n \in P_\lambda$.

Next we verify that the map Θ is continuous. By Corollary 5.3.3, it suffices to show that the composition of Θ with the canonical map $G^{\mathrm{an}} \to G^{\mathrm{an}}/\mathbf{P}$ is continuous. Therefore, it suffices to show that if $\mathscr{W}$ is an open neighborhood of the coset $\mathbf{t}_\lambda\mathbf{P}$, then there exists $\varepsilon > 0$ such that the set $\Theta^{-1}(\mathscr{W})$ contains all points $\lambda' \in \mathscr{B}(G)$ with $d(\lambda, \lambda') < \varepsilon$. Since $\mathbf{P}_\lambda\mathbf{t}_\lambda\mathbf{P} = \mathbf{t}_\lambda\mathbf{P}$, there exists an open neighborhood $\mathscr{U}$ of $\mathbf{t}_\lambda\mathbf{P}$ such that $\mathbf{P}_\lambda\mathscr{U} \subset \mathscr{W}$. From Corollary 5.3.6 it follows that the restriction of the map Θ to Λ is continuous. Hence, there exists $\varepsilon > 0$ such that $\Theta^{-1}(\mathscr{U})$ contains all points $\lambda' \in \Lambda$ with $d(\lambda, \lambda') < \varepsilon$. Let λ' be a point of $\mathscr{B}(G)$ with $d(\lambda, \lambda') < \varepsilon$. Since there exists an apartment containing λ and λ', and since the group $\mathbf{P}_\lambda(k) = \widehat{P}_\lambda$ acts transitively on the set of apartments containing λ, it follows that there exist $g \in \mathbf{P}_\lambda(k)$ and $\lambda'' \in \Lambda$ such that $\lambda' = g\lambda''$. Since $g\lambda = \lambda$, we have $d(\lambda, \lambda'') = d(g\lambda, g\lambda'') = d(\lambda, \lambda') < \varepsilon$, and therefore $\lambda'' \in \Theta^{-1}(\mathscr{U})$. It follows that $\Theta(\lambda') = g\Theta(\lambda'') \in \mathbf{P}_\lambda\mathscr{U} \subset \mathscr{W}$, i.e., $\lambda' \in \Theta^{-1}(\mathscr{W})$.

From Corollary 5.3.6 it follows that the mapping $\lambda \mapsto \mathbf{t}_\lambda * \mathbf{p}$ induces a homomorphism of Λ with a closed subset of Ω^{an}. We claim that the image of Λ in G^{an} is closed. It suffices to show that there exists an open neighborhood of the set $G^{\mathrm{an}}\backslash\Omega^{\mathrm{an}}$ which does not meet $\Theta(\Lambda)$. We know that $A(\Omega) = A(G)[d^{-1}]$ for some element $d \in A(G)$ which takes the value 1 on the identity of G. Since $A(\Omega) = k[\chi_i, \chi_i^{-1}, X_a]_{1\le i\le l, a\in\Phi}$, the image of d in $A(\Omega)$ is equal to $\chi = \prod_{i=1}^l \chi_i^{n_i}$, $n_i \in \mathbf{Z}$. The set $\mathscr{U} = \{x \in G^{\mathrm{an}} \,|\, |d(x)| < 1\}$ is an open neighborhood of $G^{\mathrm{an}}\backslash\Omega^{\mathrm{an}}$. By Corollary 5.3.6, $|\chi(\mathbf{t}_\lambda * \mathbf{p})| = 1$. Hence, $\Theta(\Lambda)\cap\mathscr{U} = \varnothing$. The theorem now follows from the following lemma.

5.4.3. LEMMA. *Suppose that the field k is local or finite. Let G^{an} act on a k-analytic space X, and let $\varphi\colon \mathscr{B}(G) \to X$ be a continuous $G(k)$-equivariant embedding such that φ induces a homeomorphism of Λ with a closed subset of an open $G(k)$-invariant set $\mathscr{U} \subset X$. Then φ induces a homeomorphism of $\mathscr{B}(G)$ with a closed subset of $\mathscr{U}$.*

PROOF. First of all we note that all of the groups $\mathbf{P}_\lambda(k)$ are compact, and $\mathscr{B}(G) = \mathbf{P}_\lambda(k)\Lambda$. Since the set $\Sigma = \varphi(\Lambda) \cup (X \setminus \mathscr{U})$ is closed in X, from Proposition 5.1.1(iii) it follows that the set $\mathbf{P}(k)\Sigma$ is also closed in X, and therefore the set $\varphi(\mathscr{B}(G)) = \mathbf{P}(k)\Sigma \cap \mathscr{U}$ is closed in $\mathscr{U}$.

Now let $\lambda \in \Lambda$ and $\varepsilon > 0$. We must show that there exists an open neighborhood $\mathscr{W}$ of $\varphi(\lambda)$ in $\mathscr{U}$ such that $\varphi^{-1}(\mathscr{W})$ is contained in the set

$D(\lambda, \varepsilon) = \{\lambda' \in \mathscr{B}(G) | d(\lambda, \lambda') < \varepsilon\}$. To do this we take an open neighborhood $\mathscr{V}$ of $\varphi(\lambda)$ in $\mathscr{U}$ such that $\varphi^{-1}(\mathscr{V}) \cap \Lambda \subset D(\lambda, \varepsilon)$, and we set

$$\mathscr{W} = \bigcap_{g \in \mathbf{P}_\lambda(k)} g\mathscr{V} .$$

The set $\mathscr{W}$ is open, because its complement in X coincides with the set $\mathbf{P}_\lambda(k)(X \setminus \mathscr{V})$, which is closed, by Proposition 5.1.1(iii). Let $\lambda' \in \varphi^{-1}(\mathscr{W})$. Then there exists $g \in \mathbf{P}_\lambda(k)$ and $\lambda'' \in \Lambda$ with $\lambda' = g\lambda''$. The point λ'' lies in $\varphi^{-1}(\mathscr{W})$, because $\mathscr{W}$ is invariant under $\mathbf{P}_\lambda(k)$. Hence, $d(\lambda, \lambda'') < \varepsilon$. We have $d(\lambda, \lambda') = d(g\lambda, g\lambda'') = d(\lambda, \lambda'') < \varepsilon$. ∎

5.4.4. THEOREM. *Suppose that G is semisimple. Then the correspondence $(g, \lambda) \mapsto g\mathbf{p}_\lambda g^{-1}$ induces a continuous embedding*

$$\theta \colon \mathscr{B}(G) \longrightarrow G^{\mathrm{an}}$$

that commutes with the action of $G(k)$ (here $G(k)$ acts on G^{an} by inner automorphisms). If the field k is local or finite, then θ is a homeomorphism of $\mathscr{B}(G)$ with a closed subset of G^{an}.

PROOF. If $\lambda' = n\lambda$ and $g^{-1}g'n \in P_\lambda$ for some $n \in N(k)$, then $g'\mathbf{p}_{\lambda'}g'^{-1} = g'n\mathbf{p}_\lambda(g'n)^{-1} = g\mathbf{p}_\lambda g^{-1}$. That is, the mapping θ is well defined. Now suppose that $g\mathbf{p}_\lambda g^{-1} = g'\mathbf{p}_{\lambda'}g'^{-1}$. Then $g\mathbf{t}_\lambda \mathbf{P}\mathbf{t}_\lambda^{-1}g^{-1} = g'\mathbf{t}_{\lambda'}\mathbf{P}\mathbf{t}_{\lambda'}^{-1}g'^{-1}$. It follows that $(\mathbf{t}_{\lambda'}^{-1}g'^{-1}g\mathbf{t}_\lambda)\mathbf{P}(\mathbf{t}_{\lambda'}^{-1}g'^{-1}g\mathbf{t}_\lambda)^{-1} = \mathbf{P}$. By Proposition 5.3.1(i), $\mathbf{t}_{\lambda'}^{-1}g'^{-1}g\mathbf{t}_\lambda \subset \mathbf{P}$, i.e., $g'^{-1}g \in \mathbf{t}_{\lambda'}\mathbf{P}\mathbf{t}_\lambda^{-1}$. From Theorem 5.3.7 it follows that there exists $n \in N(k)$ with $\lambda' = n\lambda$, and therefore $g^{-1}g'n \in \mathbf{P}_\lambda(k)$, i.e., the mapping θ is injective.

5.4.5. LEMMA. *There is a commutative diagram:*

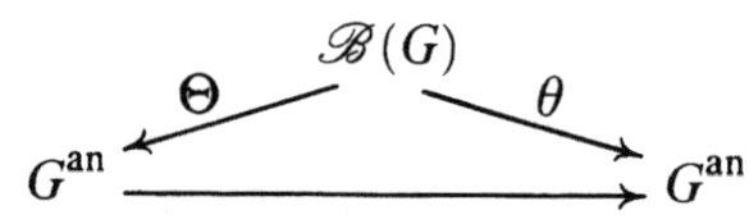

where the horizontal mapping sends a point $g \in G^{\mathrm{an}}$ to the point ${}^{g*}\mathbf{p}$ ($*$-multiplication corresponding to the action of G^{an} on G^{an} by inner automorphisms).

PROOF. It suffices to verify that ${}^{\mathbf{p}*}\mathbf{p} = \mathbf{p}$ and ${}^{\mathbf{t}_\lambda *}\mathbf{p} = \mathbf{p}_\lambda$. By Proposition 5.2.8, ${}^{\mathbf{p}}\mathbf{p} \subset \{{}^{\mathbf{p}*}\mathbf{p}\}^h$ and ${}^{\mathbf{t}_\lambda}\mathbf{p} \subset \{{}^{\mathbf{t}_\lambda *}\mathbf{p}\}^h$. The required equalities follow from the inclusions ${}^{\mathbf{p}}\mathbf{p} \subset \mathbf{P} = \{\mathbf{p}\}^h$ and ${}^{\mathbf{t}_\lambda}\mathbf{p} \subset \mathbf{P}_\lambda = \{\mathbf{p}_\lambda\}^h$. ∎

By Proposition 5.2.8(v), the horizontal map in Lemma 5.4.5 is continuous. Hence, the map θ is continuous. The fact that θ induces a homeomorphism of Λ with a closed subset of G^{an} is verified in the same way as the corresponding fact for the map Θ. Therefore, the theorem follows from Lemma 5.4.3. ∎

5.4.6. REMARK. For a point $\lambda \in \mathscr{B}(G)$ we set $\mathbf{P}_\lambda = g\mathbf{P}_{g^{-1}\lambda}g^{-1}$, where g is an element of $G(k)$ such that $g^{-1}\lambda \in \Lambda$. The affinoid subgroup $\mathbf{P}_\lambda$ is a maximal compact subgroup of G^{an}. The stabilizer of the point λ in $G(k)$ coincides with $\mathbf{P}_\lambda(k)$. According to results of Bruhat and Tits ([**BT**], 8.2.1), if the field k is maximally complete, then any relatively compact subgroup of $G(k)$ is contained in the stabilizer of some point $\lambda \in \mathscr{B}(G)$, i.e., in the subgroup $\mathbf{P}_\lambda(k)$. But it is not true that any compact subgroup of G^{an} is contained in the subgroup $\mathbf{P}_\lambda$ for some point $\lambda \in \mathscr{B}(G)$. Indeed, let K be a finite Galois extension of k such that the fixed point set of the Galois group $\mathrm{Gal}(K/k)$ in $\mathscr{B}(G, K)$ is larger than $\mathscr{B}(G)$ (see the example below). Let λ' be such a point in $\mathscr{B}(G, K)$, and let $\mathbf{P}'_{\lambda'}$ be the corresponding subgroup of $(G\widehat{\otimes}K)^{\mathrm{an}}$. Then ${}^\sigma\mathbf{P}'_{\lambda'} = \mathbf{P}'_{\lambda'}$ for any $\sigma \in \mathrm{Gal}(K/k)$, and therefore the set $H = \pi_K(\mathbf{P}'_{\lambda'})$ is a compact subgroup of G^{an}. By construction, H is not contained in any subgroup $\mathbf{P}_\lambda$ for $\lambda \in \mathscr{B}(G)$. Here is an example of such a situation (from [**Ti**], 2.6.1). Let $G = \mathrm{SL}_2$, $k = \mathbf{Q}_2$ and let K' be a totally ramified Galois extension of k different from k. Then $\mathscr{B}(G)$ (and $\mathscr{B}(G, K)$) is a tree such that all of its vertices have exactly three adjacent edges. Let λ be a point of Λ which is a vertex in $\mathscr{B}(G, K)$ but not in $\mathscr{B}(G)$. Then there exists exactly one adjacent edge of λ in $\mathscr{B}(G, K)$ not contained in $\mathscr{B}(G)$. It is clear that all points of this edge are fixed under $\mathrm{Gal}(K/k)$.

5.5. Embeddings of the building $\mathscr{B}(G)$ into compact homogeneous spaces $(G/H)^{\mathrm{an}}$. In this subsection we suppose that the group G is semisimple.

Let H be a parabolic subgroup of G which contains the Borel subgroup TU_+ and is different from G. The homogeneous space G/H is a smooth projective scheme over k. It is easy to see that the k-analytic space $(G/H)^{\mathrm{an}}$ is naturally homeomorphic to the quotient space $G^{\mathrm{an}}/H^{\mathrm{an}}$ defined in §5.2. Recall that there exists a subset Ψ of the system of simple roots in Φ (which we fixed above) such that H is generated by the Borel subgroup TU_+ and by all subgroups U_a for which a is a $\mathbf{Z}$-linear combination of simple roots from Ψ. Let Ψ_- denote the set of negative roots which are not $\mathbf{Z}$-linear combinations of simple roots from Ψ. The product

$$U_{-,H} = \prod_{a\in\Psi_-} U_a$$

is a closed subscheme in U_- (the ordering in the product corresponds to the one on Φ_-). The morphism

$$U_{-,H} \times H \longrightarrow G$$

induced by the multiplication on G is an open immersion. Its image in G/H, which will be denoted by Ω_H, is an open subset isomorphic to $U_{-,H}$. We set

$$\Omega^H = \bigcap_{g\in G(k)} g\Omega_H^{\mathrm{an}}.$$

The set Ω^H is invariant under the action of $G(k)$.

5.5.1. THEOREM. (i) *The correspondence* $(g, \lambda) \mapsto g\mathbf{t}_\lambda * \mathbf{p}H^{\mathrm{an}}$ *induces a continuous embedding*

$$\Theta_H : \mathscr{B}(G) \longrightarrow (G/H)^{\mathrm{an}}$$

that commutes with the action of $G(k)$. *Its image lies in* Ω^H.

(ii) *Suppose that the field* k *is local or finite. Then* Ω^H *is open in* $(G/H)^{\mathrm{an}}$, *and* Θ_H *is a homeomorphism of* $\mathscr{B}(G)$ *with a closed subset of* Ω^H.

PROOF. (i) We note that there is a commutative diagram

$$\begin{array}{ccc} & \mathscr{B}(G) & \\ {}^{\Theta}\swarrow & & \searrow^{\Theta_H} \\ G^{\mathrm{an}} & \longrightarrow & (G/H)^{\mathrm{an}} \end{array}$$

where the lower map is the canonical one. Therefore, Θ_H is continuous and commutes with the action of $G(k)$. Furthermore, the image of the point $\mathbf{t}_\lambda * \mathbf{p}$ under Θ_H lies in Ω_H^{an}. If $f \in A(\Omega_H) = A(U_{-,H})$ is represented by a polynomial $\sum \alpha_\nu X^\nu$, $\nu = (\nu_a)_{a \in \Psi_-}$, then

$$|f(\Theta_H(\mathbf{t}_\lambda * \mathbf{p}))| = \max |\alpha_\nu| \prod_{a \in \Psi_-} \lambda(a)^{\nu_a} .$$

In particular, Θ_H induces a homeomorphism of Λ with a closed subset of Ω_H^{an}. We claim that the stabilizer of the point $\Theta_H(\lambda)$, $\lambda \in \mathscr{B}(G)$, coincides with $\mathbf{P}_\lambda(k)$.

Indeed, we may assume that $\lambda \in \Lambda$. Let g be an element of $G(k)$ for which $g\mathbf{t}_\lambda * \mathbf{p}H^{\mathrm{an}} = \mathbf{t}_\lambda * \mathbf{p}H^{\mathrm{an}}$. It follows that $g\mathbf{p}_\lambda H^{\mathrm{an}} = \mathbf{p}_\lambda H^{\mathrm{an}}$, because $\mathbf{t}_\lambda \in H^{\mathrm{an}}$. From the Cartan decomposition $G(k) = \mathbf{P}_\lambda(k) T(k) \mathbf{P}_\lambda(k)$ it follows that $g = p' t p''$ for some p', $p'' \in \mathbf{P}_\lambda(k)$ and $t \in T(k)$. Since $p'\mathbf{p}_\lambda = p''\mathbf{p}_\lambda = \mathbf{p}_\lambda$, we have $t\mathbf{p}_\lambda H^{\mathrm{an}} = \mathbf{p}_\lambda H^{\mathrm{an}}$, and therefore $\Theta_H(\lambda + \psi(t)) = \Theta_H(\lambda)$. We already know that the map Θ_H is injective on Λ. Hence, $t \in T_1^{\mathrm{an}}(k)$, i.e., $g \in \mathbf{P}_\lambda(k)$.

Suppose that $\Theta_H(\lambda) = \Theta_H(\lambda')$ for some $\lambda, \lambda' \in \mathscr{B}(G)$. By the above assertion, $\mathbf{P}_\lambda(K) = \mathbf{P}_{\lambda'}(K)$ for all larger fields $K \supset k$. It follows that $\mathbf{P}_\lambda = \mathbf{P}_{\lambda'}$. Theorem 5.4.4 implies that $\lambda = \lambda'$. Hence, the map Θ_H is injective.

The last assertion in (i) follows from the fact that points in $\Theta_H(\mathscr{B}(G))$ cannot lie in any closed k-analytic subset of $(G/H)^{\mathrm{an}}$ different from $(G/H)^{\mathrm{an}}$.

(ii) It is easy to see that the open subscheme Ω_H is invariant under the action of the group of k-rational points of the Borel subgroup $B = U_- T$. From the Iwasawa decomposition ([**BT**], 7.3.1) it follows that $G(k) = \mathbf{P}(k)B(k)$. Therefore

$$(G/H)^{\mathrm{an}} \setminus \Omega^H = \mathbf{P}(k)((G/H)^{\mathrm{an}} \setminus \Omega_H^{\mathrm{an}}) .$$

Since the group $\mathbf{P}(k)$ is compact, the latter set is closed. Note that the image of Λ is closed in Ω^H. Hence, the theorem follows from Lemma 5.4.3. ∎

5.5.2. REMARKS. (i) Suppose that the field k is local. Then the closure of the image of $\mathscr{B}(G)$ in $(G/H)^{\mathrm{an}}$ is a compactification of the building. For example, consider the group SL_n acting on n-dimensional affine space. If H is the stabilizer of the line passing through zero and the point $(1, 0, \cdots, 0)$, then $(\mathrm{SL}_n/H)^{\mathrm{an}}$ is the projective space $\mathbf{P}^{n-1}$, and the cell $\mathbf{\Omega}_H^{\mathbf{an}}$ is the complement of the hyperplane $\{x \in \mathbf{P}^{n-1} | T_1(x) = 0\}$. The set Ω^H coincides with the set Ω^n of all points of $\mathbf{P}^{n-1}$ not lying in any hyperplane defined over k. The intersection of the closure of $\mathscr{B}(\mathrm{SL}_n)$ with a hyperplane defined over k (and so isomorphic to $\mathbf{P}^{n-2}$) coincides with the closure of $\mathscr{B}(\mathrm{SL}_{n-1})$ in $\mathbf{P}^{n-2}$.

(ii) The k-analytic space Ω^n, first introduced and studied by Drinfel'd ([**Dr**]), is an $(n-1)$-dimensional non-Archimedean analog of the upper half-plane. This space is very important in the study of uniformization by discrete subgroups of $\mathrm{GL}_n(k)$ and representations of $\mathrm{GL}_n(k)$. It would be interesting to investigate whether the k-analytic spaces Ω^H play a similar role for arbitrary reductive groups.

CHAPTER 6

The Homotopy Type of Certain Analytic Spaces

6.1. Projective space and its subsets. In this section we use the construction in §5.2 to find the homotopy type of certain analytic spaces. The idea is as follows. Let a k-analytic group G act on a k-analytic space X. Suppose that there is a map $I = [0, 1] \to G: t \mapsto g_t$ such that all of the points g_t are peaked and $g_0 = e$. Then one can construct a mapping $X \times I \to X: (x, t) \mapsto g_t * x$. If it is continuous, we have a homotopy between the identity mapping and the mapping $x \mapsto g_1 * x$. In the following examples, the image of the latter mapping can be found explicitly. Thus, first of all we are interested in the cases when the mapping $(x, t) \mapsto g_t * x$ is continuous. As in the previous section, we only consider separated analytic spaces.

For $r = (r_1, \dots, r_n)$, we set $I_r^n = [0, r_1] \times \cdots \times [0, r_n]$ and denote by ℓ the mapping $I_r^n \to \mathcal{M}(k\{r^{-1}T\})$ that sends a point $t \in I_r^n$ to the point corresponding to the norm on the algebra $k\{t^{-1}T\}$. Next, for a k-analytic space X we denote by ℓ_X the mapping $X \times I_r^n \to X \times \mathcal{M}(k\{r^{-1}T\})$ that sends a point (x, t) to the point in $p^{-1}(x, t) = \mathcal{M}(\mathcal{H}(x) \widehat{\otimes} \mathcal{H}(\ell(x)))$ corresponding to the (multiplicative) norm on the algebra $\mathcal{H}(x) \widehat{\otimes} \mathcal{H}(\ell(x))$.

6.1.1. LEMMA. *The mapping* $\ell_X: X \times I_r^n \to X \times \mathcal{M}(k\{r^{-1}T\})$ *is continuous.*

PROOF. We may assume that the space $X = \mathcal{M}(\mathcal{A})$ is k-affinoid. It suffices to show that, for any $f \in \mathcal{A}\{r^{-1}T\}$ and $c > 0$, the sets $\mathcal{W}_{>c} = \{(x, t) \mid |f(\ell_X(x, t))| > c\}$ and $\mathcal{W}_{<c} = \{(x, t) \mid |f(\ell_X(x, t))| < c\}$ are open. Let $f = \sum_\nu a_\nu T^\nu$, where $a_\nu \in \mathcal{A}$ and $||a_\nu|| r^\nu \to 0$ as $|\nu| \to \infty$. Then

$$\mathcal{W}_{>c} = \bigcup_\nu \{(x, t) \mid |a_\nu(x)| t^\nu > c\},$$

and therefore $\mathcal{W}_{>c}$ is open. We next take a sufficiently large integer N such that $||a_\nu|| r^\nu < c$ for all $|\nu| > N$. Then

$$\mathcal{W}_{<c} = \bigcap_{|\nu| \le N} \{(x, t) \mid |a_\nu(x)| t^\nu < c\},$$

and therefore $\mathcal{W}_{<c}$ is open. ∎

6.1.2. COROLLARY. *Let G be a k-analytic group, and let $I_r^n \to G: t \mapsto g_t$ be a mapping such that there exist an open neighborhood $\mathcal{U}$ of its image,* *an open immersion* $\mathcal{U} \hookrightarrow \mathcal{M}(k\{r'^{-1}T\})$, *and a continuous map* $I_r^n \to I_{r'}^{n'}$ *making the diagram*

$$\begin{array}{ccc} I_r^n & \longrightarrow & \mathcal{U} \\ \downarrow & & \downarrow \\ I_{r'}^{n'} & \xrightarrow{\ell} & \mathcal{M}(k\{r'^{-1}T\}) \end{array}$$

commutative. Then for any action of G *on a* k*-analytic space* X, *the mapping* $X \times I_r^n \to X: (x, t) \mapsto g_t * x$ *is continuous.*

PROOF. Recall that $g_t * x = \mu(w_{g_t, x})$, where $w_{g_t, x}$ is the point of $G \times X$ lying in $p^{-1}(g_t, x) = \mathcal{M}(\mathcal{H}(g_t)\widehat{\otimes}\mathcal{H}(x))$ and corresponding to the norm on the algebra $\mathcal{H}(g_t)\widehat{\otimes}\mathcal{H}(x)$. This means that the mapping $(x, t) \mapsto g_t * x$ is the composition of the mappings $X \times I_r^n \to G \times X: (x, t) \mapsto w_{g_t, x}$ and $\mu: G \times X \to X$. By hypothesis and Lemma 6.1.1, the first mapping is continuous. Hence, the mapping $(x, t) \mapsto g_t * x$ is continuous. ∎

6.1.3. REMARK. In some cases one can give an explicit formula for the homotopy in Corollary 6.1.2.

(i) Let G be the n-dimensional closed disc $\mathcal{M}(k\{r^{-1}T\})$, which is an affinoid group under addition. Let g_t, $t \in I_r^n$, denote the point of G corresponding to the norm on the algebra $k\{t^{-1}T\}$ (i.e., $g_t = \ell(t)$). Then for any $x \in G$ and $f \in k\{r^{-1}T\}$ one has

$$|f(g_t * x)| = \max_i |\partial_i f(x)| t^i,$$

where ∂_i, $i = (i_1, \ldots, i_n)$, denotes the following bounded k-linear operator $k\{r^{-1}T\} \to k\{r^{-1}T\}$:

$$\partial_i\left(\sum_\nu a_\nu T^\nu\right) = \sum_\nu \binom{\nu + i}{i} a_{\nu+i} T^\nu.$$

(ii) Let G be the split torus G_m^n acting naturally on n-dimensional affine space. For $t \in I_1^n$, we denote by g_t the maximal point of the closed disc $E(1, t)$. Then $|f(g_t * x)|$ for $x \in \mathbf{A}^n$ and $f \in k[T_1, \ldots, T_n]$ is given by the same formula but with other operators $\partial_i: k[T_1, \ldots, T_n] \to k[T_1, \ldots, T_n]$:

$$\partial_i\left(\sum_\nu a_\nu T^\nu\right) = \sum_{\nu = i}^\infty \binom{\nu}{i} a_\nu T^\nu.$$

6.1.4. THEOREM. *The n-dimensional closed disc* $X = \mathcal{M}(k\{r^{-1}T\})$ *is contractible. Moreover, any subset of* X *whose complement is a union of holomorphically convex subsets is contractible. In particular, the boundary of* X *is contractible.*

PROOF. We use the construction in Remark 6.1.3(i). The point g_r is the maximal point of X. From Proposition 5.2.10 (or from the formula

6.1.3(i)) it follows that $g_r * x = g_r$ for all $x \in X$. Hence, the homotopy $X \times I \to X: (x, t) \mapsto g_{tr} * x$ is a contraction of X to the point g_r. The more general statement follows from the fact that $x \le g_t * x$ for all $x \in X$ and $t \in I_r^n$ (we write $x \le y$ if $|f(x)| \le |f(y)|$ for all $f \in k\{r^{-1}T\}$). ∎

Now we consider the right action of the split torus $T = \mathrm{G}_\mathrm{m}^{n+1}/\mathrm{G}_\mathrm{m}$ (G_m is the diagonal in $\mathrm{G}_\mathrm{m}^{n+1}$) on n-dimensional projective space: $(x_0 : x_1 : \cdots : x_n ; t_0, \ldots, t_n) \mapsto (x_0 t_0 : \cdots : x_n t_n)$. Using the notation in Remark 6.1.3(ii), one can construct a homotopy $\mathbf{P}^n \times I \to \mathbf{P}^n : x \mapsto x * g_t$ (g_t is the $g_{(t, \ldots, t)}$ in Remark 6.1.3(ii)). The point g_1 coincides with $\mathbf{t}_0$ (the maximal point of T_1^{an}) in Remark 5.2.12. Therefore, the image of the mapping $x \mapsto x * g_1$ is homeomorphic to $\mathbf{P}^n / T_1^{\mathrm{an}}$. We claim that the correspondence $x \mapsto (|T_0(x)| : \cdots : |T_n(x)|)$ induces a homeomorphism

$$\mathbf{P}^n / T_1^{\mathrm{an}} \tilde{\to} (\mathbf{R}_+^{n+1} \backslash \{0\}) / \mathbf{R}_+^* .$$

Note that the correspondence

$$(\alpha_0 : \cdots : \alpha_n) \mapsto \left(\frac{\alpha_0}{\alpha_0 + \cdots + \alpha_n}, \ldots, \frac{\alpha_n}{\alpha_0 + \cdots + \alpha_n} \right)$$

induces a homeomorphism of the latter space with the n-dimensional simplex $\Delta^n = \{(\alpha_0, \ldots, \alpha_n) \in \mathbf{R}^{n+1} | \alpha_0 + \cdots + \alpha_n = 1\}$. Since $\mathbf{P}^n \tilde{\to} \mathbf{A}^{n+1}/\mathrm{G}_\mathrm{m}^{\mathrm{an}}$, it suffices to verify that the natural action of the torus $(\mathrm{G}_\mathrm{m}^{\mathrm{an}})^{n+1}$ on $\mathbf{A}^{n+1}$ induces a homeomorphism $\mathbf{A}^{n+1}/(\mathrm{G}_{\mathrm{m},1}^{\mathrm{an}})^{n+1} \tilde{\to} \mathbf{R}_+^{n+1} : x \mapsto (|T_0(x)|, \ldots, |T_n(x)|)$. But the latter fact is obvious.

6.1.5. THEOREM. *The projective space* $\mathbf{P}^n$ *is contractible. Moreover, any subset of* $\mathbf{P}^n$ *whose complement lies in a union of closed k-analytic subsets different from* $\mathbf{P}^n$ *is contractible.*

PROOF. The first fact has already been established. To verify the second fact, it suffices to show that, for any $x \in \mathbf{P}^n$ and $t > 0$, the point $x * g_t$ does not lie in any closed k-analytic subset. Moreover, it suffices to verify the same fact for the natural action of the torus $(\mathrm{G}_\mathrm{m}^{\mathrm{an}})^n$ on $\mathbf{A}^n$. In the notation of Remark 6.1.3(ii), this follows from the following obvious fact. For any nonzero polynomial $f \in k[T_1, \ldots, T_n]$ and any $x \in \mathbf{A}^n$, there exists i with $\partial_i f(x) \neq 0$. ∎

6.1.6. COROLLARY. *The subset* Ω^H *of* $(G/H)^{\mathrm{an}}$ *constructed in* §5.5 *for a parabolic subgroup* H *of a Chevalley group* G *is contractible. In particular, the upper half-plane* Ω^n *is contractible.* ∎

Let τ denote the mapping $\mathbf{P}^n \to \mathbf{P}^n : x \mapsto x * g_1$, and let $\Delta(\mathbf{P}^n)$ denote its image (homeomorphic to the n-dimensional simplex Δ^n). Since $g_t * g_{t'} = g_{\max(t, t')}$, it follows that for any subset $U \subset \Delta(\mathbf{P}^n)$ the homotopy $\mathbf{P}^n \times I \to \mathbf{P}^n : (x, t) \mapsto x * g_t$ induces a homotopy $\Phi_U : \tau^{-1}(U) \times I \to \tau^{-1}(U)$. Moreover, the image of the mapping $x \mapsto \Phi_U(x, 1)$ coincides with U. If

U is open in $\Delta(\mathbf{P}^n)$, $\tau^{-1}(U)$ is open in $\mathbf{P}^n$. Thus, we have the following fact.

6.1.7. THEOREM. *For any open subset U of the n-dimensional simplex Δ^n there exists a smooth n-dimensional k-analytic space that is homotopy equivalent to U.* ■

6.1.8. COROLLARY. *For any differentiable manifold V there exists a smooth k-analytic space that is homotopy equivalent to V.*

PROOF. Recall that for any differentiable manifold V there exists a proper regular embedding of V into the Euclidean space $\mathbf{R}^{2n+1}$. Moreover, there exists an open tubular neighborhood U of V such that V is a strong deformation retract of U (see **[Rh]**, §3). By the above construction, there exists an open subset of $\mathbf{A}^{2n+1}$ that is homotopy equivalent to U, and therefore to V. ■

6.1.9. EXAMPLE. The k-analytic space $X = \mathbf{A}^{n+1}\backslash\{x \in \mathbf{A}^{n+1} | |T_1(x)| = \cdots = |T_{n+1}(x)| = 1\}$ is homotopy equivalent to the n-dimensional sphere S^n.

6.2. Chevalley groups and their compact homogeneous spaces

6.2.1. THEOREM. *The k-analytic space G^{an} associated with a Chevalley group G over k is contractible. Furthermore, if H is a parabolic subgroup of G, then the k-analytic space $(G/H)^{\mathrm{an}}$ is contractible.*

PROOF. First we construct a continuous map $I \to \Omega^{\mathrm{an}} \subset G^{\mathrm{an}}: t \mapsto g_t$ which satisfies the hypothesis of Corollary 6.1.2 (we use the notation in §5.3). Recall that $\Omega^{\mathrm{an}} = U_-^{\mathrm{an}} \times T^{\mathrm{an}} \times U_+^{\mathrm{an}}$. The spaces U_-^{an} and U_+^{an} are isomorphic to affine spaces (with fixed coordinate systems). Let $U_{-,t}^{\mathrm{an}}$ and $U_{+,t}^{\mathrm{an}}$ be the closed discs of radius t with center at zero. Let T_t^{an} be the closed disc of radius t with center at the identity of T. Then the set $U_{-,t}^{\mathrm{an}} \times T_t^{\mathrm{an}} \times U_{+,t}^{\mathrm{an}}$ is isomorphic to a closed disc for $t < 1$ and coincides with $\mathbf{V}$ for $t = 1$. Let g_t be its maximal point. We have $g_0 = e$ and $g_1 = \mathbf{p}$ (see §5.3). It is clear that the mapping $t \mapsto g_t$ satisfies the hypothesis of Corollary 6.1.2.

Now we verify that the space $(G/H)^{\mathrm{an}}$ is contractible. We may assume that H contains the Borel subgroup TU_+. By the Iwasawa decomposition, $G^{\mathrm{an}} = \mathbf{P}H^{\mathrm{an}}$. In particular, the canonical mapping $G^{\mathrm{an}} \to (G/H)^{\mathrm{an}}$ induces a surjection $\varphi: \mathbf{P} \to (G/H)^{\mathrm{an}}$. Consider the homotopy $(G/H)^{\mathrm{an}} \times I \to (G/H)^{\mathrm{an}}: (x, t) \mapsto g_t * x$. We claim that $g_1 * x = \varphi(\mathbf{p})$ for all $x \in (G/H)^{\mathrm{an}}$. Indeed, $x = \varphi(y)$ for some $y \in \mathbf{P}$, and, by Proposition 5.2.10, $\mathbf{p} * y = \mathbf{p}$. From Proposition 5.2.8(ii) it follows that $g_1 * x = \varphi(\mathbf{p} * y) = \varphi(\mathbf{p})$. Thus, the space $(G/H)^{\mathrm{an}}$ is contractible.

To show that G^{an} is contractible, use the homotopy $G^{\mathrm{an}} \times I \to G^{\mathrm{an}}: (x, t) \mapsto g_t * x * g_t^{-1}$. It is easy to see that $g_1 * x * g_1^{-1} = \mathbf{p} * x * \mathbf{p}$ is the maximal point of the double coset $\mathbf{P}x\mathbf{P}$. From the Cartan decomposition it follows that the space of double cosets $\mathbf{P}\backslash G^{\mathrm{an}}/\mathbf{P}$ is naturally homeomorphic to the quotient space of Λ by the action of the Weyl group W. Hence, G^{an} is contractible. ∎

6.3. Analytic tori. Let T be an algebraic torus over k of dimension n, and let K be a finite Galois extension of k such that the torus $T' = T \otimes K$ is split over K. For $0 \leq t \leq 1$, we set

$$T_t'^{\mathrm{an}} = \{x \in T'^{\mathrm{an}} \,|\, |\chi(x) - 1| \leq t \text{ for all } \chi \in \widehat{T}\}.$$

If $0 < t < 1$, $T_t'^{\mathrm{an}}$ is an affinoid subgroup of T'^{an} (the group $T_1'^{\mathrm{an}}$ was considered above). The groups $T_t'^{\mathrm{an}}$, $0 \leq t \leq 1$, are invariant under the action of the Galois group $G = \mathrm{Gal}(K/k)$. It follows that the maximal points g_t of $T_t'^{\mathrm{an}}$ are also invariant under the action of G, and $T_t^{\mathrm{an}} = (T_t'^{\mathrm{an}})^G$ are subgroups of T^{an} (affinoid if $t > 0$). By **[Bo2]**, the correspondence $T \mapsto T_1^{\mathrm{an}}$ is an equivalence between the category of algebraic tori over k and the category of k-affinoid tori (a *k-affinoid torus* is a k-affinoid group which becomes isomorphic to $(\mathbf{G}_{\mathrm{m},1}^{\mathrm{an}})^n$ over a finite separable extension of k).

Consider the homotopy $\Phi'\colon T'^{\mathrm{an}} \times I \to T'^{\mathrm{an}} : (x, t) \mapsto g_t * x$. The image of the mapping $x \mapsto g_1 * x$, which we denote by $\Delta(T')$, is naturally homeomorphic to $\mathrm{Hom}(\widehat{T}, \mathbf{R}_+^*)$. Since the points g_t are invariant under G, Φ' induces a homotopy $\Phi\colon T^{\mathrm{an}} \times I \to T^{\mathrm{an}}$. The image of the mapping $x \mapsto \Phi(x, 1)$ coincides with $\Delta(T')/G$. It is called the *skeleton of* T and is denoted by $\Delta(T)$. In particular, the space T^{an} is contractible.

Now let Γ be a torsion-free discrete subgroup of $T(k)$. Then Γ is a free abelian group of rank $\leq \mathrm{rk}(\widehat{T}(k))$, where $\widehat{T}(k) = \widehat{T}^G$. It acts discretely and freely on the k-analytic space T^{an}. Hence, the quotient space $X_\Gamma = T^{\mathrm{an}}/\Gamma$ is a k-analytic space. (If the torus T is split and the rank of Γ is maximal, i.e., equal to $\dim(T)$, X_Γ is called a *split k-analytic torus.*) It is easy to see that $\gamma\Phi(x, t) = \Phi(\gamma x, t)$ for all $\gamma \in \Gamma$. Hence the homotopy Φ induces a homotopy $\Phi_\Gamma\colon X_\Gamma \times I \to X_\Gamma$, and $\Delta(T)$ is invariant under Γ. Moreover, the image of the mapping $x \mapsto \Phi(x, 1)$ coincides with $\Delta(X_\Gamma) := \Delta(T)/\Gamma$ (the *skeleton of* X_Γ) and is naturally homeomorphic to $\mathrm{Hom}(\widehat{T}, \mathbf{R}_+^*)/\psi(\Gamma)$ (ψ is the map $T^{\mathrm{an}} \to \mathrm{Hom}(\widehat{T}, \mathbf{R}_+^*)$ in Remark 5.2.12). Note that in this situation T^{an} is a universal covering of X_Γ.

Recall that a split k-analytic torus X_Γ is algebraizable (and, hence, is associated with an abelian variety) if and only if the transcendence degree of the field of meromorphic functions on X_Γ is equal to the dimension of X_Γ (**[Ge]**, 4.4). We shall later need the following criterion for algebraizability of X_Γ (**[Ge]**, 4.5). Let $Q\colon \Gamma \times \widehat{T} \to k^*$ be the canonical $\mathbf{Z}$-bilinear pairing (note that Γ is a subgroup of $T(k) = \mathrm{Hom}(\widehat{T}, k^*)$). Then X_Γ is algebraizable if and only if there exists a homomorphism $\varphi\colon \Gamma \to \widehat{T}$ such that the $\mathbf{Z}$-bilinear pairing $\Gamma \times \Gamma \to k^* : (\gamma, \gamma') \mapsto Q(\gamma, \varphi(\gamma'))$ is symmetric and $|Q(\gamma, \varphi(\gamma))| < 1$ for all nonzero $\gamma \in \Gamma$ (*Riemann's period relations*).

A *k-analytic torus* is a k-analytic group X for which there exists a finite Galois extension K of k such that $X\widehat{\otimes}K$ is a split K-analytic torus. The universal covering of $X\widehat{\otimes}K$ is T'^{an} for some split torus T' over K. The natural action of the Galois group $G = \mathrm{Gal}(K/k)$ extends in a unique way to an action on the k-analytic space T'^{an} (and on T') such that the identity of T' is invariant under G. Then $T = T'^{G}$ is a torus over k, and there is a unique surjective homomorphism $T^{\mathrm{an}} \to X$ which gives rise to the homomorphism $T'^{\mathrm{an}} \to X\widehat{\otimes}K$ (T is called the *algebraic torus associated with* X). Since the elements $g_t \in T'^{\mathrm{an}}$, $0 \le t \le 1$, are invariant under the action of the Galois group $G = \mathrm{Gal}(K/k)$, the homotopy $\Phi'\colon X' \times I \to X'$ induces a homotopy $\Phi\colon X \times I \to X$. The image of the mapping $x \mapsto \Phi(x, 1)$ coincides with $\Delta(X')/G$ and is denoted by $\Delta(X)$ (the *skeleton of* X). We summarize these considerations in the following theorem.

6.3.1. THEOREM. *Let X be a k-analytic torus that splits over a finite Galois extension K of k, and let T be the associated algebraic torus, so that $X\widehat{\otimes}K \widetilde{\to} (T \otimes K)^{\mathrm{an}}/\Gamma$ for a discrete subgroup $\Gamma \subset T(K)$. Then the skeleton $\Delta(X)$ is a strong deformation retract of X and is naturally homeomorphic to the quotient space of the real torus $\mathrm{Hom}(\widehat{T}, \mathbf{R}_+^*)/\psi(\Gamma)$ by the action of the finite group $G = \mathrm{Gal}(K/k)$.* ∎

In the next subsection we shall show that all abelian varieties over k admit a similar description.

6.3.2. THEOREM. *Suppose that the valuation on k is nontrivial. Further suppose that we have an action of the Galois group G of a finite Galois extension K of k on the real torus $(S^1)^n$ which comes from a linear action of G on the universal covering of $(S^1)^n$. Then there exists an abelian variety A over k such that A^{an} is a k-analytic torus whose skeleton is homeomorphic to $(S^1)^n/G$.*

PROOF. Let $(S^1)^n = V/\Gamma$, where V is a vector space over R (isomorphic to $(\mathbf{R}_+^*)^n$) and Γ is a lattice of maximal rank in V. We take a positive definite symmetric bilinear form $\langle\ ,\ \rangle\colon \Gamma \times \Gamma \to \mathbf{Z}$ invariant under the action of G and an element $q \in k^*$ with $|q| < 1$. The form $\langle\ ,\ \rangle$ induces a $\mathbf{R}$-bilinear form $(\ ,\ )\colon V \times V \to \mathbf{R}_+^*$ such that $(\gamma, \gamma') = |q|^{\langle \gamma, \gamma' \rangle}$ for all $\gamma, \gamma' \in \Gamma$. In particular, we have a G-equivariant isomorphism $V \widetilde{\to} \mathrm{Hom}(\Gamma, \mathbf{R}_+^*)$. Now we take an algebraic torus T over k whose character group $\widehat{T}$ is isomorphic to Γ as a G-module (we fix an isomorphism $\Gamma \widetilde{\to} \widehat{T}$). The form $\langle\ ,\ \rangle$ induces a G-equivariant embedding $\Gamma \hookrightarrow \mathrm{Hom}(\widehat{T}, \mathbf{Z}) \hookrightarrow \mathrm{Hom}(\widehat{T}, K) = T(K)$. The second homomorphism is induced by the embedding $\mathbf{Z} \hookrightarrow K^*$, which sends 1 to q. Thus, we have a commutative diagram of G-equivariant homomorphisms:

$$\begin{array}{ccc} \Gamma & \hookrightarrow & T(K) \\ \| & & \downarrow \\ \Gamma & \hookrightarrow & \mathrm{Hom}(\widehat{T}, \mathbf{R}_+^*) \end{array} = V.$$

Let X' be the split K-analytic torus $(T \otimes K)^{\mathrm{an}}/\Gamma$. By construction, there is a G-equivariant homeomorphism $\Delta(X') \tilde{\to} V/\Gamma$. Furthermore, Riemann's period relations are satisfied. Therefore, X' is the analytification of an abelian variety A' over K. It also follows that X' comes from a k-analytic group, i.e., $A' = A \otimes K$ for some abelian variety A over k. The skeleton of A^{an} is homeomorphic to $(V/\Gamma)/G$. ∎

6.3.3. EXAMPLE. Let E be a *twisted Tate elliptic curve* over k (i.e., E is an abelian variety of dimension one that is not a Tate elliptic curve over k but becomes such a curve over a quadratic extension of k). Then E^{an} is contractible, but $(E \times E)^{\mathrm{an}}$ is homotopy equivalent to the two-dimensional sphere S^2. This follows from Theorem 6.3.1, because $\Delta(E) \approx S^1/(\pm 1)$ (contractible) and $\Delta(E \times E) \approx (S^1)^2/(\pm 1)$ (homeomorphic to S^2).

6.4. Analytic groups with potentially good reduction. First of all we recall some definitions and facts from [**Bo3**] and [**Bo4**] which are a natural continuation of those at the beginning of §4.3 (see also Remark 4.3.5). Although in [**Bo3**] and [**Bo4**] one supposes that the valuation on k is nontrivial, everything also works in the case of trivial valuation. Indeed, this case is significantly simpler.

A *formal k-analytic space* is a k-analytic space X with a fixed equivalence class of formal coverings of X by strictly affinoid domains. For such a space there is a reduction map $X \to \widetilde{X}$, where $\widetilde{X}$ is a scheme of locally finite type over $\widetilde{k}$ (see §4.3). By [**Bo4**], 4.3, the inverse image of an open affine subscheme of $\widetilde{X}$ is a (strictly) affinoid domain in X. X is said to be *distinguished* if all of the affinoid domains in the formal covering of X are distinguished.

A morphism of formal k-analytic spaces $\varphi: X \to Y$ is a morphism of k-analytic spaces for which there exist formal coverings $\{U_i\}_{i \in I}$ of X and $\{V_j\}_{j \in J}$ of Y defining formal k-analytic space structures on X and Y such that, for any $i \in I$, there exists $j \in J$ with $\varphi(U_i) \subset V_j$. Such a morphism φ determines a morphism of reductions $\widetilde{\varphi}: \widetilde{X} \to \widetilde{Y}$ which commutes with the reduction maps. The category of formal k-analytic spaces admits direct products. Hence one can define a *formal k-analytic group* as a group object in the category of formal k-analytic spaces. For a formal k-analytic group G, the reduction $\widetilde{G}$ is a group scheme over $\widetilde{k}$. By a result of Bosch [**Bo3**], 1.1, the functor from the category of formal k-analytic groups to the category of k-analytic groups is fully faithful. In particular, a k-analytic group can have at most one formal k-analytic group structure.

A k-analytic group G is said to have *good reduction* if G is a distinguished formal k-analytic group with geometrically reduced reduction. By **[Bo3]**, 1.10, any compact geometrically reduced formal k-analytic group G has *potentially good reduction.* That is, there exists a finite separable extension K of k such that $G\widehat{\otimes}K$ has good reduction.

6.4.1. REMARK. Suppose that the valuation on k is trivial. Then the analytification X^{an} of a projective scheme X over k has a canonical formal k-analytic space structure, which is defined as follows. Every open affine subscheme of finite type $\text{Spec}(A) \subset X$ determines a strictly affinoid domain $\mathcal{M}(\mathcal{A})$ in X^{an}, where A is endowed with the trivial norm. The space X^{an} is a union of such affinoid domains, because this is true for projective space. The affinoid covering obtained in this way is formal, and $\widetilde{X^{\text{an}}} = X$. If X is reduced, then X^{an} is distinguished. The correspondence $X \mapsto X^{\text{an}}$ is a functor. In particular, the analytification of an abelian variety over k has good reduction.

6.4.2. THEOREM. *A connected compact geometrically reduced formal k-analytic group G is contractible.*

PROOF. First we consider the case when G has good reduction. Let φ denote the reduction map $G \to \widetilde{G}$. Take an arbitrary open affine subscheme $\widetilde{V} \subset \widetilde{G}$ which contains the point $\widetilde{e} = \varphi(e)$, and set $V = \varphi^{-1}(\widetilde{V})$. Then V is a distinguished affinoid domain in G whose reduction is $\widetilde{V}$. By a result of Bosch **[Bo5]**, 6.3, the fibre $\varphi^{-1}(\widetilde{e})$ is isomorphic to the open disc $D(0, 1)$ in $\mathbf{A}^n$, where n is the dimension of G. If the valuation on k is trivial, this is obvious. We assume that the point e corresponds to zero under this isomorphism. We construct a mapping $I \to G: t \mapsto g_t$ as follows. If $t < 1$, then g_t is the point of $\varphi^{-1}(\widetilde{e})$ corresponding to the maximal point of the closed disc $E(0, t)$. If $t = 1$, then g_1 is the inverse image of the generic point of $\widetilde{G}$ under φ (which is unique, by Proposition 2.4.4(ii)). It is easy to see that the mapping $t \mapsto g_t$ is continuous. We need a stronger fact.

6.4.3. LEMMA. *For any k-analytic space Y, the mapping $\mu: Y\times I \to Y\times G$ which sends a point (y, t) to the point of $p^{-1}(y, t) = \mathcal{M}(\mathcal{H}(y)\widehat{\otimes}\mathcal{H}(g_t))$ corresponding to the norm on $\mathcal{H}(y)\widehat{\otimes}\mathcal{H}(g_t)$, is continuous.*

PROOF. We may assume that the space $Y = \mathcal{M}(\mathcal{B})$ is k-affinoid, and we can replace G by the k-affinoid space $V = \mathcal{M}(\mathcal{A})$. By Lemma 6.1.1 (resp. Corollary 5.2.7), the restriction of μ to the set $Y\times[0, 1[$ (resp. $Y\times\{1\}$) is continuous. Moreover, for any $y \in Y$ the restriction of μ to the set $\{y\}\times I$ is continuous. We remark that if $t < t'$, then $|f(\mu(y, t))| \le |f(\mu(y, t'))|$ for all $y \in Y$ and $f \in \mathcal{A}\widehat{\otimes}\mathcal{B}$. To prove the lemma, it suffices to show that, for any $f \in \mathcal{A}\widehat{\otimes}\mathcal{B}$ and $c > 0$, the sets $\mathcal{W}_{<c} = \{(y, t) \mid |f(\mu(y, t))| < c\}$ and $\mathcal{W}_{>c} = \{(y, t) \mid |f(\mu(y, t))| > c\}$ are open. Suppose that the intersection of $\mathcal{W}_{<c}$ (resp. $\mathcal{W}_{>c}$) with $Y\times\{1\}$ is nonempty. Since it is open in $Y\times\{1\}$, we can decrease Y and assume that $\mathcal{W}_{<c}$ (resp. $\mathcal{W}_{>c}$) contains $Y\times\{1\}$. From the above remark it follows that $Y\times I \subset \mathcal{W}_{<c}$. We claim that $Y\times[1-\varepsilon, 1] \subset$

$\mathscr{W}_{>c}$ for some $\varepsilon > 0$. Setting $\varphi(y) = \inf\{t|(y, t) \in \mathscr{W}_{>c}\}$ for $y \in Y$, we have to show that $\sup_{y\in Y} \varphi(y) < 1$. Suppose that this is not the case. Then there exists $y_0 \in Y$ such that, for any open neighborhood $\mathscr{U}$ of y_0 and $\varepsilon > 0$, there exists $y \in \mathscr{U}$ with $\varphi(y) > 1 - \varepsilon$. Since the intersection $\mathscr{W}_{>c} \cap (\{y_0\} \times I)$ is open in $\{y_0\} \times I$, we have $\varphi(y_0) < 1$. Choose t_0 such that $\varphi(y_0) < t_0 < 1$. The point (y_0, t_0) lies in the open set $\mathscr{W}_{>c} \cap (Y \times [0, 1[)$, and therefore there exist an open neighborhood $\mathscr{U}$ of y_0 and $\varepsilon > 0$ such that $\mathscr{U} \times [t_0 - \varepsilon, t_0 + \varepsilon] \subset \mathscr{W}_{>c}$. In particular, $\varphi(y) \leq t_0 - \varepsilon$ for all $y \in \mathscr{U}$. This contradicts our assumption. ∎

From Lemma 6.4.3 it follows that the homotopy $G \times I \to G: (x, t) \mapsto g_t * x$ is continuous. We claim that $g_1 * x = g_1$ for all $x \in G$. Indeed, if $x \in G(k)$, this is obvious. If x is arbitrary, we take a valuation field K over k such that there exists $x' \in G'(K) \cap \pi^{-1}(x)$, where $G' = G \widehat{\otimes} K$. Then $\sigma(g_1)$ is the inverse image of the generic point of $\widetilde{G}'$ under the reduction map for G', and so $\sigma(g_1) * x' = \sigma(g_1)$ (σ is the map $G_p \to G'$ in §5.2). Since $\pi(\sigma(g_1) * x') = g_1 * x$, we have $g_1 * x = g_1$. Thus, the homotopy $G \times I \to G$ is a contraction of G to the point g_1.

Now we consider an arbitrary group G. Since G is compact, there exists a finite Galois extension K of k such that $G' = G \widehat{\otimes} K$ has good reduction. We claim that the homotopy $G' \times I \to G': (x, t) \mapsto g_t * x$ constructed above is invariant under the Galois group $\mathscr{G} = \mathrm{Gal}(K/k)$. Indeed, it suffices to verify that ${}^{\tau}g_t = g_t$ for all $\tau \in \mathscr{G}$. Since G' has a unique formal k-analytic group structure, the reduction map $\varphi' : G' \to \widetilde{G}'$ is invariant under $\mathscr{G}$. In particular, the open set $\varphi'^{-1}(\widetilde{e}')$ and the point g_1 are invariant under $\mathscr{G}$. The set $\varphi'^{-1}(\widetilde{e}')$ is isomorphic to the open unit disc $D' = D(0, 1)$ in $\mathbf{A}^n \widehat{\otimes} K$. It is clear that the points g_t are invariant under the automorphisms of $D' = D \widehat{\otimes} K$ (as a k-analytic space) induced by the action of $\mathscr{G}$ on the right factor. Thus, the required fact follows from the following lemma.

6.4.4. LEMMA. *Let τ be an automorphism of the open unit disc $D(0, 1) \subset \mathbf{A}^n$ such that $\tau(0) = 0$. Then $\tau(g_t) = g_t$ for all $0 \leq t < 1$ (g_t is the maximal point of the closed disc $E(0, t)$).*

PROOF. The functions $f_i = \tau^*(T_i)$, $1 \leq i \leq n$, lie in $k^\circ[[T_1, \dots, T_n]]$ and take the value 0 at zero. Let $f_i \equiv \sum_{j=1}^{n} a_{i,j} T_j \pmod{\deg \geq 2}$. It is clear that the matrix $(a_{i,j})_{1 \leq i, j \leq n}$ lies in $\mathrm{GL}_n(k^\circ)$. Now we note that the points g_t are invariant under linear transformations in $\mathrm{GL}_n(k^\circ)$. Hence, we may assume that $f_i \equiv T_i \pmod{\deg \geq 2}$. It is easy to verify that for $\varphi = \sum_\nu a_\nu T^\nu \in \mathscr{O}(D(0, 1))$ one has

$$|\varphi({}^{\tau}g_t)| = |(\sum_{\nu} a_\nu f^\nu)(g_t)| = \max_{\nu} |a_\nu| t^{|\nu|}.$$

That is, $|\varphi({}^{\tau}g_t)| = |\varphi(g_t)|$. ∎

6.4.5. COROLLARY. *Suppose that the valuation on k is trivial. Then the analytification of an abelian variety over k is contractible.* ■

6.5. Abelian varieties. Let A be an abelian variety over k. We already know that A^{an} is contractible if the valuation on k is trivial. So we suppose that the valuation on k is nontrivial. To find the homotopy type of A^{an}, we need the following results of Bosch and Lütkebohmert ([BL2], §8, [BL3], §2).

(i) There exists a unique compact subgroup N of A^{an} such that (a) N is an analytic domain; (b) N is a formal k-analytic group; (c) for some finite separable extension K of k, $N\widehat{\otimes}K$ has *semiabelian reduction* (i.e., the reduction is an extension of an abelian variety by a torus).

(ii) The group N contains a unique closed k-analytic subgroup that is a k-affinoid torus. It is isomorphic to T_1^{an} for some algebraic torus T (the *algebraic torus associated with* A).

(iii) Suppose that A is *split over* k (that is, N has semiabelian reduction, and T is split). Then the closed immersion $T_1^{an} \to N$ extends to a homomorphism of k-analytic groups $T^{an} \to A^{an}$. Furthermore, let Ω be the fibre sum of N and T^{an} over T_1^{an}, i.e., $\Omega = (N \times T^{an})/T_1^{an}$, where T_1^{an} is embedded in $N \times T^{an}$ by $x \mapsto (x, x^{-1})$. Then there is a surjective homomorphism of k-analytic groups $\Omega \to A^{an}$ which extends the homomorphisms $N \to A^{an}$ and $T^{an} \to A^{an}$. Its kernel Γ is a discrete subgroup of $\Omega(k)$, i.e., $A^{an} = \Omega/\Gamma$.

First we consider the case when A is split over k. In the proof of Theorem 6.4.2 we constructed a map $I \to N: t \mapsto g_t$ that gives rise to a homotopy $N \times I \to N: x \mapsto g_t * x$. We extend this homotopy to a homotopy $\Omega \times I \to \Omega: x \mapsto g_t * x$. The image of the mapping $x \mapsto g_1 * x$, which is denoted by $\Delta(\Omega)$, is naturally homeomorphic to $T^{an}/T_1^{an} = \mathrm{Hom}(\widehat{T}, \mathbf{R}_+^*)$. In particular, Ω is contractible. Let $\psi(\Gamma)$ denote the image of Γ in $\mathrm{Hom}(\widehat{T}, \mathbf{R}_+^*)$. By Proposition 5.2.8(iii), the homotopy $\Omega \times I \to \Omega$ commutes with the action of Γ, and so it gives rise to a homotopy $A^{an} \times I \to A^{an}: x \mapsto g_t * x$. Moreover, $\Delta(\Omega)$ is invariant under the action of Γ, and the image of the mapping $A^{an} \to A^{an}: x \mapsto g_1 * x$ coincides with $\Delta(A) := \Delta(\Omega)/\Gamma$ and is naturally homeomorphic to the real torus $\mathrm{Hom}(\widehat{T}, \mathbf{R}_+^*)/\psi(\Gamma)$. Note that in this case N is a thick compact subgroup of A^{an} (and actually is the kernel of A^{an}), and $\Delta(A)$ is a group with respect to $*$-multiplication (it is isomorphic to $\mathrm{Hom}(\widehat{T}, \mathbf{R}_+^*)/\psi(\Gamma)$).

We now consider the general case. Let K be a finite Galois extension of k such that $A' = A \otimes K$ is split over K. That is, $N' = N\widehat{\otimes}K$ has semiabelian reduction and $T' = T \otimes K$ is split over K. In particular, A'^{an} is the quotient of its universal covering Ω by a discrete subgroup $\Gamma \subset \Omega(K)$. Let $I \to N': t \mapsto g_t$ be the map constructed above for N'. By the proof of Theorem 6.4.2, the points g_t are invariant under the action of the Galois group $G = \mathrm{Gal}(K/k)$. Hence, the homotopy $A'^{\mathrm{an}} \times I \to A'^{\mathrm{an}}: (x, t) \mapsto g_t * x$ gives rise to a homotopy $\Phi: A^{\mathrm{an}} \times I \to A^{\mathrm{an}}$. Moreover, the image of the mapping $x \mapsto \Phi(x, 1)$ coincides with $\Delta(A')/G$. We denote it by $\Delta(A)$ and call it the *skeleton of* A. We summarize the above considerations in the following theorem.

6.5.1. THEOREM. *Let A be an abelian variety over k. Then (in the previous notation) the skeleton $\Delta(A)$ is a strong deformation retract of A^{an} and is naturally homeomorphic to the quotient space of the real torus* $\mathrm{Hom}(\widehat{T}, \mathbf{R}_+^*)/\psi(\Gamma)$ *by the action of the finite group G.* ■

Let τ denote the retraction $A^{\mathrm{an}} \to \Delta(A): x \mapsto g_1 * x$. Since $\Phi(\Phi(x, t), t') = \Phi(x, \max(t, t'))$ for any open subset $U \subset \Delta(A)$, the homotopy Φ induces a homotopy $\Phi_U: \tau^{-1}(U) \times I \to \tau^{-1}(U)$. Therefore A^{an} can be covered by a finite number of open contractible subsets, because the space $\Delta(A)$ is locally contractible. We obtain the following fact.

6.5.2. COROLLARY. *The analytification A^{an} of an abelian variety A over k has a universal covering which is a k-analytic group. Moreover, the fundamental group of A^{an} is a free abelian group whose rank is equal to the dimension of the maximal split subtorus of the algebraic torus associated with A.* ■

6.5.3. COROLLARY. *Let A be an abelian variety over k. The first Betti number of A^{an} is equal to the dimension of A if and only if A^{an} is a split k-analytic torus.* ■

6.6. Weak local contractibility of the two-dimensional closed disc. A topological space X is said to be *weakly locally contractible* if, for any point $x \in X$ and for any open neighborhood $\mathscr{U}$ of x, there exists an open neighborhood $\mathscr{V}$ of x in $\mathscr{U}$, which can be contracted in $\mathscr{U}$ to a point.

6.6.1. THEOREM. *The two-dimensional closed disc $X = \mathscr{M}(k\{r^{-1}T\})$ is weakly, locally contractible.*

PROOF. Let $x \in X$. If x is the maximal point of X (i.e., x corresponds to the norm on $k\{r^{-1}T\}$), then every open neighborhood of x contains an open neighborhood of the form $\{y \in X \,|\, |f_i(y)| > r_i,\ 1 \le i \le n\}$. By Theorem 6.1.4, the latter set is contractible. So we assume that x is not the maximal point of X. Consider the homotopy $\Phi: X \times I \to X$ for which $|f(\Phi(y, t))| = \max_i\{|\partial_i f(y)| t^{|i|}\}$ (see Remark 6.1.3(i)). It contracts X to its maximal point. Let $\mathscr{U}$ be an open neighborhood of the point x. We take $0 < \alpha < 1$ such that $\Phi(x, \alpha) \ne x$ and $\Phi(x, t) \in \mathscr{U}$ for all $t \le \alpha$. Let f be a function on $k\{r^{-1}T\}$ with $|f(x)| < \beta := |f(\Phi(x, \alpha))|$, and let $\mathscr{V}$ be the connected component of the open set $\{y \in \mathscr{U} \,|\, |f(y)| < \beta\}$, which contains the point x. Consider the map ${}^*f: X \to \mathbf{A}^1$ corresponding to the function f, and set $\Sigma = \{y \in X \,|\, {}^*f(y) = {}^*f(x)\}$. From Theorem 4.2.1 it follows that for any point $y \in \mathscr{V}$ there exists a unique $0 < \gamma < 1$ such that $\Phi(y, \gamma) \in \Sigma$. It is easy to see that the mapping $\mathscr{V} \to \Sigma: y \mapsto \Phi(y, \gamma)$ is continuous. We now note that the set Σ is the spectrum of a one-dimensional $\mathscr{H}({}^*f(x))$-affinoid algebra and so is locally contractible, by Corollary 4.3.3. Hence, we can take a smaller open neighborhood $\mathscr{W}$ of x in $\mathscr{V}$ such that the image of $\mathscr{W}$ under the mapping $y \mapsto \Phi(y, \gamma)$ is contractible. The required statement follows. ∎

CHAPTER 7

Spectral Theory

In this section all homomorphisms between algebras with identity are assumed to send the identity to the identity.

7.1. The spectrum and the resolvent. Let B be a Banach space over k, and let $X = \mathcal{M}(\mathcal{A})$ be a k-affinoid space. For an open subset $\mathcal{U} \subset X$, we set:

$$\Gamma(\mathcal{U}, \mathcal{O}_X(B)) = \varprojlim B\widehat{\otimes}\mathcal{A}_V,$$

where the limit is taken over all special subsets $V \subset \mathcal{U}$. The correspondence $\mathcal{U} \mapsto \Gamma(\mathcal{U}, \mathcal{O}_X(B))$ is a sheaf. Indeed, it suffices to verify that, if $\mathcal{V} = \{V_i\}_{i \in I}$ is a finite affinoid covering of X, then the complex $C^{\cdot}(\mathcal{V})\widehat{\otimes}B$ is exact. This follows from the fact that the Čech complex $C^{\cdot}(\mathcal{V})$ is exact and admissible (Proposition 2.2.5), and the tensor product functor is exact on the category of Banach spaces (with admissible linear operators as morphisms). Every element $f \in \Gamma(\mathcal{U}, \mathcal{O}_X(B))$ induces a function which associates to a point $x \in X$ the element $f(x)$ of the Banach space $B\widehat{\otimes}\mathcal{H}(x)$ over $\mathcal{H}(x)$. In the case when X is reduced, the elements of $\Gamma(\mathcal{U}, \mathcal{O}_X(B))$ can be characterized as functions f of the above type which satisfy the following condition. For any point $x \in \mathcal{U}$, there exist an affinoid neighborhood V of x in $\mathcal{U}$ and an element $f_V \in B\widehat{\otimes}\mathcal{A}_V$ such that $f_V(y) = f(y)$ for all $y \in V$. The sheaf $\mathcal{O}_X(B)$ is naturally defined now for an arbitrary k-analytic space X; it is called the *sheaf of analytic functions on* X *with values in* B.

7.1.1. DEFINITION. (i) Let $\mathcal{A}$ be a Banach k-algebra with identity. The *spectrum* of an element $f \in \mathcal{A}$ is the set Σ_f of points $x \in \mathbf{A}^1$ such that $1 \otimes T(x) - f \otimes 1$ is noninvertible in the Banach algebra $\mathcal{A}\widehat{\otimes}\mathcal{H}(x)$. The function $x \mapsto R_f(x) = (1 \otimes T(x) - f \otimes 1)^{-1} \in \mathcal{A}\widehat{\otimes}\mathcal{H}(x)$ on $\mathbf{A}^1 \backslash \Sigma_f$ is called the *resolvent* of f.

(ii) Let B be a Banach space over k, and let $A\colon B \to B$ be a bounded linear operator. By the *spectrum* Σ_A and the *resolvent* R_A of A we mean the spectrum and resolvent of A considered as an element of the Banach k-algebra $\mathrm{End}(B)$ of bounded linear operators on B.

7.1.2. THEOREM. *Let* $\mathcal{A}$ *be a Banach* k*-algebra with identity, and let* $f \in \mathcal{A}$. *Then*:

(i) *the spectrum* Σ_f *is nonempty and compact in* $\mathbf{A}^1$;

(ii) *the radius of the minimal closed disc with center at zero which contains* Σ_f *is equal to* $\rho(f)$;

(iii) *the resolvent* $x \mapsto R_f(x)$ *is an analytic function on the set* $\mathbf{P}^1 \backslash \Sigma_f$ *which is equal to zero at infinity.*

PROOF. Let $x \in \mathbf{A}^1 \backslash \Sigma_f$. Then there exists an element $\varphi \in \mathscr{A} \widehat{\otimes} \mathscr{H}(x)$ such that $(1 \otimes T(x) - f \otimes 1)\varphi = 1$. Since the subfield of $\mathscr{H}(x)$ generated by the image of $k[T]$ is everywhere dense, and $\mathscr{A} \otimes \mathscr{H}(x)$ is everywhere dense in $\mathscr{A} \widehat{\otimes} \mathscr{H}(x)$, it follows that there exist an affinoid neighborhood V of x and an element $\psi \in \mathscr{A} \widehat{\otimes} \mathscr{A}_V$ such that $\|\psi(x) - \varphi\| < \|1 \otimes T(x) - f \otimes 1\|^{-1}$. Then $\|(1 \otimes T(x) - f \otimes 1)\psi(x) - 1\| = \|(1 \otimes T(x) - f \otimes 1)(\psi(x) - \varphi)\| < 1$. Therefore, we can decrease the affinoid neighborhood V so that $\|(1 \otimes T - f \otimes 1)\psi - 1\|_V < 1$. From this it follows that the element $(1 \otimes T - f \otimes 1)\psi$ is invertible in the algebra $\mathscr{A} \widehat{\otimes} \mathscr{A}_V$, and hence the element $1 \otimes T - f \otimes 1$ is invertible in this algebra. In particular, the set Σ_f is closed in $\mathbf{A}^1$, and the resolvent $x \mapsto R_f(x)$ is analytic on $\mathbf{A}^1 \backslash \Sigma_f$.

Now let $V = \mathbf{P}^1 \backslash D(0, r)$. If $r > \rho(f)$, then the *Neumann series* $g = \sum_{j=1}^{\infty} f^{j-1} \otimes T^{-j}$ belongs to the algebra $\mathscr{A} \widehat{\otimes} \mathscr{A}_V$ and satisfies the equality $(1 \otimes T - f \otimes 1)g = 1$. From this it follows that Σ_f is compact in $\mathbf{A}^1$, and the resolvent is analytic on $\mathbf{P}^1 \backslash \Sigma_f$ and equal to zero at infinity. This implies that Σ_f is nonempty, because analytic functions on $\mathbf{P}^1$ are constant. The statement (iii) follows from the fact that the radius of convergence of the Neumann series is equal to $\rho(f)^{-1}$. ∎

7.1.3. EXAMPLE. We consider the operator with "empty" spectrum constructed by Vishik in [Vi], §9. Suppose that k is algebraically closed. Let B be the Banach space of sequences $\{a_i\}_{i \in \mathbf{Z}}$, $a_i \in k$, with $|a_i| = O(1)$ as $i \to \infty$ and $|a_i| \to 0$ as $i \to -\infty$, endowed with the norm $\|\{a_i\}\| = \sup |a_i|$. Let $A\colon B \to B$ be the operator sending a sequence $\{a_i\}$ to $\{b_i\}$, where $b_i = a_{i+1}$. We claim that $\Sigma_A = \{x\}$, where $x = p(E(0, 1))$. Indeed, since $\|A\| = 1$, the resolvent R_A is analytic on $\mathbf{P}^1 \backslash E(0, 1)$. Next, let $a \in k$ with $|a| \le 1$. The operator $(A - a)^{-1}$ sends a sequence $\{a_i\}$ to $\{b_i\}$, where $b_i = a_{i-1} + a \cdot a_{i-2} + a^2 \cdot a_{i-3} + \cdots$. Hence, $\|(A - a)^{-1}\| = 1$. From this it follows that the resolvent R_A is analytic on $D(0, 1)$. Thus, $R_A \in \Gamma(\mathbf{P}^1 \backslash \{x\}, \mathscr{O}_{\mathbf{P}^1}(\mathrm{End}(B)))$, and therefore $\Sigma_A = \{x\}$.

Note that if $\varphi\colon \mathscr{A} \to \mathscr{B}$ is a bounded homomorphism of Banach k-algebras with identity, then $\Sigma_{\varphi(f)} \subset \Sigma_f$ for all $f \in \mathscr{A}$.

7.1.4. PROPOSITION. *Let* $\mathscr{A}$ *be a commutative Banach* k*-algebra with identity.*

(i) *The spectrum* Σ_f *of an element* $f \in \mathscr{A}$ *coincides with the image of the map* ${}^*f\colon \mathscr{M}(\mathscr{A}) \to \mathbf{A}^1$ *induced by the homomorphism* $k[T] \to \mathscr{A}$ *which sends* T *to* f.

(ii) *The canonical map*

$$\mathcal{M}(\mathcal{A}) \to \prod_{f\in\mathcal{A}} \Sigma_f : x \mapsto \{{}^*f(x)\}_{f\in\mathcal{A}}$$

induces a homeomorphism of $\mathcal{M}(\mathcal{A})$ *with its image.*

PROOF. We set $\Sigma'_f = {}^*f(\mathcal{M}(\mathcal{A}))$. Let $x' \in \mathcal{M}(\mathcal{A})$ and $x = {}^*f(x')$. Then there is a commutative diagram:

$$\begin{array}{ccc} k[T] & \xrightarrow{T\mapsto f} & \mathcal{A} \\ \downarrow & & \downarrow \\ \mathcal{H}(x) & \longrightarrow & \mathcal{H}(x') \end{array}$$

This implies that the image of the element $1 \otimes T(x) - f \otimes 1$ under the induced character $\mathcal{A}\widehat{\otimes}\mathcal{H}(x) \to \mathcal{H}(x')$ is zero. In particular, this element is noninvertible in $\mathcal{A}\widehat{\otimes}\mathcal{H}(x)$, i.e., $\Sigma'_f \subset \Sigma_f$. Conversely, if $x \in \Sigma_f$, then we have a commutative diagram of the same type, and we find that $x \in \Sigma'_f$.

(ii) Since both spaces are Hausdorff and compact, it suffices to verify the injectivity of the map. Suppose that the images of two characters $x, y \in \mathcal{M}(\mathcal{A})$ coincide. Then $\Sigma_{f(x)} = \Sigma_{f(y)}$ for all $f \in \mathcal{A}$. By Theorem 7.1.2(ii), the radius of the minimal closed disc with center at zero which contains $\Sigma_{f(x)}$ (resp. $\Sigma_{f(y)}$) is equal to $\rho(f(x)) = |f(x)|$ (resp. $\rho(f(y)) = |f(y)|$). Hence, $|f(x)| = |f(y)|$ for all $f \in \mathcal{A}$, i.e., $x = y$. ∎

7.1.5. COROLLARY. *Let* Σ *be a compact subset of* $\mathbf{A}^1$, *and let* f *be the image of the coordinate function on* $\mathbf{A}^1$ *in the Banach* k*-algebra* $\mathcal{P}(\Sigma) \subset \mathcal{R}(\Sigma) \subset \mathcal{H}(\Sigma)$. *Then* $\Sigma_f(\mathcal{H}(\Sigma)) = \Sigma$, $\Sigma_f(\mathcal{R}(\Sigma)) = \Sigma^m$, *and* $\Sigma_f(\mathcal{P}(\Sigma)) = \Sigma^h$. ∎

7.1.6. PROPOSITION. *Let* $\mathcal{A}$ *be a Banach* k*-algebra with identity, and let* $f \in \mathcal{A}$. *Let* K *be a non-Archimedean field over* k, *and let* π *be the canonical map* $\mathbf{A}^1_K = \mathbf{A}^1\widehat{\otimes}K \to \mathbf{A}^1$. *Then the spectrum* $\Sigma_{f,K}$ *of the element* $f \otimes 1$ *of* $\mathcal{A}\widehat{\otimes}K$ *coincides with* $\pi^{-1}(\Sigma_f)$.

We say that a subalgebra B of an algebra A with identity is *saturated* if B contains the identity of A and if any element of B invertible in A is invertible in B. For example, maximal commutative subalgebras of A are saturated.

7.1.7. LEMMA. *In the situation of Proposition* 7.1.6, $\mathcal{A}$ *is a saturated subalgebra of* $\mathcal{A}\widehat{\otimes}K$.

PROOF. First of all, we reduce the statement to the case when K is of countable type. Let f be an element of $\mathcal{A}$ which is invertible in $\mathcal{A}\widehat{\otimes}K$. Since $\mathcal{A} \otimes K$ is everywhere dense in $\mathcal{A}\widehat{\otimes}K$, there exists an element $\varphi \in \mathcal{A} \otimes K$ with $\|f\varphi - 1\| < 1$. Take a representation $f\varphi - 1 = \sum_{i=1}^n f_i \otimes \lambda_i$,

$f_i \in \mathscr{A}$, $\lambda_i \in K$, with $\max\{\|f_i\| \cdot |\lambda_i|\} < 1$. Let L be the closure of the subfield of K generated over k by the elements $\lambda_1, \dots, \lambda_n$. Then L is of countable type, and f is invertible in $\mathscr{A}\widehat{\otimes}L$. If $\mathscr{A}$ is saturated in $\mathscr{A}\widehat{\otimes}L$, then f is invertible in $\mathscr{A}$.

Thus, we may assume that K is of countable type. Note that the lemma is obviously true for K finite over k and for $K = K_r$, $r \notin \sqrt{|k^*|}$. We assume that K is infinite over k. Suppose that the lemma is true in the case of nontrivial valuation, and suppose that the valuation on k is trivial. Choose a positive number $r \neq 1$, and let x be a point in $\mathscr{M}(K_r\widehat{\otimes}K)$. Then the field $L = \mathscr{H}(x)$ is of countable type over k (and hence over K_r). Therefore, $\mathscr{A}\widehat{\otimes}K_r$ is a saturated subalgebra of $(\mathscr{A}\widehat{\otimes}K_r)\widehat{\otimes}_{K_r}L = \mathscr{A}\widehat{\otimes}L$. Let f be an element of $\mathscr{A}$ which is invertible in $\mathscr{A}\widehat{\otimes}K$. Then it is invertible in $\mathscr{A}\widehat{\otimes}L$. It follows that f is invertible in $\mathscr{A}\widehat{\otimes}K_r$. Since $\mathscr{A}$ is a saturated subalgebra of $\mathscr{A}\widehat{\otimes}K_r$, f is invertible in $\mathscr{A}$.

Suppose that the valuation on k is nontrivial. Let $\{v_i\}_{1\le i<\infty}$ be a linearly independent family of elements of K which generate an everywhere dense subspace. We may assume that $v_1 = 1$. From **[BGR]**, 2.6.2/4, it follows that for any $\alpha > 1$ there exists a family $\{y_i\}_{1\le i<\infty}$ of elements of K such that: (a) $y_1 = v_1 = 1$; (b) $\sum_{i=1}^n kv_i = \sum_{i=1}^n ky_i$; (c) $\max_{1\le i\le n}\{|a_i| \cdot |y_i|\} \le \alpha|\sum_{i=1}^n a_i y_i|$, $a_i \in k$. It follows that the family of elements of the form $\sum_{i=1}^n f_i \otimes y_i$, $f_i \in \mathscr{A}$, is everywhere dense in $\mathscr{A}\widehat{\otimes}K$, and $\max_{1\le i\le n}\{\|f_i\| \cdot |y_i|\} \le \alpha\|\sum_{i=1}^n f_i \otimes y_i\|$. Let f be an element of $\mathscr{A}$ which is invertible in $\mathscr{A}\widehat{\otimes}K$. Then there exists an element $\varphi = \sum_{i=1}^n f_i \otimes y_i \in \mathscr{A}\widehat{\otimes}K$ such that $\|f\varphi - 1\| < \alpha^{-1}$. We have $\|ff_1 - 1\| \le \alpha\|f\varphi - 1\| < 1$. Hence f is invertible in $\mathscr{A}$. ∎

Proof of Proposition 7.1.6. The inclusion $\Sigma_{f,K} \subset \pi^{-1}(\Sigma_f)$ is trivial. Let $x \in \Sigma_f$, and let $x' \in \pi^{-1}(x)$. Then the element $1 \otimes T(x) - f \otimes 1$ is noninvertible in $\mathscr{A}\widehat{\otimes}\mathscr{H}(x)$. Since $\mathscr{A}\widehat{\otimes}\mathscr{H}(x)$ is a saturated subalgebra of $(\mathscr{A}\widehat{\otimes}\mathscr{H}(x))\widehat{\otimes}_{\mathscr{H}(x)}\mathscr{H}(x') = \mathscr{A}\widehat{\otimes}\mathscr{H}(x')$, it follows that $1 \otimes T(x') - f \otimes 1$ is noninvertible in $\mathscr{A}\widehat{\otimes}\mathscr{H}(x') = (\mathscr{A}\widehat{\otimes}K)\widehat{\otimes}_K\mathscr{H}(x')$. This means that $x' \in \Sigma_{f,K}$. ∎

7.1.8. Remark. Suppose that k is algebraically closed and is not maximally complete. Let x be a point of type (4) in $\mathbf{A}^1$, and let f be the image of the coordinate function in the field $\mathscr{H}(x)$. By Corollary 7.1.5, $\Sigma_f = \{x\}$. Hence, $R_f \in \Gamma(\mathbf{P}^1\backslash\{x\}, \mathscr{O}_{\mathbf{P}^1}(\mathscr{H}(x)))$. Since the field $\mathscr{H}(x)$ is of countable type over k, there exists a nontrivial bounded linear functional $\mu: \mathscr{H}(x) \to k$ (**[BGR]**, 2.8.2/3). Then the function $\mu(R_f)$ belongs to $\Gamma(\mathbf{P}^1\backslash\{x\}, \mathscr{O}_{\mathbf{P}^1})$. Its restriction to k is a nonzero bounded function on k which is Krasner analytic and equal to zero at infinity (see **[Ro]**, §10). Note also that since the image of the polynomial ring $k[T]$ is everywhere dense in $\mathscr{H}(x)$, the operator of multiplication by f on $\mathscr{H}(x)$ has no nontrivial invariant subspaces.

7.2. The spectrum with respect to a subalgebra

7.2.1. PROPOSITION. *Let $\mathscr{A}$ be a Banach k-algebra with identity, and let $\mathscr{B}$ be a closed subalgebra of $\mathscr{A}$ containing the identity of $\mathscr{A}$. Then for any $f \in \mathscr{B}$ we have inclusions $\Sigma_f(\mathscr{A}) \subset \Sigma_f(\mathscr{B})$ and $\partial(\Sigma_f(\mathscr{B})) \subset \partial(\Sigma_f(\mathscr{A}))$.*

PROOF. The first inclusion is trivial. Let $x \in \partial(\Sigma_f(\mathscr{B}))$. It suffices to show that $x \in \Sigma_f(\mathscr{A})$. Suppose that $x \notin \Sigma_f(\mathscr{A})$. Then there exists a connected affinoid neighborhood V of x such that $1 \otimes T - f \otimes 1$ is invertible in $\mathscr{A}\widehat{\otimes}\mathscr{A}_V$. In particular, setting $\varphi = R_f|_V \in \mathscr{A}\widehat{\otimes}\mathscr{A}_V$, we have $(1 \otimes T(y) - f \otimes 1)\varphi(y) = 1$ for all $y \in V$. But V contains infinitely many points in the complement of $\Sigma_f(\mathscr{B})$. For such points y we have $\varphi(y) \in \mathscr{B}\widehat{\otimes}\mathscr{H}(y)$. This implies that $\varphi(y) \in \mathscr{B}\widehat{\otimes}\mathscr{H}(y)$ for all $y \in V$. Indeed, φ induces an element of $(\mathscr{A}/\mathscr{B})\widehat{\otimes}\mathscr{A}_V$ which is equal to zero at infinitely many points of V. Hence it is equal to zero everywhere on V. It follows that $\varphi \in \mathscr{B}\widehat{\otimes}\mathscr{A}_V$. This is impossible, because $x \in \Sigma_f(\mathscr{B})$. ∎

7.2.2. COROLLARY. *The spectrum $\Sigma_f(\mathscr{B})$ is the union of $\Sigma_f(\mathscr{A})$ and a set of bounded connected components of the complement of $\Sigma_f(\mathscr{A})$.*

PROOF. Let $\mathscr{U}$ be a connected component of $\mathbf{A}^1\backslash\Sigma_f(\mathscr{A})$. By Proposition 7.2.1, each point of the boundary of $\Sigma_f(\mathscr{B}) \cap \mathscr{U}$ in $\mathscr{U}$ lies in $\partial(\Sigma_f(\mathscr{A}))$. It follows that $\Sigma_f(\mathscr{B}) \cap \mathscr{U}$ has no boundary points in $\mathscr{U}$. Since $\mathscr{U}$ is connected, we have $\Sigma_f(\mathscr{B}) \cap \mathscr{U} = \emptyset$, i.e., $\mathscr{U} \subset \Sigma_f(\mathscr{B})$. ∎

7.2.3. COROLLARY. *The polynomially convex envelope of a compact subset $\Sigma \subset \mathbf{A}^1$ coincides with the union of Σ and all the bounded connected components of the set $\mathbf{A}^1\backslash\Sigma$. In particular, Σ is polynomially convex if and only if its complement is connected.*

PROOF. Let Σ' denote the union of Σ and all the bounded connected components of $\mathbf{A}^1\backslash\Sigma$. It is compact, because its complement is the only unbounded connected component of $\mathbf{A}^1\backslash\Sigma$. From the Maximum Modulus Principle it follows that $\Sigma' \subset \Sigma^h$. So we must verify that the set Σ' is polynomially convex. Let f be the image of the coordinate function in the algebra $\mathscr{P}(\Sigma') \subset \mathscr{H}(\Sigma')$. By Corollary 7.2.2, $\Sigma_f(\mathscr{P}(\Sigma')) = \Sigma_f(\mathscr{H}(\Sigma')) = \Sigma'$. The required fact now follows from Corollary 7.1.5. ∎

7.2.4. PROPOSITION. *Let $\mathscr{A}$ be a Banach k-algebra with identity. If $\mathscr{B}$ is a maximal commutative subalgebra of $\mathscr{A}$, then $\Sigma_f(\mathscr{B}) = \Sigma_f(\mathscr{A})$ for all $f \in \mathscr{B}$. If the valuation on k is nontrivial, the same is true for any closed saturated subalgebra of $\mathscr{A}$.*

PROOF. First suppose that the valuation on k is nontrivial, and let $\mathscr{B}$ be a maximal commutative subalgebra of $\mathscr{A}$. Let $x \in \mathbf{A}^1\backslash\Sigma_f(\mathscr{A})$. Then there exist a strictly affinoid neighborhood V of x and $\varphi \in \mathscr{A}\widehat{\otimes}\mathscr{A}_V$ such that $(1 \otimes T - f \otimes 1)\varphi = 1$. We claim that $\varphi \in \mathscr{B}\widehat{\otimes}\mathscr{A}_V$. Indeed, if $y \in V_0 =$

$\mathrm{Max}(\mathscr{A}_V)$, then $(1\otimes T(y)-f\otimes 1)\varphi(y)=1$ in $\mathscr{A}\widehat{\otimes}\mathscr{H}(y)$. Since $\mathscr{H}(y)$ is finite over k, $\mathscr{B}\widehat{\otimes}\mathscr{H}(y)=\mathscr{B}\otimes\mathscr{H}(y)$ is a maximal commutative subalgebra of $\mathscr{A}\widehat{\otimes}\mathscr{H}(y)=\mathscr{A}\otimes\mathscr{H}(y)$. Hence, $\varphi(y)\in\mathscr{B}\otimes\mathscr{H}(y)$ for all $y\in V_0$. Since V_0 is everywhere dense in V, it follows that $\varphi\in\mathscr{B}\widehat{\otimes}\mathscr{A}_V$. Note that the second statement can be reduced to the case of commutative Banach algebras. In this case it is verified by the same reasoning, using the following lemma.

7.2.5. LEMMA. *Let $\mathscr{A}$ be a commutative Banach k-algebra with identity, let $\mathscr{B}$ be a closed subalgebra of $\mathscr{A}$ containing the identity, and let K be a finite extension of k. Then $\mathscr{B}$ is a saturated subalgebra of $\mathscr{A}$ if and only if $\mathscr{B}\otimes K$ is a saturated subalgebra of $\mathscr{A}\otimes K$.*

PROOF. The converse implication follows from the fact that $\mathscr{A}$ and $\mathscr{B}$ are saturated subalgebras of $\mathscr{A}\otimes K$ and $\mathscr{B}\otimes K$, respectively.

Suppose that $\mathscr{B}$ is a saturated subalgebra of $\mathscr{A}$. Using the converse implication, we can increase the field K. So we must consider the following two cases: K is a Galois extension of k, or K is purely inseparable over k. Let the element $f\in\mathscr{B}\otimes K$ be invertible in $\mathscr{A}\otimes K$. In the first case the element ${}^{\sigma}f$ is also invertible in $\mathscr{A}\otimes K$ for any $\sigma\in\mathrm{Gal}(K/k)$. So the element $\prod{}^{\sigma}f\in\mathscr{B}$ is invertible in $\mathscr{A}$. It follows that f is invertible in $\mathscr{B}\otimes K$. In the second case let n be an integer such that $\alpha^{p^n}\in k$ for all $\alpha\in K$, $p=\mathrm{char}(k)$. Then the element f^{p^n} is invertible in $\mathscr{A}$. But $f^{p^n}\in\mathscr{B}$. It follows that f is invertible in $\mathscr{B}\otimes K$. ∎

We now remark that if $\mathscr{B}$ is a maximal commutative subalgebra of $\mathscr{A}$ and $r\notin\sqrt{|k^*|}$, then $\mathscr{B}\widehat{\otimes}K_r$ is a maximal commutative subalgebra of $\mathscr{A}\widehat{\otimes}K_r$. Therefore, by Proposition 7.1.6, the general case can be reduced to the case of nontrivial valuation. ∎

7.2.6. REMARK. If the valuation on k is trivial, the second assertion of Proposition 7.2.4 is not true. This is connected with the fact that there exist nonrational affinoid domains in $\mathbf{A}^1$ (see Remark 2.6.4).

7.3. Holomorphic functional calculus. Let $f_1,\dots,f_n$ be elements of a commutative Banach k-algebra $\mathscr{B}$ with identity. Consider the homomorphism $k[T_1,\dots,T_n]\to\mathscr{B}$ sending T_i to f_i. The image of the induced map ${}^*f:\mathscr{M}(\mathscr{B})\to\mathbf{A}^n$ is called the *joint spectrum* of the set of elements $f=(f_1,\dots,f_n)$, and it is denoted by $\Sigma_{f_1,\dots,f_n}$. As in the proof of Proposition 7.1.4(i), one verifies that $\Sigma_{f_1,\dots,f_n}$ coincides with the set of points $x\in\mathbf{A}^n$ such that the ideal of the algebra $\mathscr{B}\widehat{\otimes}\mathscr{H}(x)$ generated by the elements $1\otimes T_i(x)-f_i\otimes 1$, $1\le i\le n$, is not equal to the algebra itself. The notion of the joint spectrum is a special case of the notion of the spectrum of a bounded homomorphism $f:\mathscr{A}\to\mathscr{B}$ from an affinoid k-algebra $\mathscr{A}$ to a commutative Banach k-algebra $\mathscr{B}$ with identity, which was defined in §2.2 as the image of $\mathscr{M}(\mathscr{B})$ in $\mathscr{M}(\mathscr{A})$. We will work with the more general notion.

Let $f:\mathscr{A}\to\mathscr{B}$ be a bounded homomorphism from an affinoid k-algebra

$\mathscr{A}$ to a commutative Banach k-algebra $\mathscr{B}$ with identity. It extends naturally to a homomorphism $\mathscr{A}_{(\Sigma_f)} \to \mathscr{B}$. (Recall that $\mathscr{A}_{(\Sigma_f)}$ is the localization of $\mathscr{A}$ with respect to the elements which do not vanish on Σ_f.)

7.3.1. PROPOSITION. *Let $\mathscr{B}'$ be the closed k-subalgebra of $\mathscr{B}$ containing the identity and generated by the image of $\mathscr{A}$ (resp. $\mathscr{A}_{(\Sigma_f)}$) in $\mathscr{B}$, and let f' denote the induced homomorphism $\mathscr{A} \to \mathscr{B}'$. Then $\Sigma_{f'} = \Sigma_f^h$ (resp. $\Sigma_{f'} = \Sigma_f^m$).*

PROOF. First of all, we note that $||g||_{\Sigma_f} = ||g||_{\Sigma_{f'}}$ for all elements $g \in \mathscr{A}$. Indeed,

$$||g||_{\Sigma_f} = \max\{|g(x)| \; |x \in \Sigma_f\} = \max\{|f(g)(\chi)| \; |\chi \in \mathscr{M}(\mathscr{B})\} = \rho_{\mathscr{B}}(f(g)) .$$

Similarly, $||g||_{\Sigma_{f'}} = \rho_{\mathscr{B}'}(f(g))$. But $\rho_{\mathscr{B}}(f(g)) = \rho_{\mathscr{B}'}(f(g))$. It follows that $\Sigma_{f'}^h = \Sigma_f^h$. So we must show that $\Sigma_{f'}$ is holomorphically convex in $X = \mathscr{M}(\mathscr{A})$. Let $x \in \Sigma_{f'}^h$. Then $|g(x)| \le \rho(f(g)) \le ||f(g)||$ for all $g \in \mathscr{A}$. Hence, the character χ_x can be uniquely extended to the completion of $\mathscr{A}$ with respect to the seminorm $g \mapsto ||f(g)||$. But this completion coincides with $\mathscr{B}'$. The second statement is verified similarly. ∎

If Σ is a closed subset of $X = \mathscr{M}(\mathscr{A})$, a homomorphism $\Gamma(\Sigma, \mathscr{O}_X) \to \mathscr{B}$ to a Banach k-algebra $\mathscr{B}$ is said to be *bounded* if, for any special neighborhood V of Σ, the induced homomorphism $\mathscr{A}_V \to \mathscr{B}$ is bounded.

7.3.2. THEOREM (Holomorphic Functional Calculus). *There exists a unique map which to every bounded homomorphism $f: \mathscr{A} \to \mathscr{B}$ from an affinoid k-algebra $\mathscr{A}$ to a commutative Banach k-algebra $\mathscr{B}$ with identity assigns a bounded homomorphism $\theta_f: \Gamma(\Sigma_f, \mathscr{O}_X) \to \mathscr{B}$, $X = \mathscr{M}(\mathscr{A})$, which extends f and has the following properties.*

(i) (*Composition Property*) *For any bounded homomorphism $\varphi: \mathscr{B} \to \mathscr{B}'$, the induced diagram*

$$\begin{array}{ccc} \Gamma(\Sigma_f, \mathscr{O}_X) & \xrightarrow{\theta_f} & \mathscr{B} \\ \downarrow & & \downarrow \varphi \\ \Gamma(\Sigma_{\varphi(f)}, \mathscr{O}_X) & \xrightarrow{\theta_{\varphi(f)}} & \mathscr{B}' \end{array}$$

is commutative.

(ii) (*Superposition Property*) *Let $g: \mathscr{U} \to X'$ be a morphism from an open neighborhood $\mathscr{U}$ of Σ_f to an affinoid space $X' = \mathscr{M}(\mathscr{A}')$ over k which gives rise to a bounded homomorphism $f': \mathscr{A}' \to \Gamma(\Sigma_f, \mathscr{O}_X) \xrightarrow{\theta_f} \mathscr{B}$. Then $\Sigma_{f'} = g(\Sigma_f)$, and the induced diagram*

$$\begin{array}{ccc} \Gamma(\Sigma_{f'}, \mathscr{O}_{X'}) & \xrightarrow{\theta_{f'}} & \mathscr{B} \\ \downarrow & & \| \\ \Gamma(\Sigma_f, \mathscr{O}_X) & \xrightarrow{\theta_f} & \mathscr{B} \end{array}$$

is commutative.

7.3.3. PROPOSITION (Arens-Calderon Lemma). *Let* $f: \mathscr{A} \to \mathscr{B}$ *be a bounded homomorphism from a* k*-affinoid algebra* $\mathscr{A}$ *to a commutative Banach* k*-algebra* $\mathscr{B}$ *with identity. Then for any open neighborhood* $\mathscr{U}$ *of the spectrum* Σ_f, *there exists a bounded homomorphism*

$$f': \mathscr{A}\{r_1^{-1}T_1, \dots, r_n^{-1}T_n\} \to \mathscr{B}$$

extending f *such that* $\pi(\Sigma_{f'}^h) \subset \mathscr{U}$, *where* π *is the canonical map* $X' = \mathscr{M}(\mathscr{A}\{r_1^{-1}T_1, \dots, r_n^{-1}T_n\}) \to X = \mathscr{M}(\mathscr{A})$.

PROOF. It suffices to show that, for any point $x \in X \backslash \Sigma_f$, there exists a homomorphism $f': \mathscr{A}\{r_1^{-1}T_1, \dots, r_n^{-1}T_n\} \to \mathscr{B}$ extending f such that $x \notin \pi(\Sigma_{f'}^h)$. Let $\mathbf{b}$ be the ideal of $\mathscr{B}\widehat{\otimes}\mathscr{H}(x)$ generated by the elements $f(a) \otimes 1 - 1 \otimes a(x)$, $a \in \mathscr{A}$. We claim that $\mathbf{b} = \mathscr{B}\widehat{\otimes}\mathscr{H}(x)$. Indeed, suppose that $\mathbf{b} \neq \mathscr{B}\widehat{\otimes}\mathscr{H}(x)$. Then $\bar{\mathbf{b}} \neq \mathscr{B}\widehat{\otimes}\mathscr{H}(x)$. Setting $\mathscr{B}' = \mathscr{B}\widehat{\otimes}\mathscr{H}(x)/\bar{\mathbf{b}}$, we obtain a commutative diagram:

$$\begin{array}{ccc} \mathscr{A} & \xrightarrow{f} & \mathscr{B} \\ \downarrow \chi_x & & \downarrow \\ \mathscr{H}(x) & \longrightarrow & \mathscr{B}' \end{array}$$

It follows that the image of $\mathscr{M}(\mathscr{B}')$ in X coincides with the point x, and it lies in Σ_f. But $x \notin \Sigma_f$.

We now choose an element $b = \sum_{i=1}^m (f(a_i) \otimes 1 - 1 \otimes a_i(x)) g_i \otimes \lambda_i$ lying in $\mathbf{b}$ such that $\|1 - b\| < 1$ (here $a_i \in \mathscr{A}$, $g_i \in \mathscr{B}$ and $\lambda_i \in \mathscr{H}(x)$). Let r_i be numbers with $r_i > \|g_i\|$, $1 \le i \le n$, and let $\mathscr{B}'$ denote the closed subalgebra of $\mathscr{B}$ generated over $\mathscr{A}$ by the elements g_i, $1 \le i \le n$. We consider the homomorphism $f': \mathscr{A}\{r_1^{-1}T_1, \dots, r_n^{-1}T_n\} \to \mathscr{B}$ which extends f and sends T_i to g_i. The image of f' is everywhere dense in $\mathscr{B}'$. By Proposition 7.3.1, the spectrum $\Sigma_{f'}$ is holomorphically convex in X'. Since the element b lies in $\mathscr{B}'\widehat{\otimes}\mathscr{H}(x)$, the ideal $\mathscr{B}'\widehat{\otimes}\mathscr{H}(x)$ generated by the elements $f(a) \otimes 1 - 1 \otimes a(x)$, $a \in \mathscr{A}$, coincides with $\mathscr{B}'\widehat{\otimes}\mathscr{H}(x)$. From this it follows that the character $\chi_x: \mathscr{A} \to \mathscr{H}(x)$ does not extend to a character of $\mathscr{B}'$, i.e., $x \notin \pi(\Sigma_{f'})$. ∎

PROOF OF THEOREM 7.3.2. Consider a bounded homomorphism $f: \mathscr{A} \to \mathscr{B}$ of commutative Banach k-algebras with identity, where $\mathscr{A}$ is an affinoid k-algebra. Let V be a special neighborhood of Σ_f, and let

$$f': \mathscr{A}\{r_1^{-1}T_1, \dots, r_n^{-1}T_n\} \to \mathscr{B}$$

be an extension of f such that $\pi(\Sigma_{f'}^h) \subset \mathrm{Int}(V/X)$. We have a homomorphism $\pi^*: \mathscr{A}_V \to \Gamma(\Sigma_{f'}^h, \mathscr{O}_{X'})$, and, by Corollary 2.5.16, f' extends uniquely to a bounded homomorphism $\theta_{f', \Sigma_{f'}^h}: \Gamma(\Sigma_{f'}^h, \mathscr{O}_{X'}) \to \mathscr{B}$. We set $\theta_{f,V} = \theta_{f', \Sigma_{f'}^h} \circ \pi^*: \mathscr{A}_V \to \mathscr{B}$. It is easy to verify that $\theta_{f,V}$ does not depend on the choice of f' and that all of the $\theta_{f,V}$ are compatible. Hence they induce a bounded homomorphism $\theta_f\colon \Gamma(\Sigma_f, \mathcal{O}_X) \to \mathcal{B}$. All of the properties of the homomorphisms θ_f are verified by standard arguments (see **[Bou2]**, §4). ∎

In the case of one variable, one can also consider noncommutative Banach algebras.

7.3.4. COROLLARY. *There exists a unique map which to every element f of a Banach k-algebra $\mathcal{B}$ with identity assigns a bounded homomorphism $\theta_f\colon \Gamma(\Sigma_f, \mathcal{O}_{\mathbf{A}^1}) \to \mathcal{B}$ which sends the coordinate function of $\mathbf{A}^1$ to f and has the following properties.*

(i) *For any bounded homomorphism $\varphi\colon \mathcal{B} \to \mathcal{B}'$, the induced diagram*

$$\begin{array}{ccc} \Gamma(\Sigma_f, \mathcal{O}_{\mathbf{A}^1}) & \xrightarrow{\theta_f} & \mathcal{B} \\ \downarrow & & \downarrow \varphi \\ \Gamma(\Sigma_{\varphi(f)}, \mathcal{O}_{\mathbf{A}^1}) & \xrightarrow{\theta_{\varphi(f)}} & \mathcal{B}' \end{array}$$

is commutative.

(ii) *Let $g \in \Gamma(\Sigma_f, \mathcal{O}_{\mathbf{A}^1})$, and set $f' = \theta_f(g)$. Then $\Sigma_{f'} = g(\Sigma_f)$, and the induced diagram*

$$\begin{array}{ccc} \Gamma(\Sigma_{f'}, \mathcal{O}_{\mathbf{A}^1}) & \xrightarrow{\theta_{f'}} & \mathcal{B} \\ \downarrow & & \| \\ \Gamma(\Sigma_f, \mathcal{O}_{\mathbf{A}^1}) & \xrightarrow{\theta_f} & \mathcal{B} \end{array}$$

is commutative.

PROOF. To construct the homomorphism θ_f, it is enough to replace $\mathcal{B}$ by a maximal commutative subalgebra containing the element f and then apply Proposition 7.2.4. ∎

The value of the homomorphism θ_f on an element $g \in \Gamma(\Sigma_f, \mathcal{O}_X)$ will be denoted by $g(f)$.

7.4. Partitions of the spectrum of an algebra

7.4.1. THEOREM (Shilov Idempotent Theorem). *Let $\mathcal{A}$ be a commutative Banach k-algebra with identity, and let $\mathcal{U}$ be an open and closed subset of $\mathcal{M}(\mathcal{A})$. Then there exists a unique idempotent $e \in \mathcal{A}$ which is equal to 1 on $\mathcal{U}$ and 0 outside $\mathcal{U}$.*

PROOF. By Proposition 7.1.4(ii), $\mathcal{M}(\mathcal{A})$ is identified with a closed subset of the compact space $\prod \Sigma_f$. From this it follows that there exist $f_1, \dots, f_n \in \mathcal{A}$ such that the projections of $\mathcal{U}$ and $\mathcal{M}(\mathcal{A})\backslash\mathcal{U}$ on $\prod_{i=1}^n \Sigma_{f_i}$ do not intersect. Thus, denoting by φ the canonical map $\mathcal{M}(\mathcal{A}) \to \Sigma_{f_1,\dots,f_n} \subset \mathbf{A}^n$, we have $\varphi(\mathcal{U}) \cap \varphi(\mathcal{M}(\mathcal{A})\backslash\mathcal{U}) = \varnothing$. Let $\mathcal{U}'$ and $\mathcal{U}''$ be disjoint neighborhoods of the sets $\varphi(\mathcal{U})$ and $\varphi(\mathcal{M}(\mathcal{A})\backslash\mathcal{U})$, respectively, and let g be the analytic function on $\mathcal{U}' \cup \mathcal{U}''$ which is equal to 1 on $\mathcal{U}'$ and 0 on $\mathcal{U}''$. We set $e = g(f)$. Since $g^2 = g$, we have $e^2 = e$. Moreover, e is equal to 1 on $\mathcal{U}$ and 0 outside $\mathcal{U}$. Thus, the existence of e is established. Let e' be an idempotent in $\mathcal{A}$ which has the same property. Then the element $1 - e' - e$, which is equal to -1 on $\mathcal{U}$ and 1 outside $\mathcal{U}$, is invertible in $\mathcal{A}$. But $(e' - e)(1 - e' - e) = 0$. From this it follows that $e' = e$. ∎

7.4.2. Corollary. *The following properties of a commutative Banach k-algebra $\mathcal{A}$ with identity are equivalent:*

(a) *$\mathcal{A}$ is not isomorphic to a direct product of two nonzero Banach k-algebras;*

(b) *0 and 1 are the only idempotents in $\mathcal{A}$;*

(c) *$\mathcal{M}(\mathcal{A})$ is connected.* ∎

7.4.3. Remark. One can define the spectrum $\mathcal{M}(\mathcal{A})$ of a commutative Banach k-algebra $\mathcal{A}$ (not necessarily with identity) as the space of all nonzero bounded multiplicative seminorms. If $\mathcal{A}'$ is obtained from $\mathcal{A}$ by introducing an identity element, then $\mathcal{M}(\mathcal{A}') = \mathcal{M}(\mathcal{A}) \cup \{x_0\}$, where $\{x_0\}$ corresponds to the multiplicative seminorm of $\mathcal{A}'$ which is equal to zero on $\mathcal{A}$ and is equal to 1 on the identity. In particular, the space $\mathcal{M}(\mathcal{A})$ is locally compact. From Theorem 7.4.1 it follows that, if $\mathcal{A}$ is a commutative Banach k-algebra having no nonzero quasinilpotent elements, then the spectrum $\mathcal{M}(\mathcal{A})$ is compact if and only if $\mathcal{A}$ has identity.

7.5. Partitions of the spectrum of an element. Let $\mathcal{A}$ be a Banach k-algebra with identity. Given an element $f \in \mathcal{A}$, let $\mathcal{S}_f$ denote the family of subsets of Σ_f which are open and closed simultaneously. Let $S \in \mathcal{S}_f$. There exists a unique element $\varphi_S \in \Gamma(\Sigma_f, \mathcal{O}_{\mathbf{A}^1})$ which is equal to 1 in a neighborhood of S and is equal to 0 in a neighborhood of $S' = \Sigma_f \backslash S$. The element $e_S = \varphi_S(f)$ is an idempotent of $\mathcal{A}$ called the *idempotent associated with f and S*. One has $e_\varnothing = 0$, $e_{\Sigma_f} = 1$, $e_S e_{S'} = e_{S'} e_S = 0$ and $1 = e_S + e_{S'}$. Furthermore, let ψ_S be the element of $\Gamma(\Sigma_f, \mathcal{O}_{\mathbf{A}^1})$ which is equal to the coordinate function in a neighborhood of S and is equal to 0 in a neighborhood of S', and set $f_S = \psi_S(f)$. Since $\psi_S = \varphi_S T = T\varphi_S$, we have $f_S = e_S f = f e_S$.

For $S \in \mathcal{S}_f$, we set $\mathcal{A}_S = e_S \mathcal{A} e_S$. This is a closed subalgebra of $\mathcal{A}$ with identity e_S, and $f_S \in \mathcal{A}_S$. We denote by $R_{f,S}$ the resolvent of the element f_S with respect to the algebra $\mathcal{A}_S$.

7.5.1. Proposition. *Let $S \in \mathcal{S}_f$ and $S \neq \varnothing$. Then:*

(i) *$\Sigma_{f_S}(\mathcal{A}_S) = S$; in particular, $e_S \neq 0$;*

(ii) *if $S' = \Sigma_f \backslash S$, then $R_f = R_{f,S} + R_{f,S'}$; in particular, the function $R_f - R_{f,S}$ is analytic in a neighborhood of S.*

Proof. For $x \in \mathbf{A}^1$, we denote by π_x the canonical map $\mathbf{A}^1_{\mathcal{H}(x)} \to \mathbf{A}^1$.

By Proposition 7.1.6, the spectrum $\Sigma_{f,x}$ of the element $f\otimes 1$ with respect to $\mathscr{A}\widehat{\otimes}\mathscr{H}(x)$ coincides with $\pi_x^{-1}(\Sigma_f)$. The sets $S_x = \pi_x^{-1}(S)$ and $S'_x = \pi_x^{-1}(S')$ lie in $\mathscr{S}_{f\otimes 1}$.

Let $x \in \mathbf{A}^1\backslash S$. We denote by $\delta_{S,x}$ the element of $\Gamma(\Sigma_{f,x}, \mathscr{O}_{\mathbf{A}^1_{\mathscr{H}(x)}})$ that is equal to $(1\otimes T(x) - T\otimes 1)^{-1}$ in a neighborhood of S_x and is equal to 0 in a neighborhood of S'_x, and we set $R'_{f,S}(x) = \delta_{S,x}(f\otimes 1)$. Since $\delta_{S,x} = (\varphi_S\otimes 1)\delta_{S,x} = \delta_{S,x}(\varphi_S\otimes 1)$, we have $R'_{f,S}(x)\in \mathscr{A}_S\widehat{\otimes}\mathscr{H}(x)$. Furthermore, since $(1\otimes T(x) - \psi_S\otimes 1)\delta_{S,x} = \varphi_S\otimes 1$, it follows that $(1\otimes T(x) - f_S\otimes 1)R'_{f,S}(x) = e_S\otimes 1$. In particular, $\Sigma_{f_S}(\mathscr{A}_S)\subset S$ and $R'_{f,S}(x) = R_{f,S}(x)$ for all $x\notin S$.

Suppose that for a point $x\in S$ the element $1\otimes T(x) - f_S\otimes 1$ is invertible in $\mathscr{A}_S\widehat{\otimes}\mathscr{H}(x)$. We claim that the element $1\otimes T(x) - f\otimes 1$ is invertible in $\mathscr{A}\widehat{\otimes}\mathscr{H}(x)$. Indeed, setting $g_x = (1\otimes T(x) - f_S\otimes 1)^{-1}$, we have:

$$\begin{aligned} 1 &= e_S\otimes 1 + e_{S'}\otimes 1 \\ &= (1\otimes T(x) - f_S\otimes 1)g_x + (1\otimes T(x) - f_{S'}\otimes 1)R_{f,S'}(x) \\ &= (1\otimes T(x) - f\otimes 1)(g_x + R_{f,S'}(x)). \end{aligned}$$

It follows that $1\otimes T(x) - f\otimes 1$ is invertible in $\mathscr{A}\widehat{\otimes}\mathscr{H}(x)$, but this is impossible.

The equality in (ii) follows from the identity $\delta_{S,x} + \delta_{S',x} = (1\otimes T(x) - T\otimes 1)^{-1}$ in $\Gamma(\Sigma_{f,x}, \mathscr{O}_{\mathbf{A}^1_{\mathscr{H}(x)}})$, where $x\notin \Sigma_f$. ∎

For a point $a\in \mathbf{A}^1_0$, we denote by Q_a the monic polynomial which generates the maximal ideal of a in $k[T]$. The *order of pole at* a of a function φ which is meromorphic in a neighborhood of a is defined to be the minimal integer n such that the function $Q_a^n\varphi$ is analytic. In this case the number $-n$ is called the *order of zero at* a.

7.5.2. PROPOSITION. *Suppose that a point* $a\in \mathbf{A}^1_0$ *is an isolated point of the spectrum* Σ_f. *Then*:

(i) *the element* $Q_a(f)e_{\{a\}}$ *is quasinilpotent*;

(ii) *the point* a *is a pole of order* $n>0$ *of the resolvent* R_f *if and only if* $Q_a(f)^{n-1}e_{\{a\}}\neq 0$ *and* $Q_a(f)^n e_{\{a\}} = 0$.

PROOF. We may consider the algebra $\mathscr{A}_{\{a\}}$ instead of $\mathscr{A}$, and so we assume that $\Sigma_f = \{a\}$. If $g = Q_a(f)$, then, by Corollary 7.3.4(ii), we have $\Sigma_g = g(\Sigma_f) = \{0\}$, and hence the element g is quasinilpotent.

The statement (ii) is easily reduced to the case $a\in k$. In this case it follows from the identity $R_f = \sum_{i=1}^{\infty}(T-a)^{-i}(f-a)^{i-1}$. ∎

We apply the above consideration to the case when $\mathscr{A}$ is the algebra $\mathrm{End}(B)$ of bounded linear operators on a Banach space B over k. Let $A\in\mathscr{A}$ and $S\in\mathscr{S}_A$. Then the idempotent e_S associated with A and S is a projection onto a closed subspace $B_S \subset B$ (the *Riesz projection*). Let B'_S be the kernel of the projection. Then B is the direct sum of the subspaces B_S and B'_S invariant under the action of A. The algebra $\mathcal{A}_S$ consists of all operators which send B_S to B_S and B'_S to zero. One has $\Sigma_{A|B_S} = S$ and $\Sigma_{A|B'_S} = \Sigma_A \backslash S$. In this situation Proposition 7.5.2 gives the following facts.

7.5.3. PROPOSITION. *Let B be a Banach space over k, and let $A: B \to B$ be a bounded linear operator. Suppose that $a \in \mathbf{A}_0^1$ is an isolated point of the spectrum Σ_A, and let $P = e_{\{a\}}$ be the Riesz projection.*

(i) *The subspaces $N = \mathrm{Im}(P)$ and $W = \mathrm{Ker}(P)$ are invariant under A, and the operator $Q_a(A)$ is quasinilpotent on N and invertible on W.*

(ii) *The point a is a pole of order $n > 0$ of the resolvent R_A if and only if $Q_a(A)^{n-1}|_N \neq 0$ and $Q_a(A)^n|_N = 0$.* ■

In the situation of Proposition 7.5.3, N is called the *root subspace* of the operator $Q_a(A)$ (or of the point a). The *discrete spectrum* Σ_A^d of the operator A is the set of points $a \in \Sigma_A \cap \mathbf{A}_0^1$ whose root subspace has finite dimension. This dimension is denoted by $\nu_A(a)$. The complement of Σ_A^d in Σ_A is called the *essential spectrum* of A and is denoted by Σ_A^e. The set Σ_A^e is compact. It is nonempty if $\dim(B) = \infty$. From Proposition 7.5.3 it follows that the resolvent R_A is meromorphic outside of the essential spectrum. It is convenient to set the root subspace of a point $a \in \mathbf{A}_0^1 \backslash \Sigma_A^e$ equal to zero, and to set $\nu_A(a) = 0$.

CHAPTER 8

Perturbation Theory

8.1. The Cauchy transform of a distribution with compact support. Let Σ be a nonempty compact subset of $\mathbf{P}^1$, and let B be a Banach space over k. A *B-valued distribution with support in* Σ is a bounded linear operator $\Gamma_0(\Sigma, \mathcal{O}_{\mathbf{P}^1}) \to B$. That is, for any special neighborhood V of Σ the induced operator $(\mathcal{A}_V)_0 \to B$ is bounded. The subscript zero means that if the set contains infinity, then it is the subspace of functions equal to zero at infinity that is considered. Let $M: \Gamma_0(\Sigma, \mathcal{O}_{\mathbf{P}^1}) \to B$ be a distribution. For each point $x \in \mathbf{P}^1$ it induces a distribution $M_x: \Gamma_0(\pi_x^{-1}(\Sigma), \mathcal{O}_{\mathbf{P}^1_{\mathscr{H}(x)}}) \to B\widehat{\otimes}\mathscr{H}(x)$. If $x \in \mathbf{A}^1\backslash\Sigma$, then the function $(1 \otimes T(x) - T \otimes 1)^{-1}$ belongs to $\Gamma_0(\pi_x^{-1}(\Sigma), \mathcal{O}_{\mathbf{P}^1_{\mathscr{H}(x)}})$. Its image under M_x gives an element of $B\widehat{\otimes}\mathscr{H}(x)$, which we denote by $\mathscr{K}M(x)$. The function $x \mapsto \mathscr{K}M(x)$ is called the *Cauchy transform* of the distribution M.

8.1.1. THEOREM. *The Cauchy transform induces a linear isomorphism*:

$$\mathscr{K}: \mathrm{Hom}(\Gamma_0(\Sigma, \mathcal{O}_{\mathbf{P}^1}), B) \tilde{\to} \Gamma_0(\mathbf{P}^1\backslash\Sigma, \mathcal{O}_{\mathbf{P}^1}(B)).$$

8.1.2. REMARKS. (i) The inverse map

$$\mathscr{E}: \Gamma_0(\mathbf{P}^1\backslash\Sigma, \mathcal{O}_{\mathbf{P}^1}(B)) \to \mathrm{Hom}(\Gamma_0(\Sigma, \mathcal{O}_{\mathbf{P}^1}), B)$$

constructed below plays the role of the Cauchy integral along a contour around the set Σ. We shall later explain what plays the role of the contour.

(ii) If $B = \mathcal{A}$ is a Banach k-algebra with identity, then, by Corollary 7.3.4, the distributions that are induced by bounded homomorphisms $\Gamma(\Sigma, \mathcal{O}_{\mathbf{P}^1}) \to \mathcal{A}$ of algebras with identity correspond to elements $f \in \mathcal{A}$ with $\Sigma_f \subset \Sigma$. The Cauchy transform sends such an element f to its resolvent R_f, and, for any $g \in \Gamma_0(\Sigma, \mathcal{O}_{\mathbf{P}^1})$, one has $g(f) = \mathscr{E}R_f(g)$.

(iii) If the valuation on k is nontrivial, then $\mathscr{K}$ is an isomorphism of topological vector spaces over k.

(iv) The space $\Gamma(\Sigma, \mathcal{O}_{\mathbf{P}^1})$ maps canonically to the Banach space $\mathscr{H}(\Sigma)$ and has everywhere dense image in it. Theorem 8.1.3 describes the functions in $\Gamma_0(\mathbf{P}^1\backslash\Sigma, \mathcal{O}_{\mathbf{P}^1}(B))$ that correspond to distributions factoring through $\mathscr{H}_0(\Sigma)$.

Given a point $x \in \mathbf{A}^1\backslash\Sigma$, we introduce the *distance* $d_\Sigma(x)$ between x and Σ as follows:

$$d_\Sigma(x) = \|(1 \otimes T(x) - T \otimes 1)^{-1}\|^{-1}_{\pi_x^{-1}(\Sigma)}.$$

Let $\mathscr{U} = \mathbf{P}^1\backslash\Sigma$. We let $\mathscr{H}(\mathscr{U}, B)$ ($\mathscr{H}(\mathscr{U})$ if $B = k$) denote the space of functions $\varphi \in \Gamma(\mathbf{P}^1\backslash\Sigma, \mathscr{O}_{\mathbf{P}^1}(B))$ such that

$$\|\varphi\| := \sup_{x \in \mathscr{U}\backslash\infty} d_\Sigma(x)\|\varphi(x)\| < \infty.$$

It is easy to verify that $\mathscr{H}(\mathscr{U}, B)$ is complete with respect to this norm and is contained in $\Gamma_0(\mathscr{U}, \mathscr{O}_{\mathbf{P}^1}(B))$.

8.1.3. THEOREM. *The Cauchy transform induces an isometric isomorphism:*

$$\mathrm{Hom}(\mathscr{H}_0(\Sigma), B) \tilde{\to} \mathscr{H}(\mathbf{P}^1\backslash\Sigma, B).$$

8.1.4. REMARKS. (i) From Theorem 8.1.3 it follows that the following properties of an element f of a Banach k-algebra $\mathscr{A}$ with identity, are equivalent:

(a) the homomorphism $\Gamma(\Sigma_f, \mathscr{O}_{\mathbf{A}^1}) \to \mathscr{A} : g \mapsto g(f)$ extends to a bounded homomorphism of Banach algebras $\mathscr{H}(\Sigma_f) \to \mathscr{A}$.

(b) $R_f \in \mathscr{H}(\mathbf{P}^1\backslash\Sigma_f, \mathscr{A})$.

An element $f \in \mathscr{A}$ with either of the equivalent properties (a), (b) is said to be *special*. If f is special, then the homomorphism $\mathscr{H}(\Sigma_f) \to \mathscr{A}$ is admissible. If the norm on $\mathscr{A}$ is equivalent to the spectral norm (for example, if $\mathscr{A}$ is uniform), then all elements of $\mathscr{A}$ are special.

(ii) The results formulated above generalize results obtained by Vishik ([**Vi**]) in the case $\Sigma \subset k$ (and under the assumption that k is algebraically closed and has nontrivial valuation). In the construction of the map $\mathscr{E}$, Vishik used the Shnirelman integral. The special elements of the algebra of bounded linear operators (called operators of scalar type in [**Vi**]) are analogs of the normal operators in a Hilbert space, because in that case $\mathscr{H}(\Sigma) = \mathscr{C}(\Sigma)$ (the space of continuous functions $\Sigma \to k$).

To construct the map $\mathscr{E}$ we first need some notation.

A closed disc E in $\mathbf{P}^1$ is a set of the form $E(a, r)$ or $\mathbf{P}^1\backslash D(a, r)$, $a \in \mathbf{A}^1_0$. Its radius $r(E)$ is equal to r. If E and E' are closed discs, we write $E \prec E'$ if they are of the same type (i.e., they either do or do not contain infinity simultaneously), $E \subset E'$ and $E \neq E'$. The family of finite unions of closed discs which are not $\mathbf{P}^1$ is denoted by S^h.

Let S_c denote the family of connected affinoid domains in $\mathbf{P}^1$. Each $V \in S_c$ can be represented as an intersection of closed discs: if $V = E\backslash\bigcup_{i=1}^n D_i$, $n \geq 0$, then $V = E \cap (\bigcap_{i=1}^n \mathbf{P}^1\backslash D_i)$; and if $V = \mathbf{P}^1\backslash\bigcup_{i=1}^n D_i$, $n \geq 1$, then $V = \bigcap_{i=1}^n \mathbf{P}^1\backslash D_i$. This set of discs is denoted by $E(V)$. There is an isometric isomorphism of Banach spaces:

$$(\mathscr{A}_V)_0 \tilde{\to} \bigoplus_{E \in E(V)} (\mathscr{A}_E)_0.$$

If $V \subset V'$, then each disc $E' \in E(V')$ contains exactly one disc $E \in E(V)$. We write $V \prec V'$, if $E(V') \tilde{\to} E(V)$: $E' \mapsto E$ and $E \prec E'$ for all $E' \in E(V')$.

Let S denote the family of affinoid domains in $\mathbf{P}^1$. Each $V \in S$ is a disjoint union $\bigcup_{i=1}^n V_i$ of certain $V_i \in S_c$. We set $E(V) = \bigcup_{i=1}^n E(V_i)$. The relation $V \prec V'$ obviously extends to the entire family S. Since $\mathscr{A}_V \tilde{\to} \bigoplus_{i=1}^n \mathscr{A}_{V_i}$ and $(\mathscr{A}_{V_i})_0 \tilde{\to} \bigoplus(\mathscr{A}_E)_0$, $E \in E(V_i)$, we obtain an isometric isomorphism of Banach spaces:

$$(8.1.5) \qquad (\mathscr{A}_V)_0 \tilde{\to} \bigoplus_{E \in E(V)} (\mathscr{A}_E)_0 : f \mapsto \{f_E\}_{E \in E(V)}.$$

Two closed discs E and E' are said to be *complementary* if they are of different types, $E \cup E' = \mathbf{P}^1$, and $r(E) \neq r(E')$. Two sets $V, V' \in S$ are said to be *complementary* if there is a bijection $E(V) \tilde{\to} E(V')$ sending each disc $E \in E(V)$ to a complementary disc $E' \in E(V')$. In this case the bijection is unique, and $V \cup V' = \mathbf{P}^1$. The family of elements of S having a complementary set is denoted by S'. We further set $S'_c = S' \cap S_c$. It is easy to see that $V \in S'$ if and only if there exists $V' \in S$ with $V' \prec V$. Moreover, for any closed subset $\Sigma \subset \mathbf{P}^1$, $\Sigma \neq \varnothing, \mathbf{P}^1$, the intersection of its affinoid neighborhoods that belong to S' coincides with Σ.

Let $V \in S$. We represent V as an intersection of elements in S^h as follows. Let U be a set which is complementary to some V' with $V \prec V'$. Then $E(V) \tilde{\to} E(V') \tilde{\to} E(U)$: $E \mapsto E'$. If U is a disjoint union $\bigcup_{i=1}^n U_i$ of certain $U_i \in S_c$, then:

$$V_i = \bigcup_{E' \in E(U_i)} E \in S^h \quad \text{and} \quad V = \bigcap_{i=1}^n V_i.$$

This representation does not depend on the choice of U. Since $V_i \cup V_j = \mathbf{P}^1$ for $i \neq j$, there is an isomorphism $(\mathscr{A}_V)_0 \tilde{\to} \bigoplus_{i=1}^n (\mathscr{A}_{V_i})_0$. We take its composition with the isomorphism $(\mathscr{A}_{V_i})_0 \tilde{\to} \bigoplus(\mathscr{A}_E)_0$, $E \in E(V_i)$, and obtain an isometric isomorphism of Banach spaces:

$$(8.1.6) \qquad (\mathscr{A}_V)_0 \tilde{\to} \bigoplus_{E \in E(V)} (\mathscr{A}_E)_0 : f \mapsto \{f^E\}_{E \in E(V)}.$$

The isomorphisms 8.1.5 and 8.1.6 are connected as follows: $f_E = \sum f^{E'}$, where the sum is taken over all $E' \in E(V)$ with $E \subset E'$. If V belongs to S_c or to S^h, then $f_E = f^E$ for all $E \in E(V)$. The isomorphisms 8.1.5 and 8.1.6 extend naturally to the tensor products with a given Banach space.

Given a closed disc E, we define the *residue operator* $\mathrm{Res}_E : \mathscr{A}_E \to k$ as follows. If $E \subset \mathbf{A}^1$, then we set $\mathrm{Res}_E = 0$. Let $E = \mathbf{P}^1 \backslash D(a, r)$, $a \in \mathbf{A}_0^1$. If $a \in k$, any $f \in \mathscr{A}_E$ can be written as $\sum_{i=0}^{\infty} f_i (T - a)^{-i}$, and we set $\mathrm{Res}_E(f) = f_1$. This definition does not depend on the choice of a, and we have $\|\mathrm{Res}_E\| = r(E)$. In the general case we take a disc $D(a', r')$ containing $D(a, r)$ for which $a' \in k$ and set $\mathrm{Res}_E = \mathrm{Res}_{E'}|_{\mathscr{A}_E}$, where $E' = \mathbf{P}^1\backslash D(a', r')$. This also does not depend on the choice of $D(a', r')$. If B is a Banach space, then the operator we have constructed induces an operator $\mathrm{Res}_E \colon B\widehat{\otimes}\mathscr{A}_E \to B$.

Now let Σ be a closed subset of $\mathbf{P}^1$, $\Sigma \neq \varnothing, \mathbf{P}^1$. A *contour around* Σ is a pair (U, V) of complementary sets such that U is a neighborhood of Σ and $V \subset \mathbf{P}^1\backslash\Sigma$. Let $\gamma = (U, V)$ be a contour around Σ. Then:

$$U \cap V = \bigcup_{E \in E(V)} (E \cap E'),$$

where E' is the disc in $E(u)$ complementary to E. Since $(\mathscr{A}_{E\cap E'})_0 \xrightarrow{\sim} (\mathscr{A}_E)_0 \oplus (\mathscr{A}_{E'})_0$, the operators Res_E and $\mathrm{Res}_{E'}$ are defined on $B\widehat{\otimes}\mathscr{A}_{E\cap E'}$. (Only one of them is not equal to zero.)

Let $\varphi \in \Gamma_0(\mathbf{P}^1\backslash\Sigma, \mathscr{O}_{\mathbf{P}^1}(B))$. Given a function $f \in \Gamma_0(\Sigma, \mathscr{O}_{\mathbf{P}^1})$, we define an element $\mathscr{E}\varphi(f) \in B$ as follows. Let $\gamma = (U, V)$ be a contour around Σ such that $f \in \mathscr{A}_U$ and $\varphi \in B\widehat{\otimes}\mathscr{A}_V$, and set:

$$\mathscr{E}\varphi(f) = \sum_{E \in E(V)} (\mathrm{Res}_E - \mathrm{Res}_{E'})(f^{E'}\varphi_E). \tag{8.1.7}$$

(The element $f^{E'}\varphi_E$ belongs to $B\widehat{\otimes}\mathscr{A}_{E\cap E'}$.) Note that if γ is fixed, the linear operators $\mathscr{E}\varphi \colon (\mathscr{A}_U)_0 \to B$ and $\mathscr{E} \colon B\widehat{\otimes}(\mathscr{A}_V)_0 \to \mathrm{Hom}((\mathscr{A}_U)_0, B)$ are bounded.

8.1.8. LEMMA. *The element $\mathscr{E}\varphi(f)$ does not depend on the choice of contour γ.*

PROOF. We must verify the following fact. Let (U, V) and (U', V') be pairs of complementary sets with $U \subset U'$ and $V' \subset V$. Then the following diagram is commutative:

$$\begin{array}{ccc} B\widehat{\otimes}(\mathscr{A}_V)_0 & \xrightarrow{\mathscr{E}_V} & \mathrm{Hom}((\mathscr{A}_U)_0, B) \\ \downarrow & & \downarrow \\ B\widehat{\otimes}(\mathscr{A}_{V'}) & \xrightarrow{\mathscr{E}_{V'}} & \mathrm{Hom}((\mathscr{A}_{U'})_0, B). \end{array}$$

Let $V = \bigcup_{i=1}^n V_i$, $V_i \in S_c$, and $U = \bigcap_{i=1}^n U_i$, $U_i \in S^h$ (as above). First suppose that $V' = V_i$ and $U' = U_i$ for some $1 \le i \le n$. Let $\varphi \in B\widehat{\otimes}(\mathscr{A}_V)_0$ and $f \in (\mathscr{A}_{U_i})_0$. Then:

$$\begin{aligned} \mathscr{E}_V\varphi(f) &= \sum_{E \in E(V)} (\mathrm{Res}_E - \mathrm{Res}_{E'})(f^{E'}\varphi_E) \\ &= \sum_{E \in E(V_i)} (\mathrm{Res}_E - \mathrm{Res}_{E'})(f^{E'}\varphi_E) = \mathscr{E}_{V_i}\varphi(f) \end{aligned}$$

because $f^{E'} = 0$ for $E \notin E(V_i)$. Since

$$\mathscr{A}_V \xrightarrow{\sim} \bigoplus_{i=1}^{n} \mathscr{A}_{V_i} \quad \text{and} \quad (\mathscr{A}_U)_0 \xrightarrow{\sim} \bigoplus_{i=1}^{n} (\mathscr{A}_{U_i})_0,$$

the statement is reduced to the case $V, V' \in S_c$.

Now suppose that $V' = E_0 \in E(V)$ and $U' = E'_0 \in E(U)$. Let $\varphi \in B\widehat{\otimes}(\mathscr{A}_{E_0})_0$ and $f \in (\mathscr{A}_U)_0$. Then:

$$\mathscr{E}_V \varphi(f) = (\mathrm{Res}_{E_0} - \mathrm{Res}_{E'_0})(f^{E'_0}\varphi_{E_0}) = \mathscr{E}_{E_0}\varphi(f)$$

because $\varphi_E = 0$ for $E \neq E_0$. So the statement is reduced to the case when all of the sets under consideration are discs. Moreover, passing to a finite extension of k, we may assume that the discs have centers in k.

If V and V' are of the same type, the required fact is trivial. So let $V = \mathbf{P}^1 \backslash D(a_1, r_1)$ and $V' = E(a_2, r_2)$, where $\max(r_1, r_2) < |a_1 - a_2|$. Then $U = E(a_1, r'_1)$ and $U' = \mathbf{P}^1 \backslash D(a_2, r'_2)$, where $r'_1 > r_1$ and $r'_2 < r_2$. We must show that, for all $\varphi \in B\widehat{\otimes}(\mathscr{A}_V)_0$ and $f \in (\mathscr{A}_{U'})_0$, one has $\mathrm{Res}_V(f\varphi) = -\mathrm{Res}_{U'}(f\varphi)$. Since the families of functions of the form $\varphi = b \otimes (T - a_2)^{-j}$, $j \geq 1$, $b \in B$, and $f = (T - a)^{-i}$, $i \geq 1$, generate everywhere dense subspaces in $B\widehat{\otimes}(\mathscr{A}_V)_0$ and $(\mathscr{A}_{U'})_0$, respectively, and the operators Res_V and $\mathrm{Res}_{U'}$ are bounded, it suffices to verify the equality:

$$\mathrm{Res}_{T=a_1}((T - a_1)^{-i}(T - a_2)^{-j}) = -\mathrm{Res}_{T=a_2}((T - a_1)^{-i}(T - a_2)^{-j})$$

for all $i, j \geq 1$. But it is easy to show that both sides are equal to

$$(-1)^{i-1}\binom{i+j-2}{i-1}(a_1 - a_2)^{-(i+j-1)}. \qquad \blacksquare$$

By Lemma 8.1.8, we have a linear map:

$$\mathscr{E} : \Gamma_0(\mathbf{P}^1 \backslash \Sigma, \mathscr{O}_{\mathbf{P}^1}(B)) \to \mathrm{Hom}(\Gamma_0(\Sigma, \mathscr{O}_{\mathbf{P}^1}), B).$$

PROOF OF THEOREM 8.1.1. It suffices to show that the maps $\mathscr{K}$ and $\mathscr{E}$ are inverse to one another. This follows from the following lemma.

8.1.9. LEMMA. *Let $U, U', V, V' \in S$ with $U \prec U'$, $V' \prec V$, $U \cap V = \varnothing$, $U' \cap V' = \varnothing$. Suppose that the sets U' and V are complementary. Then the following diagram is commutative*:

$$\begin{array}{ccc} \mathrm{Hom}((\mathscr{A}_U)_0, B) & \xrightarrow{\mathscr{K}} & B\widehat{\otimes}(\mathscr{A}_V)_0 \\ \downarrow & \swarrow_{\mathscr{E}} & \downarrow \\ \mathrm{Hom}((\mathscr{A}_{U'})_0, B) & \xrightarrow{\mathscr{K}} & B\widehat{\otimes}(\mathscr{A}_{V'})_0. \end{array}$$

PROOF. As in the proof of Lemma 8.1.8, the situation can be reduced to the case when all of the sets are discs with centers in k. We assume that $U \subset \mathbf{A}^1$ (the case $\infty \in U$ is similar). Let $U = E(a, r_1)$, $U' = E(a, r'_1)$,

$V = \mathbf{P}^1 \backslash D(a, r_2)$, $V' = \mathbf{P}^1 \backslash D(a, r_2')$ where $a \in k$ and $r_1 < r_2 < r_1' < r_2'$. We remark that for $x \in \mathbf{A}^1 \backslash U$, one has $(1 \otimes T(x) - T \otimes 1)^{-1} = \sum_{j=1}^{\infty} (T - a)^{j-1} \otimes (T(x) - a)^{-j}$. Hence, for a bounded operator $M : \mathscr{A}_U \to B$ one has:

$$\mathscr{K} M = \sum_{j=1}^{\infty} M((T-a)^{j-1}) \otimes (T-a)^{-j}.$$

The upper triangle. We have to show that $\mathscr{E}\mathscr{K} M(f) = M(f)$ for all bounded operators $M : \mathscr{A}_U \to B$ and $f \in \mathscr{A}_{U'}$. It is enough to consider elements of the form $f = (T-a)^i$, $i \geq 0$, which generate an everywhere dense subspace in $\mathscr{A}_{U'}$. One has:

$$\begin{aligned} \mathscr{E}\mathscr{K} M(f) &= \operatorname{Res}_V\Big(\sum_{j=1}^{\infty} M((T-a)^{j-1}) \otimes (T-a)^{i-j}\Big) \\ &= M((T-a)^i) = M(f). \end{aligned}$$

The lower triangle. Again it is enough to verify commutativity on elements of the form $\varphi = b \otimes (T-a)^{-i}$, $b \in B$, $i \geq 1$, which generate an everywhere dense subspace in $B \widehat{\otimes} (\mathscr{A}_V)_0$. One has:

$$\begin{aligned} \mathscr{K}\mathscr{E} \varphi &= \sum_{j=1}^{\infty} \mathscr{E}\varphi((T-a)^{j-1}) \otimes (T-a)^{-j} \\ &= \sum_{j=1}^{\infty} \operatorname{Res}_V(b \otimes (T-a)^{-i+j-1}) \otimes (T-a)^{-j} \\ &= b \otimes (T-a)^{-i} = \varphi. \end{aligned}$$ ■

PROOF OF THEOREM 8.1.3. It suffices to show that if $M \in \operatorname{Hom}(\mathscr{H}_0(\Sigma), B)$, then $\mathscr{K} M \in \mathscr{H}(\mathbf{P}^1 \backslash \Sigma, B)$ and $\|\mathscr{K} M\| \leq \|M\|$, and if $\varphi \in \mathscr{H}(\mathbf{P}^1 \backslash \Sigma, B)$, then $\mathscr{E}\varphi \in \operatorname{Hom}(\mathscr{H}_0(\Sigma), B)$ and $\|\mathscr{E}\varphi\| \leq \|\varphi\|$. We remark that since these are metric properties, they are invariant under extension of the ground field. So we may assume that k is algebraically closed.

Let $M \in \operatorname{Hom}(\mathscr{H}_0(\Sigma), B)$. Then for $x \in \mathbf{A}^1 \backslash \Sigma$ one has:

$$\begin{aligned} \|\mathscr{K} M(x)\| &= \|M_x((1 \otimes T(x) - T \otimes 1)^{-1})\| \\ &\leq \|M\| \cdot \|(1 \otimes T(x) - T \otimes 1)^{-1}\|_{\pi_x^{-1}(\Sigma)} = \|M\| d_\Sigma(x)^{-1}. \end{aligned}$$

Hence, $\mathscr{K} M \in \mathscr{H}(\mathbf{P}^1 \backslash \Sigma, B)$ and $\|\mathscr{K} M\| \leq \|M\|$.

Now let $\varphi \in \mathscr{H}(\mathbf{P}^1 \backslash \Sigma, B)$. It suffices to verify that, if U is an affinoid neighborhood of Σ, then the norm $\|\mathscr{E}\varphi\|_U$ of the operator $(\mathscr{A}_U)_0 \to \Gamma_0(\Sigma, \mathscr{O}_{\mathbf{P}^1}) \xrightarrow{\mathscr{E}\varphi} B$ does not exceed $\|\varphi\|$. Let $\gamma = (U, V)$ be a contour around Σ. By formula (8.1.7), we have:

$$\mathscr{E}\varphi(f) = \sum_{E \in E_\infty(V)} \operatorname{Res}_E(f^{E'} \varphi_E) - \sum_{E \in E_f(V)} \operatorname{Res}_{E'}(f^{E'} \varphi_E),$$

where $E_\infty(V)$ (resp. $E_f(V)$) denotes the set of discs in $E(V)$ which contain (resp. do not contain) infinity. Since $||\mathrm{Res}_E|| = r(E)$, $||\mathrm{Res}_{E'}|| = r(E')$ and $r(E') < r(E)$ for $E \in E_f(V)$, we have

$$||\mathscr{E}\varphi(f)|| \leq \max\{\max_{E\in E_\infty(V)}\{r(E)||\varphi_E||\}\cdot||f||_U,\ \max_{E\in E_f(V)}\{r(E')||\varphi_E||\}\cdot||f||_U\}$$
$$\leq \max_{E\in E(V)}\{r(E)||\varphi_E||\}\cdot||f||_U.$$

Hence, the required inequality $||\mathscr{E}\varphi||_U \leq ||\varphi||$ follows from the following lemma.

8.1.10. LEMMA. *For all $V \in S'$ lying in $\mathbf{P}^1\backslash\Sigma$ and $\varphi \in B\widehat{\otimes}(\mathscr{A}_V)_0$, one has the inequality*

$$\max_{E\in E(V)} r(E)||\varphi_E|| \leq \sup_{x\in V} d_\Sigma(x)||\varphi(x)||.$$

PROOF. We may assume that V is connected. Suppose that the inequality is wrong, i.e., for some $E \in E(V)$ one has

$$r(E)||\varphi_E|| > \sup_{x\in V} d_\Sigma(x)||\varphi(x)||.$$

For example, let $E = \mathbf{P}^1\backslash D(a, r)$ (the case $E \subset \mathbf{A}^1$ is similar). Choose $r' > r$ so that for $E' = E(a, r')$ the set $E\cap E'$ is contained in V. Then we may assume that $V = E\cap E'$. For $x \in V$ we have $d_\Sigma(x) = |T(x) - a|$, and hence

$$\sup_{x\in V} d_\Sigma(x)||\varphi(x)|| = \sup_{x\in V}||(T-a)\varphi(x)|| = ||(T-a)\varphi||.$$

Let $\varphi_E = \sum_{i=1}^\infty b_i\otimes(T-a)^{-i}$ and $\varphi_{E'} = \sum_{i=0}^\infty b_i'\otimes(T-a)^i$, in which case:

$$||(T-a)\varphi|| = \max\{\sup_{i\geq1}||b_{i+1}||r^{-i},\ ||b_1||,\ \sup_{i\geq1}||b_{i-1}'||r'^i\}$$
$$= \max\{r\sup_{i\geq1}||b_i||r^{-i},\ r'\sup_{i\geq0}||b_i'||r'^i\}$$
$$= \max\{r||\varphi_E||,\ r'||\varphi_{E'}||\}.$$

We obtain the inequality $r||\varphi_E|| > \max\{r||\varphi_E||, r'||\varphi_{E'}||\}$, which is impossible. ∎

8.2. Fredholm Theory. In this subsection we give a survey of the non-Archimedean Fredholm Theory developed in **[Se]** and **[Gr]**. Although in **[Se]** and **[Gr]** it was assumed that the valuation on the ground field is nontrivial, all of the results can be extended in a straightforward way to the case of trivial valuation.

Let B be a Banach space over k, let $\mathrm{End}(B)$ be the algebra of bounded linear operators on B (we consider only such operators), let $\mathrm{End}^f(B)$ be the subspace of operators of finite rank, and let $\mathrm{End}^c(B)$ be the closure of $\mathrm{End}^f(B)$ in $\mathrm{End}(B)$. Elements of $\mathrm{End}^c(B)$ are called *completely continuous operators*. The subspaces $\mathrm{End}^f(B)$ and $\mathrm{End}^c(B)$ are two-sided ideals in $\mathrm{End}(B)$. We note that, if $L \in \mathrm{End}^f(B)$ (resp. $\mathrm{End}^c(B)$), then $L \otimes 1 \in \mathrm{End}^f(B\widehat{\otimes}K)$ (resp. $\mathrm{End}^c(B\widehat{\otimes}K)$) for any non-Archimedean field K over k.

8.2.1. REMARK. One might expect that $\mathrm{End}^f(B) = \mathrm{End}^c(B)$ if the valuation on k is trivial. For example, this is the case if the norm on B is also trivial. But it is not the case for arbitrary B. Let $B = k[[T]]$ endowed with the norm $||\sum_{i=0}^{\infty} a_i T^i|| = \varepsilon^n$, where $a_n \neq 0$ and $a_0 = \cdots = a_{n-1} = 0$. The operator $L: B \to B$ sending $\sum_{i=0}^{\infty} a_i T^i$ to $\sum_{i=0}^{\infty} a_i T^{2i}$ is completely continuous and of infinite rank.

Every operator $L: B \to B$ induces an operator $\bigwedge^q L: \bigwedge^q B \to \bigwedge^q B$ ($\bigwedge^q B$ is the q-exterior power of B). Completely continuous operators have the following property.

8.2.2. *If* $L \in \mathrm{End}^c(B)$ *and* $\lambda > 0$, *then* $\lambda^q ||\bigwedge^q L|| \to 0$ *as* $q \to \infty$.

The trace map $\mathrm{tr}: \mathrm{End}^f(B) \to k$ is a bounded linear functional of norm 1, and so extends naturally to a bounded linear functional $\mathrm{tr}: \mathrm{End}^c(B) \to k$ of norm 1. By 8.2.2, for $L \in \mathrm{End}^c(B)$ one can set

$$\det(1 + L) = \sum_{i=0}^{\infty} \mathrm{tr}(\bigwedge{}^q L).$$

The function $L \mapsto \det(1 + L)$ is continuous on $\mathrm{End}^c(B)$, and on operators of finite rank it coincides with the usual determinant. One has:

$$\det(1 + AL) = \det(1 + LA), \quad A \in \mathrm{End}(B),$$

$$\det(1 + L' + L'' + L'L'') = \det(1 + L')\det(1 + L'').$$

Furthermore, from 8.2.2 it follows that the series $\sum_{i=0}^{\infty}(-1)^q \mathrm{tr}(\bigwedge^q L)T^q$ is convergent on the affine line. This series is called the *Fredholm determinant* of the operator L and is denoted by $\det(1 - LT)$. The main result of Fredholm Theory is the following.

8.2.3. *Let* $L \in \mathrm{End}^c(B)$. *The operator* $1 - L$ *is noninvertible if and only if* $\det(1 - L) = 0$. *Furthermore, in this case the dimension of the root subspace of* $1 - L$ *is finite and equal to the order of zero of the function* $\det(1 - LT)$ *at* $T = 1$.

8.2.4. COROLLARY. *Let* $L \in \mathrm{End}^c(B)$.

(i) *The spectrum* Σ_L *consists of at most a countable set of points in* $\mathbf{A}_0^1$. *If this set is infinite, then its only limit point is zero.*

(ii) *All nonzero points of the spectrum* Σ_L *belong to the discrete spectrum* Σ_L^d. *If* $\dim(B) = \infty$, *then* $\Sigma_L^e = \{0\}$.

(iii) *For all* $a \in \mathbf{A}_0^1\backslash\{0\}$, *one has*:

$$\nu_L(a) = \deg_{a^{-1}}(\det(1 - LT)). \qquad \blacksquare$$

Here $\nu_L(a)$ is the dimension of the root subspace of the point a (defined at the end of §7.5), and the right hand side is the order of zero at a^{-1} multiplied by $[\mathscr{H}(a^{-1})\colon k]$ (a^{-1} is the image of a under the map $\mathbf{P}^1 \to \mathbf{P}^1$ induced by the function T^{-1}).

As in classical Fredholm Theory, the proof of 8.2.3 given in **[Gr]** uses a family of continuous maps:

$$P_n\colon \mathrm{End}^c(B) \overset{L\mapsto \bigwedge L}{\longrightarrow} \sum_{q=0}^{\infty} \mathrm{End}^c(\textstyle\bigwedge^q B) \overset{\sum_{q=0}^{\infty} h_{n,q}}{\longrightarrow} \mathrm{End}(\textstyle\bigwedge^n B),$$

where the $h_{n,q}\colon \mathrm{End}^c(\bigwedge^q B) \to \mathrm{End}(\bigwedge^n B)$ are certain bounded linear operators of norm 1. (The direct sum $\sum_{q=0}^{\infty} B_q$ of Banach spaces B_q is the space of all families $\{b_q\}_{q\geq 0}$ with $\|b_q\| \to 0$ as $q \to \infty$, endowed with the norm $\|\{b_q\}\| = \sup \|b_q\|$.) The maps P_n satisfy the following *Fredholm formulas*:

$$\textstyle\bigwedge^n(1+L)P_n(L) = P_n(L)\bigwedge^n(1+L) = \det(1+L),$$

$$P_n(L' + L'' + L'L'') = P_n(L')P_n(L'').$$

If $n = 0$, then $h_{0,q} = \mathrm{tr}\colon \mathrm{End}^c(\bigwedge^q B) \to k$, and so $P_0(L) = \det(1+L)$. The function $P_1(-LT)\colon x \mapsto P_1(-L\otimes T(x))$ is called the *Fredholm resolvent* of the operator L. It is an entire function on $\mathbf{A}^1$ with values in $\mathrm{End}(B)$:

$$P_1(-LT) = \sum_{q=0}^{\infty} (-1)^q h_{1,q}(\textstyle\bigwedge^q L)T^q.$$

For $n = 1$ the first Fredholm formula gives

(8.2.5) $$(1-L)P_1(-L) = P_1(-L)(1-L) = \det(1-L).$$

From (8.2.5) one deduces the following recurrence relation between the coefficients of $P_1(-LT)$:

(8.2.6) $$h_{1,q}(\textstyle\bigwedge^q L) + Lh_{1,q-1}(\bigwedge^{q-1} L) = \mathrm{tr}(\bigwedge^q L).$$

8.2.7. LEMMA. *Let $\mathscr{U}$ be an open subset of $\mathbf{P}^1$, $\varphi \in \Gamma(\mathscr{U}, \mathscr{O}_{\mathbf{P}^1}(\mathrm{End}(B)))$ and $L \in \mathrm{End}^c(B)$. Then the function $x \mapsto P_n((L\otimes 1)\varphi(x))$ belongs to $\Gamma(\mathscr{U}, \mathscr{O}_{\mathbf{P}^1}(\mathrm{End}(\bigwedge^n X)))$.*

PROOF. Since P_n is the composition of the map $L \mapsto \bigwedge L$ and a bounded linear operator, it suffices to verify that the function $x \mapsto \bigwedge((L\otimes 1)\varphi(x))$ belongs to $\Gamma(\mathscr{U}, \mathscr{O}_{\mathbf{P}^1}(\sum_{q=0}^{\infty} \mathrm{End}^c(\bigwedge^q B)))$. First we show that the function $x \mapsto \bigwedge^q((L\otimes 1)\varphi(x))$ belongs to $\Gamma(\mathscr{U}, \mathscr{O}_{\mathbf{P}^1}(\bigwedge^q B))$. Indeed, if L is of finite rank, this is obvious. If L is arbitrary, then for any compact subset $\Sigma \subset \mathscr{U}$ and any $\varepsilon > 0$, one can choose an operator L' of finite rank such that $\|\bigwedge^q((L\otimes 1)\varphi(x)) - \bigwedge^q((L'\otimes 1)\varphi(x))\| < \varepsilon$ for all $x \in \Sigma$. This gives the required fact. Now, by 8.2.2, for any compact subset $\Sigma \subset \mathscr{U}$ one has

$$\sup_{x\in\Sigma} \|\textstyle\bigwedge^q((L\otimes 1)\varphi(x))\| \leq \|\bigwedge^q L\| \cdot \|\varphi\|_\Sigma^q \to 0 \quad \text{as } q \to \infty.$$

The required statement follows. ∎

8.3. Perturbations of operators. The perturbation theory considered here concerns properties of the perturbation $A + L$ of an operator $A \in \operatorname{End}(B)$ by a completely continuous operator $L \in \operatorname{End}^c(B)$.

8.3.1. LEMMA. *Let* $f \in \Gamma(\Sigma_{A+L} \cup \Sigma_A, \mathscr{O}_{\mathbf{A}^1})$. *Then the operator* $f(A+L) - f(A)$ *is completely continuous.*

PROOF. Let $\gamma = (U, V)$ be a contour around the set $\Sigma_{A+L} \cup \Sigma_A$ for which $f \in \mathscr{A}_U$. By the formula (8.1.7), we have:

$$f(A+L) - f(A) = \sum_{E \in E(V)} (\operatorname{Res}_E - \operatorname{Res}_{E'})(f^{E'} \varphi_E),$$

where $\varphi = R_{A+L} - R_A$. Since $R_{A+L} - R_A = R_{A+L} L R_A$, it follows that $\varphi \in \mathscr{A}_V \widehat{\otimes} \operatorname{End}^c(B)$ and $\varphi_E \in \mathscr{A}_E \widehat{\otimes} \operatorname{End}^c(B)$ for all $E \in E(V)$. Hence, the right side of the formula belongs to $\operatorname{End}^c(B)$. ■

The *perturbation determinant of* $A+L/A$ is the function $\Delta_{A+L/A} = \det(1 - LR_A)$, which belongs to $\Gamma(\mathbf{P}^1 \backslash \Sigma_A, \mathscr{O}_{\mathbf{P}^1})$, by Lemma 8.2.7. This function satisfies the following simple identities:

$$\Delta_{A/A+L} = \Delta^{-1}_{A+L/A},$$

$$\Delta_{A+L'+L''/A} = \Delta_{A+L'+L''/A+L'} \Delta_{A+L'/A}.$$

8.3.2. THEOREM. (i) $\Sigma^e_{A+L} = \Sigma^e_A$, *i.e., the essential spectrum of an operator does not change under perturbations by completely continuous operators;*

(ii) *the perturbation determinant* $\Delta_{A+L/A}$ *is meromorphic outside* Σ^e_A, *and for all* $a \in \mathbf{A}^1_0 \backslash \Sigma^e_A$ *one has*

$$\deg_a(\Delta_{A+L/A}) = \nu_{A+L}(a) - \nu_A(a).$$

PROOF. (i) First of all, we claim that one has the formula:

$$R_{A+L} = \Delta^{-1}_{A+L/A} R_A P_1(-LR_A). \tag{8.3.3}$$

Indeed, let $x \in \mathbf{A}^1 \backslash (\Sigma_{A+L} \cup \Sigma_A)$. Applying the formula (8.2.5) to the operator $(L \otimes 1)R_A(x) \in \operatorname{End}^c(B \widehat{\otimes} \mathscr{H}(x))$, we have:

$$\det(1 - (L \otimes 1)R_A(x)) = (1 - (L \otimes 1)R_A(x))P_1(-(L \otimes 1)R_A(x)).$$

The required formula now follows from the identity:

$$1 - (L \otimes 1)R_A(x) = (1 \otimes T(x) - (A+L) \otimes 1)(1 \otimes T(x) - A \otimes 1)^{-1}.$$

By Lemma 8.2.7, the function $x \mapsto P_1(-(L \otimes 1)R_A(x))$ is analytic outside Σ_A. From the formula (8.3.3) it follows that the resolvent R_{A+L} is analytic outside $\Sigma_A \cup \{\text{zeros of } \Delta_{A+L/A}\}$, and hence $\Sigma_{A+L} \subset \Sigma_A \cup \{\text{zeros of } \Delta_{A+L/A}\}$. We will show that, if $a \in \mathbf{A}^1_0 \backslash \Sigma^e_A$, then $a \in \mathbf{A}^1_0 \backslash \Sigma^e_{A+L}$. This will give the inclusion $\Sigma^e_{A+L} \subset \Sigma^e_A$, and, by symmetry, the inclusion $\Sigma^e_A \subset \Sigma^e_{A+L}$.

Let $a \in \mathbf{A}_0^1 \backslash \Sigma_A^e$. First we consider the case $a \notin \Sigma_A$. Suppose that $a \in \Sigma_{A+L}$. Then a is a zero of the function $\Delta_{A+L/A}$. Since all of the zeros of $\Delta_{A+L/A}$ are isolated, it follows that a is an isolated point of the spectrum Σ_{A+L}. Let P be the corresponding Riesz projection, and let f be the germ of analytic functions which are equal to 1 in a neighborhood of the point a and to 0 in a neighborhood of the set $(\Sigma_{A+L} \cup \Sigma_A) \backslash \{a\}$. Then $P = f(A+L)$. Since $a \notin \Sigma_A$, we have $f(A) = 0$. By Lemma 8.3.1, the operator $P = f(A+L) - f(A)$ is completely continuous. It follows that P is of finite rank, i.e., $a \in \Sigma_{A+L}^d$.

Now consider the case $a \in \Sigma_A^d$. Let P be the Riesz projection, let $N = \operatorname{Im}(P)$ be the root subspace, and let $W = \operatorname{Ker}(P)$. The operator $Q_a(A)$ is invertible on W and nilpotent on N. First suppose that $a \neq 0$. We have $A + L = A' + L'$, where $A' = A - PA$ and $L' = L + PA$. The operator L' is completely continuous. The operator $Q_a(A')$ coincides with $Q_a(A)$ on W and with multiplication by $Q_a(0) \neq 0$ on N. Hence, $Q_a(A')$ is invertible. It follows that $a \notin \Sigma_{A'}$, and the situation is reduced to the case considered previously. Now suppose that $a = 0$. Let $A^n = 0$ on N, and denote by C the operator which coincides with the identity on N and with A^{-n} on W. Then $A + L = A' + L'$, where $A' = -1 + A + CA^n$ and $L' = L + 1 - CA^n$. The operator $1 - CA^n$ is a projection onto N, and so L' is completely continuous. The operator A' coincides with A on W and with $-1 + A$ on N, and so is invertible. Therefore, $a \notin \Sigma_{A'}$, and again the situation is reduced to the case considered above.

(ii) Let $a \in \mathbf{A}_0^1 \backslash \Sigma_A^e$. By (i), we have $a \in \mathbf{A}_0^1 \backslash \Sigma_{A+L}^e$. It suffices to verify the formula (ii) over a finite extension of k, and so we may assume that $a \in k$. Since $\Delta_{A+b+L/A+b}(T) = \Delta_{A+L/A}(T-b)$ for any $b \in k$, we may assume that $a \neq 0$. Let P (resp. Q) be the Riesz projection of the operator $a - A - L$ (resp. $a - L$). We have:

$$\Delta_{A+L/A} = \Delta_{A+L/(1-P)(A+L)} \Delta_{(1-P)(A+L)/(1-Q)A} \Delta_{(1-Q)A/A}.$$

Since the point a is isolated from $\Sigma_{A+L} \backslash \{a\}$ and $\Sigma_A \backslash \{a\}$, all three factors on the right are analytic outside a in a small neighborhood of a. Furthermore, since $a \neq 0$, it follows that the operators $a - (1-P)(A+L)$ and $a - (1-Q)A$ are invertible. Hence, the middle factor is defined and is not equal to zero at the point a. Finally, since the operator $a - A$ is nilpotent on its root space, we have:

$$\Delta_{(1-Q)A/A} = \det(1 + QAR_A) = \det(1 + AR_A Q)$$
$$= \left(1 + \frac{a}{T-a}\right)^{\nu_A(a)} = \frac{T^{\nu_A(a)}}{(T-a)^{\nu_A(a)}}.$$

Similarly, we have:

$$\Delta_{A+L/(1-P)(A+L)} = \Delta^{-1}_{(1-P)(A+L)/A+L} = \frac{(T-a)^{\nu_{A+L}(a)}}{T^{\nu_{A+L}(a)}}.$$

The required formula (ii) follows. ∎

8.4. The relative multiplicity of perturbation, and the trace formula. We first remark that, if φ is an analytic function on an open subset $\mathscr{U} \subset \mathbf{P}^1$ that is not identically equal to zero on any connected component of $\mathscr{U}$, then the logarithmic derivative φ'/φ defined on $\mathscr{U}' = \mathscr{U} \cup \{\text{zeros of } \varphi\}$ belongs to $\mathscr{H}(\mathscr{U}')$. By Theorem 8.1.3, the corresponding distribution $\mathscr{E}(\varphi'/\varphi)$ comes from a bounded linear functional on the Banach space $\mathscr{H}_0(\Sigma')$, where $\Sigma' = \mathbf{P}^1 \backslash \mathscr{U}'$.

Now let $A \in \operatorname{End}(B)$ and $L \in \operatorname{End}^c(B)$. By Theorem 8.3.2(ii), the zeros of the perturbation determinant $\Delta_{A+L/A}$ lie in the set $\Sigma_A \cup \Sigma_{A+L} = \Sigma_A \cup \Sigma^d_{A+L}$. The *relative multiplicity of perturbation* is the distribution:

$$M_{A+L/A} = \mathscr{E}(\Delta'_{A+L/A}/\Delta_{A+L/A}).$$

As was just remarked, this distribution comes from a bounded linear functional on the Banach space $\mathscr{H}(\Sigma_A \cup \Sigma^d_{A+L})$.

8.4.1. THEOREM (Trace Formula). *For $f \in \Gamma(\Sigma_A \cup \Sigma^d_{A+L}, \mathscr{O}_{\mathbf{A}^1})$, one has:*

$$M_{A+L/A}(f) = \operatorname{tr}(f(A+L) - f(A)).$$

If the operators A and L are special, the same formula holds for all $f \in \mathscr{H}(\Sigma_A \cup \Sigma^d_{A+L})$.

PROOF. Using the construction of the map $\mathscr{E}$, the trace formula is easily deduced from the following equality:

$$\Delta'_{A+L/A}/\Delta_{A+L/A} = \operatorname{tr}(R_{A+L} - R_A).$$

This equality is proved in Vishik's paper **[Vi]** (Lemma 15.6). We sketch a proof which uses the facts from §8.2.

By the definiticn of the determinant, one has:

$$\Delta_{A+L/A} = \sum_{q=0}^{\infty} (-1)^q \operatorname{tr}(\textstyle\bigwedge^q (LR_A)).$$

So we have to consider the action of differentiation on functions of the form $x \mapsto \operatorname{tr}(\bigwedge^q \varphi(x))$, where $\varphi \in \Gamma(\mathscr{U}, \mathscr{O}_{\mathbf{P}^1}(\operatorname{End}^c(B)))$, $\mathscr{U} \subset \mathbf{P}^1$. It is easy to see that $(\bigwedge^q \varphi)' = b_{q-1}(\varphi, \varphi')$, where $b_q(P, Q)$, $P, Q \in \operatorname{End}^c(B)$, is the operator on $\bigwedge^q B$ which acts as follows:

$$b_q(P, Q)(x_1 \textstyle\bigwedge \cdots \bigwedge x_{q+1})$$
$$= \sum_{i=1}^{q} Px_1 \textstyle\bigwedge \cdots \bigwedge Px_{i-1} \bigwedge Qx_i \bigwedge Px_{i+1} \bigwedge \cdots \bigwedge Px_{q+1}.$$

(This is the operator $b_{q,1}(P, Q)$ in **[Gr]**, §2.1, and $\Lambda_{q+1}(P, Q)$ in **[Vi]**, §13.) It turns out that the trace of $b_q(P, Q)$ can be expressed in terms of the trace of the operator $h_{1,q}$, which enters into the decomposition of the Fredholm resolvent. Namely (**[Vi]**, 13.4):

$$\operatorname{tr}(b_q(P, Q)) = \operatorname{tr}(h_{1,q}(\textstyle\bigwedge^q P)Q).$$

This equality is proved by induction, using the formula (8.2.6) and the following formula:

$$\operatorname{tr}(b_q(P, Q)) = \operatorname{tr}(\textstyle\bigwedge^q P)\operatorname{tr}(Q) - \operatorname{tr}(b_{q-1}(P, QP)),$$

which one first proves in the finite dimensional case and then easily deduces for arbitrary spaces.

Using the above formulas and the equality $R'_A = -R_A^2$, we have:

$$\begin{aligned}
\Delta'_{A+L/A} &= \sum_{q=0}^{\infty}(-1)^q(\operatorname{tr}(\textstyle\bigwedge^q(LR_A)))' \\
&= -\sum_{q=1}^{\infty}(-1)^q \operatorname{tr}(b_{q-1}(LR_A, LR_A^2)) \\
&= \sum_{q=0}^{\infty}(-1)^q \operatorname{tr}(h_{1,q}(\textstyle\bigwedge^q(LR_A))LR_A^2) \\
&= \operatorname{tr}((\sum_{q=0}^{\infty}(-1)^q h_{1,q}(\textstyle\bigwedge^q(LR_A)))LR_A^2) \\
&= \operatorname{tr}(P_1(-LR_A)LR_A^2) = \operatorname{tr}(R_A P_1(-LR_A)LR_A).
\end{aligned}$$

Using the formula (8.3.3) and the identity $R_{A+L}LR_A = R_{A+L} - R_A$, we have:

$$\Delta'_{A+L/A} = \Delta_{A+L/A}\operatorname{tr}(R_{A+L} - R_A). \qquad \blacksquare$$

CHAPTER 9

The Dimension of a Banach Algebra

9.1. Dimension and stable dimension. For a non-Archimedean field K over k, we denote by $s(K/k)$ the transcendence degree of $\widetilde{K}$ over $\widetilde{k}$ and by $t(K/k)$ the dimension of the $\mathbf{Q}$-vector space $\sqrt{|K^*|}/\sqrt{|k^*|}$, and we set $d(K/k) = s(K/k) + t(K/k)$. If $k \subset K \subset L$, then $d(L/k) = d(L/K) + d(K/k)$, and similarly for the functions s and t. For example, $d(K/k) = 0$ if and only if K can be embedded in an immediate extension of $\widehat{k}^a$. In particular, if $\widehat{k}^a$ is maximally complete, then the zero-dimensional fields over k are (up to isomorphism) the closed subfields of $\widehat{k}^a$.

Let $\mathscr{A}$ be a commutative Banach k-algebra with identity. For $x \in \mathscr{M}(\mathscr{A})$ we set $s(x) = s(\mathscr{H}(x)/k)$, $t(x) = t(\mathscr{H}(x)/k)$ and $d(x) = d(\mathscr{H}(x)/k)$. We call the *dimension of $\mathscr{A}$ over k* the number:

$$d(\mathscr{A}/k) = \sup_{x \in \mathscr{M}(\mathscr{A})} d(x).$$

One similarly introduces the numbers $s(\mathscr{A}/k)$ and $t(\mathscr{A}/k)$.

9.1.1. LEMMA. *If π is the canonical map $\mathscr{M}(\mathscr{A}\widehat{\otimes}\widehat{k}^a) \to \mathscr{M}(\mathscr{A})$ and if $x' \in \pi^{-1}(x)$, then $s(x'/\widehat{k}^a) = s(x)$, $r(x'/\widehat{k}^a) = r(x)$ and $d(x'/\widehat{k}^a) = d(x)$.*

PROOF. The field $\mathscr{H}(x)$ is naturally embedded in $\mathscr{H}(x')$. The set of elements of $\mathscr{H}(x')$ which are algebraic over $\mathscr{H}(x)$ is everywhere dense in $\mathscr{H}(x')$. Hence, $d(\mathscr{H}(x')/\mathscr{H}(x)) = 0$. The required statements follow. ∎

9.1.2. COROLLARY. $d(\mathscr{A}/k) = d(\mathscr{A}\widehat{\otimes}\widehat{k}^a/\widehat{k}^a) = d(\mathscr{A}\widehat{\otimes}\widehat{k}^a/k)$. ∎

The functions s, t, and d are naturally defined on points of k-analytic spaces. For example, for points of the affine line $\mathbf{A}^1$ one has: $d(x) = 0$ if x is of type (1) or (4), $s(x) = 1$ and $t(x) = 0$ if x is of type (2), $s(x) = 0$ and $t(x) = 1$ if x is of type (3). We note that morphisms of k-analytic spaces do not increase the values of s, t and d on points.

9.1.3. PROPOSITION. *If $\mathscr{A}$ is a strictly k-affinoid algebra, then $d(\mathscr{A}/k)$ is equal to the Krull dimension of $\mathscr{A}$.*

PROOF. First of all, if $\mathscr{B}$ is a closed k-affinoid subalgebra of $\mathscr{A}$ such that $\mathscr{A}$ is a finite Banach $\mathscr{B}$-module, then $d(\mathscr{A}/k) = d(\mathscr{B}/k)$. So it suffices to establish the required equality for the algebra $\mathscr{A} = k\{T_1, \dots, T_n\}$. If x is the point of $\mathscr{M}(\mathscr{A})$ corresponding to the norm, then $\widetilde{\mathscr{H}(x)} = \widetilde{k}(T_1, \dots, T_n)$

and $|\mathscr{H}(x)^*| = |k^*|$. Hence, $d(x) = n$, i.e., $d(\mathscr{A}/k) \geq n$. The reverse inequality is verified by induction. ■

9.1.4. PROPOSITION. *Let $\mathscr{A}$ be a commutative Banach k-algebra with identity. Then $d(\mathscr{A}\widehat{\otimes}K/K) \geq d(\mathscr{A}/k)$ for any non-Archimedean field K over k.*

PROOF. We first remark that, if x is a point of $\mathbf{A}^1$ with $d(\mathscr{H}(x)/k) = 1$ (i.e., x is of type (2) or (3)), then there exists a point $x' \in \pi^{-1}(x)$ with $d(\mathscr{H}(x')/K) = 1$ (π denotes the canonical map $\mathbf{A}^1_K \to \mathbf{A}^1$). Indeed, if $x = p(E(a, r))$, one can take $x' = p(E(a, r)_K)$.

To prove the proposition, it is enough to assume that $\mathscr{A} = L$ is a field. If $d(L/k) = 0$, the statement is trivial. Let $d(L/k) \geq 1$, and suppose that the proposition holds for fields of dimension at most $d(L/k) - 1$. We choose an element $f \in L$ with $d(\mathscr{H}(x)/k) = 1$, where $\Sigma_f = \{x\}$. Then $L\widehat{\otimes}K = L\widehat{\otimes}_{\mathscr{H}(x)}\mathscr{B}$, where $\mathscr{B} = \mathscr{H}(x)\widehat{\otimes}K$. Hence:

$$\begin{aligned} d(L\widehat{\otimes}K/K) &= \max_{y\in\mathscr{M}(\mathscr{B})} d(L\widehat{\otimes}_{\mathscr{H}(x)}\mathscr{H}(y)/K) \\ &= \max_{y\in\mathscr{M}(\mathscr{B})} \{d(L\widehat{\otimes}_{\mathscr{H}(x)}\mathscr{H}(y)/\mathscr{H}(y)) + d(\mathscr{H}(y)/K)\}. \end{aligned}$$

By induction, $d(L\widehat{\otimes}_{\mathscr{H}(x)}\mathscr{H}(y)/\mathscr{H}(y)) \geq d(L/\mathscr{H}(x)) = d(L/k) - 1$, and, by the above remark, there exists $y \in \mathscr{M}(\mathscr{B})$ with $d(\mathscr{H}(y)/K) = 1$. Therefore, $d(\mathscr{A}\widehat{\otimes}K/K) \geq d(L/k)$. ■

9.1.5. COROLLARY. $d(\mathscr{A}\widehat{\otimes}\mathscr{B}/k) \geq d(\mathscr{A}/k) + d(\mathscr{B}/k)$. ■

We call the following number the *stable dimension* of a commutative Banach k-algebra $\mathscr{A}$ with identity:

$$d^{st}(\mathscr{A}/k) = \sup_{K/k} d(\mathscr{A}\widehat{\otimes}K/K),$$

where K runs through all non-Archimedean fields over k. For example, the non-Archimedean fields of stable dimension zero are (up to isomorphism) the closed subfields of $\widehat{k^a}$. If $x \in \mathscr{M}(\mathscr{A})$, we set $d^{st}(x) = d^{st}(\mathscr{H}(x)/k)$. The following statements are easily deduced from the previous facts.

9.1.6. PROPOSITION. (i) $d^{st}(\mathscr{A}\widehat{\otimes}\mathscr{B}/k) = d^{st}(\mathscr{A}/k) + d^{st}(\mathscr{B}/k)$.

(ii) *The stable dimension of a k-affinoid algebra is equal to the dimension of the corresponding k-affinoid space.* ■

9.1.7. REMARK. If x is a point of a k-affinoid space $\mathscr{M}(\mathscr{A})$, then $d^{st}(x) \leq \dim(\mathscr{M}(\mathscr{A}/\wp_x))$. The inequality might be strict. For example, let $\mathscr{A} = k\{r^{-1}T_1, T_2\}$, where $0 < r < 1$, and let x_1 be the point of $\mathscr{M}(k\{r^{-1}T_1\})$ corresponding to the norm on $k\{r^{-1}T_1\}$. Then $[\mathscr{H}(x_1)^a : \mathscr{H}(x_1)] = \infty$. Since the valuation on $\mathscr{H}(x_1)$ is nontrivial, there exists an element $\alpha \in \widehat{\mathscr{H}(x_1)^a}\backslash\mathscr{H}(x_1)^a$ with $|\alpha| \leq 1$. Consider the point $x \in \mathscr{M}(k\{r^{-1}T_1, T_2\})$

corresponding to the character

$$\chi: k\{r^{-1}T_1, T_2\} \to \widehat{\mathscr{H}(x_1)}^a: f(T_1, T_2) \mapsto f(T_1, \alpha) .$$

It is clear that χ is injective, i.e., $\wp_x = 0$. We obtain $d^{st}(x) = 1$ and $\dim(\mathscr{M}(\mathscr{A}/\wp_x)) = 2$.

9.2. Zero-dimensional Banach algebras

9.2.1. PROPOSITION. *The following properties of a non-Archimedean field K over k are equivalent:*

(a) $d(K/k) = 0$;

(b) *all closed k-subalgebras of K are fields;*

(c) *there exists an everywhere dense subset $M \subset K$ such that for every $f \in M$ there is a polynomial $P(T) \in k[T]$ with $|1 - fP(f)| < 1$.*

PROOF. (a) $\Rightarrow$ (b). We show that every nonzero element $f \in K$ is invertible in the closed k-subalgebra generated by f. By assumption, $|f|^n = |a|$ for some $n \geq 1$ and $a \in k$. Setting $g = f^n/a$, one obtains $|g| = 1$. Since $\widetilde{K}$ is algebraic over $\widetilde{k}$, we have $|Q(g)| < 1$ for some polynomial $Q(T) \in k^\circ[T]$ with constant term 1. If $Q(T) = 1 - TP(T)$, where $P(T) \in k^\circ[T]$, then $|1 - gP(g)| < 1$. Hence the element $gP(g)$ is invertible in the closed subalgebra generated by f. Therefore, the element f is also invertible in this subalgebra.

(b) $\Rightarrow$ (c). Since a nonzero element $f \in K$ is invertible in the closed k-subalgebra generated by f, there exists a polynomial $P(T) \in k[T]$ with $|f^{-1} - P(f)| < |f|^{-1}$. From this it follows that $|1 - fP(f)| < 1$.

(c) $\Rightarrow$ (a). Suppose that $\widetilde{f} \in \widetilde{K}$, where $f \in M \cap K^\circ$ is transcendental over $\widetilde{k}$. Then the homomorphism $k\{T\} \to K: F \mapsto F(f)$ is an isometry. By assumption, there exists a polynomial $P(T) \in k[T]$ with $|1 - fP(f)| < 1$. It follows that $\|P\| = 1$ and $1 - \widetilde{f}\widetilde{P}(\widetilde{f}) = 0$. This contradicts our supposition. We now show that for any nonzero $f \in K$ we have $|f| \in \sqrt{|k^*|}$. We may assume that $f \in M$. There exists $P(T) \in k[T]$ with $|1 - fP(f)| < 1$. Let $P(T) = -\sum_{i=1}^n a_i T^{i-1}$. Then $|\sum_{i=0}^n a_i f^i| < 1$, where $a_0 = 1$. Hence, one has $|a_i f^i| = |a_j f^j|$ for some $0 \leq i < j \leq n$ with $a_i \neq 0$ and $a_j \neq 0$. It follows that $|f| \in \sqrt{|k^*|}$. ∎

9.2.2. PROPOSITION. *Let $\mathscr{A}$ be a commutative Banach k-algebra with identity, and let $X = \mathscr{M}(\mathscr{A})$. Then the set $X_t = \{x \in X | t(x) = 0\}$ is totally disconnected. If the valuation on k is nontrivial, then the set $X_s = \{x \in X | s(x) = 0\}$ is totally disconnected.*

PROOF. First we remark that the set $\mathbf{A}_t^1 = \{x \in \mathbf{A}^1 | t(x) = 0\}$ (resp. $\mathbf{A}_s^1 = \{x \in \mathbf{A}^1 | s(x) = 0\}$ if the valuation on k is nontrivial) is totally disconnected. Since this is obvious in the case of trivial valuation, we assume that the valuation on k is nontrivial. The family of open subsets of the form

$\mathscr{U} = D(a, r)\backslash\bigcup_{i=1}^{n} E(a_i, r_i)$, where $r, r_i \in \sqrt{|k^*|}$ (resp. $r, r_i \notin \sqrt{|k^*|}$), forms a basis of the topology of $\mathbf{A}^1$. For such $\mathscr{U}$ one has $\mathscr{U} \cap \mathbf{A}_t^1 = \overline{\mathscr{U}} \cap \mathbf{A}_t^1$ (resp. $\mathscr{U} \cap \mathbf{A}_s^1 = \overline{\mathscr{U}} \cap \mathbf{A}_s^1$). Hence, $\mathbf{A}_t^1$ and $\mathbf{A}_s^1$ are totally disconnected.

The required statement now follows from the following remark. If Σ is a subset of X such that for any $f \in \mathscr{A}$ the image of Σ in $\mathbf{A}^1$ under the induced map ${}^*f: X \to \mathbf{A}^1$ is totally disconnected, then Σ is totally disconnected. Indeed, this is a consequence of Proposition 7.1.4(ii). ∎

9.2.3. COROLLARY. *If $t(\mathscr{A}/k) = 0$ (resp. $s(\mathscr{A}/k) = 0$ and the valuation on k is nontrivial), then $\mathscr{M}(\mathscr{A})$ is totally disconnected.* ∎

9.2.4. THEOREM. *Let $\mathscr{A}$ be a commutative Banach k-algebra with identity and let $\mathscr{B}$ be a closed subalgebra containing the identity. If $d(\mathscr{A}/k) = 0$, then the canonical map $\mathscr{M}(\mathscr{A}) \to \mathscr{M}(\mathscr{B})$ is surjective, and $d(\mathscr{B}/k) = 0$.*

PROOF. We first remark that the complement of a point $x \in \mathbf{A}^1$ with $d(x) = 0$ is connected. Hence the complement of any compact subset of $\mathbf{A}^1$ consisting of such points is also connected. From Corollary 7.2.2 it follows that the spectra of f with respect to $\mathscr{B}$ and $\mathscr{A}$ coincide for any element $f \in \mathscr{B}$. By Proposition 9.2.2, Σ_f is totally disconnected for all $f \in \mathscr{A}$. From Proposition 7.1.4(ii) it follows that $\mathscr{M}(\mathscr{B})$ is totally disconnected. Suppose that the map $\mathscr{M}(\mathscr{A}) \to \mathscr{M}(\mathscr{B})$ is not surjective. We can choose an open-closed subset $\mathscr{U} \subset \mathscr{M}(\mathscr{B})$ which does not meet the image of $\mathscr{M}(\mathscr{A})$. By the Shilov Idempotent Theorem 7.4.1, there exists an element $f \in \mathscr{B}$ which is equal to 1 on $\mathscr{U}$ and equal to 0 outside $\mathscr{U}$. It follows that $\Sigma_f = \{0\}$, but this is impossible. Hence, the map $\mathscr{M}(\mathscr{A}) \to \mathscr{M}(\mathscr{B})$ is surjective and $d(\mathscr{B}/k) = 0$. ∎

Let X be a compact Hausdorff space, and let K be a non-Archimedean field over k. The space $\mathscr{C}(X, K)$ (resp. $\mathscr{C}(X)$ if $K = k$) of continuous functions $X \to K$ is a commutative Banach k-algebra with respect to the maximum modulus norm. Let $\mathscr{A}$ be a closed k-subalgebra of $\mathscr{C}(X, K)$ containing the identity. Two points $x, y \in X$ are said to be *$\mathscr{A}$-equivalent* if $f(x) = f(y)$ for all $f \in \mathscr{A}$. The set $X/\mathscr{A}$ of $\mathscr{A}$-equivalence classes endowed with the obvious topology is a compact Hausdorff space. There is a natural embedding $X/\mathscr{A} \hookrightarrow \mathscr{M}(\mathscr{A})$. Note that for every point $x \in X$ the field $\mathscr{H}(x)$ is naturally embedded in K.

9.2.5. THEOREM. *Suppose that $d(K/k) = 0$. Then,*

(i) *$d(\mathscr{A}/k) = 0$ and $X/\mathscr{A} \xrightarrow{\sim} \mathscr{M}(\mathscr{A})$;*

(ii) *the algebra $\mathscr{A}$ consists of all functions $f \in \mathscr{C}(X, K)$ which are constant on $\mathscr{A}$-equivalence classes and for which $f(x) \in \mathscr{H}(x)$ for all $x \in X$.*

PROOF. Without loss of generality we may assume that the space X is totally disconnected. It is easy to see that the natural map $X \to \mathscr{M}(\mathscr{C}(X, K))$ is bijective, and hence $d(\mathscr{C}(X, K)/k) = d(K/k)$. Since $d(K/k) = 0$, the statement (i) follows from Theorem 9.2.4.

To prove (ii) it suffices to assume that $X \widetilde{\to} X/\mathscr{A}$. Let $\mathscr{B}$ denote the k-subalgebra of all functions $f \in \mathscr{C}(X, K)$ such that $f(x) \in \mathscr{H}(x)$ for all $x \in X$. It is clear that $\mathscr{A} \subset \mathscr{B}$. We claim that for any $f \in \mathscr{B}$ and any $\varepsilon > 0$ there exists a function $\varphi \in \mathscr{A}$ with $\|f - \varphi\| < \varepsilon$. By Proposition 9.2.1, the image of $\mathscr{A}$ is everywhere dense in $\mathscr{H}(x)$. Therefore, for every $x \in X$ there exists a function $\varphi_x \in \mathscr{A}$ with $|f(x) - \varphi_x(x)| < \varepsilon$. From this it easily follows that there exist a finite covering $X = \bigcup_{i=1}^n \mathscr{U}_i$ by pairwise disjoint open-closed subsets $\mathscr{U}_i$ and functions $\varphi_i \in \mathscr{A}$ such that $|f(x) - \varphi_i(x)| < \varepsilon$ for all $x \in \mathscr{U}_i$, $1 \le i \le n$. Let g_i, $1 \le i \le n$, be the characteristic function of the set $\mathscr{U}_i$. Since $g_i \in \mathscr{A}$, we have $\varphi = \sum_{i=1}^n g_i \varphi_i \in \mathscr{A}$ and $\|f - \varphi\| < \varepsilon$. ∎

9.2.6. COROLLARY (Stone-Weierstrass Theorem ([**Ka**])). *Let X be a totally disconnected compact Hausdorff space, and let $\mathscr{A}$ be a k-subalgebra of $\mathscr{C}(X)$ which satisfies the following conditions:*

(a) *the elements of $\mathscr{A}$ separate points of X;*

(b) *for each $x \in X$ there exists $f \in \mathscr{A}$ with $f(x) \neq 0$.*

Then $\mathscr{A}$ is everywhere dense in $\mathscr{C}(X)$.

PROOF. We may assume that $\mathscr{A}$ is closed in $\mathscr{C}(X)$. From (a) and (b) it follows that X is naturally embedded in $\mathscr{M}(\mathscr{A})$. It suffices to show that $\mathscr{A}$ contains the identity. Suppose that this is not the case. Let $\mathscr{A}'$ be the closed k-subalgebra of $\mathscr{C}(X)$ generated by $\mathscr{A}$ and 1. By Theorem 9.2.5, $\mathscr{A}'$ coincides with $\mathscr{C}(X)$. In particular, $\mathscr{M}(\mathscr{A}') = X$. On the other hand, $\mathscr{M}(\mathscr{A}') = \mathscr{M}(\mathscr{A}) \bigcup \{x_0\}$, where x_0 corresponds to the multiplicative seminorm on $\mathscr{A}'$ which vanishes on $\mathscr{A}$. The required statement follows. ∎

9.2.7. COROLLARY. *Let $\mathscr{A}$ be a uniform commutative Banach k-algebra with identity. Suppose that all of the characters of $\mathscr{A}$ take values in k. Then:*

(i) *the space $\mathscr{M}(\mathscr{A})$ is totally disconnected;*

(ii) *the Gel′fand transform gives an isomorphism $\mathscr{A} \widetilde{\to} \mathscr{C}(\mathscr{M}(\mathscr{A}))$.*

PROOF. It suffices to verify that the Gel′fand transform sends elements of $\mathscr{A}$ to continuous functions on $\mathscr{M}(\mathscr{A})$. Let $f \in \mathscr{A}$, and let $D(a, r)$ be an open disc in k, $a \in k$. Setting $g = f - a$, we see that the inverse image of $D(a, r)$ under f coincides with the set $\mathscr{U} = \{x \in \mathscr{M}(\mathscr{A}) \,|\, |g(x)| < r\}$, which is open in $\mathscr{M}(\mathscr{A})$. ∎

9.2.8. THEOREM. *The following properties of a uniform commutative Banach k-algebra $\mathscr{A}$ with identity are equivalent:*

(a) $d^{st}(\mathscr{A}/k) = 0$;

(b) *there exists an isometric embedding of Banach k-algebras with identity $\mathscr{A} \hookrightarrow \mathscr{C}(\mathscr{M}(\mathscr{A}), \widehat{k}^a)$.*

PROOF. The implication (b) ⇒ (a) follows from Theorem 9.2.5. Since $d^{st}(\mathscr{A}/k) = 0$, every character of the $\widehat{k}^a$-algebra $\mathscr{A} \widehat{\otimes} \widehat{k}^a$ takes values in $\widehat{k}^a$.

By Corollary 9.2.7, the uniform completion of $\mathscr{A}\widehat{\otimes}\widehat{k}^a$ is naturally isomorphic to $\mathscr{C}(\mathscr{M}(\mathscr{A}\widehat{\otimes}\widehat{k}^a), \widehat{k}^a)$. It is clear that under this embedding elements of $\mathscr{A}$ go to functions which are constant on orbits of the Galois group $G = \mathrm{Gal}(k^a/k)$. Since $\mathscr{M}(\mathscr{A}\widehat{\otimes}\widehat{k}^a)/G \tilde{\to} \mathscr{M}(\mathscr{A})$, we obtain the required isometric embedding $\mathscr{A} \hookrightarrow \mathscr{C}(\mathscr{M}(\mathscr{A}), \widehat{k}^a)$. ∎

9.2.9. PROPOSITION. *Let Σ be a closed subset of a k-affinoid space $X = \mathscr{M}(\mathscr{A})$ consisting of points x with $d^{st}(x) = 0$. Then Σ is holomorphically convex in X. If $\Sigma \subset X(k)$, then any continuous function $\Sigma \to k$ can be approximated by elements of $\mathscr{A}$.*

PROOF. First suppose that $\Sigma \subset X(k)$. Every element of $\mathscr{A}$ (and of $\mathscr{H}(\Sigma)$) induces a continuous function $\Sigma \to k$. By the Stone-Weierstrass Theorem 9.2.6, the image of $\mathscr{A}$ is everywhere dense in $\mathscr{C}(\Sigma)$. Hence, $\mathscr{P}(\Sigma) = \mathscr{H}(\Sigma) = \mathscr{C}(\Sigma)$. It follows that $\Sigma^h = \mathscr{M}(\mathscr{P}(\Sigma)) = \mathscr{M}(\mathscr{C}(\Sigma)) = \Sigma$.

In the general case, consider the map $X' = X\widehat{\otimes}\widehat{k}^a \to X$. The set $\Sigma' = \pi^{-1}(\Sigma)$ lies in $X'(\widehat{k}^a)$, and so is holomorphically convex in X'. Since $\mathscr{P}(\Sigma)$ is a closed subalgebra in $\mathscr{P}(\Sigma')$, Theorem 9.2.5(i) implies that the map $\Sigma' = \mathscr{M}(\mathscr{P}(\Sigma')) \to \mathscr{M}(\mathscr{P}(\Sigma)) = \Sigma^h$ is surjective. Therefore, $\Sigma = \Sigma^h$. ∎

9.3. Dimension of a subalgebra

9.3.1. THEOREM. *Let $\mathscr{A}$ be a commutative Banach k-algebra with identity whose norm is multiplicative, and let $\mathscr{B}$ be a closed subalgebra containing the identity. Then $d(\mathscr{B}/k) \le d(\mathscr{A}/k)$.*

PROOF. First, we may assume that $\mathscr{A} = K$ is a field with $d(K/k) < \infty$. Next, we may assume that the valuation on k is nontrivial. Indeed, otherwise we take a positive number $r \notin \sqrt{|K^*|}$ and replace K by the field $K\widehat{\otimes}K_r$ and $\mathscr{B}$ by its subalgebra $\mathscr{B}\widehat{\otimes}K_r$ (because $d(K\widehat{\otimes}K_r/K_r) = d(K/k)$ and, by Proposition 9.1.4, $d(\mathscr{B}\widehat{\otimes}K_r/K_r) \ge d(\mathscr{B}/k)$). Finally, we may assume that the fraction field $Q(\mathscr{B})$ of $\mathscr{B}$ is everywhere dense in K.

We set $s = s(K/k)$, $t = t(K/k)$ and $n = s + t$. Let $x \in \mathscr{M}(\mathscr{B})$. We choose elements $f_1, \dots, f_n \in Q(\mathscr{B})$ such that $f_1, \dots, f_s \in K^\circ$, $\widetilde{f}_1, \dots, \widetilde{f}_s$ are algebraically independent over $\widetilde{k}$, and $|f_{s+1}|, \dots, |f_n|$ are linearly independent in the $\mathbf{Q}$-vector space $\sqrt{|K^*|}/\sqrt{|k^*|}$. Let $f_i = g_i/g$, where $g_i, g \in \mathscr{B}$, $1 \le i \le n$. Replacing the f_i by elements of the form $g_i/(g+a)$, where $a \in k$ with $|a|$ sufficiently small, we may assume that $g(x) \ne 0$. In this case the seminorm corresponding to the point x can be extended to the closure of the localization of $\mathscr{B}$ with respect to powers of g. So we may assume that $f_i \in \mathscr{B}$, $1 \le i \le n$.

Let $\mathscr{C}$ denote the closed subalgebra of $\mathscr{B}$ generated by the elements $f_1, \dots, f_n$. It is easy to see that for $P = \sum_{i_1, \dots, i_n=0}^m a_{i_1 \dots i_n} T_1^{i_1} \dots T_n^{i_n} \in k[T_1, \dots, T_n]$ one has the equality

$$|P(f_1, \dots, f_n)| = \max |a_{i_1 \dots i_n}| |f_{s+1}|^{i_1} \dots |f_n|^{i_n} .$$

It follows that $\mathscr{C}$ is isomorphic to the algebra $\mathscr{H}(E)$, where E is the n-dimensional closed disc of radius $(1, \dots, 1, |f_{s+1}|, \dots, |f_n|)$. It follows that $d(\mathscr{C}/k) = n$. In particular, $d(\mathscr{H}_{\mathscr{C}}(x)/k) \le n$. We will show that the extension $\mathscr{H}_{\mathscr{B}}(x)/\mathscr{H}_{\mathscr{C}}(x)$ has the property (c) of Proposition 9.2.1, where one can take M to be the image of $\mathscr{B}_{(x)}$ in $\mathscr{H}_{\mathscr{B}}(x)$ ($\mathscr{B}_{(x)}$ is the localization of $\mathscr{B}$ with respect to all elements which do not vanish at x).

Let $f \in \mathscr{B}_{(x)}$. Since $d(K/\widehat{Q(\mathscr{C})}) = 0$, there exist elements $\alpha_0, \dots, \alpha_m \in \widehat{Q(\mathscr{C})}$ with $|1 - f\sum_{i=0}^m \alpha_i f^i| < 1$. We may assume that $\alpha_i \in Q(\mathscr{C})$, $1 \le i \le n$. Let $\alpha_i = \frac{\beta_i}{\gamma}$, where $\beta_i, \gamma \in \mathscr{C}$. Replacing γ by an element of the form $\gamma + a$, where $a \in k$ with $|a|$ sufficiently small, we may assume that $\gamma(x) \ne 0$, i.e., $\alpha_i \in \mathscr{C}_{(x)}$. We obtain the inequality $|1 - f(x)\sum_{i=0}^m \alpha_i(x) f(x)^i| < 1$. By Proposition 9.2.1, $d(\mathscr{H}_{\mathscr{B}}(x)/\mathscr{H}_{\mathscr{C}}(x)) = 0$. ∎

9.3.2. Corollary. *Let $\mathscr{A}$ be a commutative Banach k-algebra with identity, and let x', x'' be points in $\mathscr{M}(\mathscr{A})$ such that $|f(x')| \le |f(x'')|$ for all $f \in \mathscr{A}$. Then $d(x') \le d(x'')$.* ∎

References

[I] V. G. Berkovich, *Spectral theory and analytic geometry over non-Archimedean fields I* (in Russian), Preprint, Moscow, April 1986.

[II] V. G. Berkovich, *Spectral theory and analytic geometry over non-Archimedean fields II*, Preprint IHES/M/88/43, September 1988.

[III] V. G. Berkovich, *Non-Archimedean analytic spaces and buildings of semi-simple groups*, Preprint IHES/M/89/11, February 1989.

[Ba] W. Bartenwerfer, *Einige Fortsetzungssätze in der p-adischen analysis*, Math. Ann. **185** (1970), 191–210.

[BKKN] R. Berger, R. Kiehl, E. Kunz, and H. J. Nastold, *Differentialrechnung in der analytischen geometrie*, Lecture Notes in Math. vol. 38, Springer, Berlin-Heidelberg-New York, 1967.

[Bo1] S. Bosch, *Orthonormalbasen in der nichtarchimedean Funktionentheorie*, Manuscripta Math. **1** (1969), 35–57.

[Bo2] S. Bosch, *k-affinoide tori*, Math. Ann. **192** (1971), 1–16.

[Bo3] S. Bosch, *Rigid analytische Gruppen mit guter Reduction*, Math. Ann. **223** (1976), 193–205.

[Bo4] S. Bosch, *Zur Kohomologietheorie rigid analytischer Räume*, Manuscripta Math. **20** (1977), 1–27.

[Bo5] S. Bosch, *Eine bemerkenswerte Eigenschaft der formellen Fasern affinoider Räume*, Math. Ann. **229** (1977), 25–45.

[BGR] S. Bosch, U. Güntzer, and R. Remmert, *Non-Archimedean analysis. A systematic approach to rigid analytic geometry*, Grundlehren der Mathematischen Wissenschaften, Bd. 261, Springer, Berlin-Heidelberg-New York, 1984.

[BL1] S. Bosch and W. Lütkebohmert, *Stable reduction and uniformization of abelian varieties I*, Math. Ann. **270** (1985), 349–79.

[BL2] S. Bosch and W. Lütkebohmert, *Stable reduction and uniformization of abelian varieties II*, Invent. Math. **78** (1984), 257–97.

[BL3] S. Bosch and W. Lütkebohmert, *Néron models from the rigid analytic viewpoint*, J. Reine Angew. Math. **364** (1986), 69–84.

[Bou1] N. Bourbaki, *Algèbre commutative*, Hermann, Paris, 1961.

[Bou2] N. Bourbaki, *Thèories spectrales*, Hermann, Paris, 1967.

[BT] F. Bruhat and J. Tits, *Groupes réductifs sur un corps local. I. Donnée radicielles valuée*, Publ. Math. I.H.E.S. **41** (1972), 5–251.

[Ch] C. Chevalley, *Certains schémas en groupes semi-simples*, Séminaire Bourbaki **13** (1960–61), n° 219, 16 pages.

[DeG] M. Demazure and A. Grothendieck, *Schémas en groupes III*, Lecture Notes in Math. vol. 153, Springer, Berlin-Heidelberg-New York, 1970.

[Dr] V. G. Drinfel'd, *Elliptic modules*, Math. USSR-Sb. **23** (1974), 561–92.

[En] R. Engelking, *General topology*, Warszawa, 1977.

[Es] A. Escassut, *Algèbres de Banach ultramétriques et algèbres de Krasner-Tate*, Astérisque, **10** (1973), 1–107.

[Fi] K.-H. Fieseler, *Zariski's Main Theorem für affinoide Kurven*, Math. Ann. **251** (1980), 97–110.

[GRS] I. Gel′fand, D. Raikov, and G. Shilov, *Commutative normed rings*, Chelsea Co., New York, 1964.

[Ge] L. Gerritzen, *On non-Archimedean representations of abelian varieties*, Math. Ann. **196** (1972), 323–46.

[GP] L. Gerritzen and M. van der Put, *Schottky Groups and Mumford curves*, Lecture Notes in Math. vol. 817, Springer, Berlin-Heidelberg-New-York, 1980.

[GR1] H. Grauert and R. Remmert, *Nichtarchimedische Funktionentheorie*, Weierstraß-Festschrift, Wissenschaftl. Abh. Arbeitsgemeinschaft für Forschung des Landes Nordrhein-Westfalen **33** (1966), 393–476.

[GR2] H. Grauert and R. Remmert, *Coherent analytic sheaves*, Springer, Berlin-Heidelberg-New York, 1984.

[SGA] A. Grothendieck, *Séminaire de géométrie algébrique. I. Revêtements étales et groupe fondemental*, Lecture Notes in Math. vol. 224, Springer, Berlin-Heidelberg-New York, 1971.

[EGA] A. Grothendieck and J. Dieudonné, *Eléments de géométrie algébrique. IV. Étude local des schémas et des morphismes de schémas*, Publ. Math. I.H.E.S. **20** (1964), **25** (1965), **28** (1966), **32** (1967).

[Gr] L. Gruson, *Théorie de Fredholm p-adique*, Bull. Soc. Math. France **94** (1966), 67–95.

[H] R. Hartshorne, *Algebraic geometry*, Springer, Berlin-Heidelberg-New York, 1977.

[Ka] I. Kaplansky, *The Weierstrass theorem in fields with valuations*, Proc. Amer. Math. Soc. **1** (1950), 356–57.

[Ki1] R. Kiehl, *Der Endlichkeitssatz für eigentliche Abbildungen in der nichtarchimedischen Funktionentheorie*, Invent. Math. **2** (1967), 191–214.

[Ki2] R. Kiehl, *Theorem A und Theorem B in der nichtarchimedischen Funktionentheorie*, Invent. Math. **2** (1967), 256–73.

[Kr] M. Krasner, *Prolongement analytique uniforme et multiforme dans les corps valués complets*, Colloque Int. CNRS **143** (1964), 97–142.

[Li] Q. Liu, *Ouverts analytiques d'une courbe algébrique en géométrie rigide*, Ann. Inst. Fourier (Grenoble) **37** (1987), 39–64.

[Lu] W. Lütkebohmert, *Der Satz von Remmert-Stein in der nichtarchime-*

dischen Funktionentheorie, Math. Z. **139** (1974), 69–84.

[Ly] Yu. I. Lyubich, *Introduction to the theory of Banach representations of groups* (in Russian), Vischa Shkola, Harkov, 1985.

[M1] D. Mumford, *Abelian varieties*, Bombay, 1968.

[M2] D. Mumford, *An analytic construction of degenerating curves over complete local fields*, Compositio Math. **24** (1972), 129–74.

[P1] M. van der Put, *Non-Archimedean function algebras*, Indag. Math. **33** (1971), 60–77.

[P2] M. van der Put, *The class group of a one-dimensional affinoid space*, Ann. Inst. Fourier (Grenoble) **30** (1980), 155–64.

[P3] M. van der Put, *Cohomology on affinoid spaces*, Compositio Math. **45** (1982), 165–98.

[P4] M. van der Put, *A note on p-adic uniformization*, Indag. Math. **49** (1987), 313–18.

[Rh] G. de Rham, *Variétés différentiables*, Hermann, Paris, 1955.

[Ro] P. Robba, *Fonctions analytiques sur les corps valués complets*, Astérisque **10** (1973), 109–218.

[Se] J.-P. Serre, *Endomorphismes complètement continus d'espaces de Banach p-adiques*, Publ. Math. IHES **12** (1962), 69–85.

[Ta] J. Tate, *Rigid analytic spaces*, Invent. Math. **12** (1971), 257–89.

[Ti] J. Tits, *Reductive groups over local fields*, Proc. Sympos. Pure Math. **33** (1979), 29–69.

[Vi] M. M. Vishik, *Non-Archimedean spectral theory*, J. Soviet Math. **30** (1985), 2513–54.

Index